Piping and Pipeline
Calculations Manual

Piping and Pipeline Calculations Manual

Construction, Design Fabrication and Examination

Second Edition

J. PHILLIP ELLENBERGER

AMSTERDAM • BOSTON • HEIDELBERG • LONDON
NEW YORK • OXFORD • PARIS • SAN DIEGO
SAN FRANCISCO • SINGAPORE • SYDNEY • TOKYO

Butterworth-Heinemann is an imprint of Elsevier

Acquiring Editor: Ken McCombs
Editorial Project Manager: Jeff Freeland
Project Manager: Mohana Natarajan
Designer: Matthew Limbert

Butterworth-Heinemann is an imprint of Elsevier
225 Wyman Street, Waltham, MA 02451, USA
The Boulevard, Langford Lane, Kidlington, Oxford, OX5 1GB, UK

Notices

Knowledge and best practice in this field are constantly changing. As new research and experience broaden our understanding, changes in research methods or professional practices, may become necessary. Practitioners and researchers must always rely on their own experience and knowledge in evaluating and using any information or methods described here in. In using such information or methods they should be mindful of their own safety and the safety of others, including parties for whom they have a professional responsibility.

To the fullest extent of the law, neither the Publisher nor the authors, contributors, or editors, assume any liability for any injury and/or damage to persons or property as a matter of products liability, negligence or otherwise, or from any use or operation of any methods, products, instructions, or ideas contained in the material herein.

Library of Congress Cataloging-in-Publication Data
Ellenberger, J. Phillip, author.
  Piping and pipeline calculations manual: construction, design fabrication and examination / J. Phillip Ellenberger. -- Second edition.
    pages cm
  Includes bibliographical references and index.
  ISBN 978-0-12-416747-6
1. Pipelines--Design and construction--Handbooks, manuals, etc. 2. Piping--Design and construction--Handbooks, manuals, etc. I. Title.
  TJ930.E438 2014
  621.8'672--dc23
                        2014002255

British Library Cataloguing-in-Publication Data
A catalogue record for this book is available from the British Library

ISBN: 978-0-12-416747-6

For information on all Butterworth-Heinemann publications
visit our website at store.elsevier.com

Working together
to grow libraries in
developing countries

www.elsevier.com • www.bookaid.org

The book is dedicated to all those unnamed members of the cloud.

CONTENTS

PREFACE TO THE FIRST EDITION

What are the necessary requirements to move from a piping or pipeline system idea to its completion? The basic premise of this book is that at the heart of those requirements are a series of calculations, which cover a wide range of subjects.

In any pipeline system, the core of the system itself is the piping, which is its skeleton. However, as with any skeleton, there must be other elements to include before the system can become the final entity that was the original idea.

Pipe is basically a transport structure. To determine what that structure requires would involve what it is intended to transport. While it is important to have knowledge of how the medium to be transported is generated, this book does not address that area. Generation of that comes from another field of expertise.

A pipe system has a beginning, an ending, and a path between the two points. To transport the medium—liquid or gas—some definition of temperatures, pressures, amount to be transported per unit of time, and the energy required to accomplish the transport need to be, at least partially, established. Many of these will be considered as a given in this book and the methods of calculating the other elements are discussed and explained.

The base codes for the design of a new system, and the ones used in this book as the reference source, are the B31 piping codes of the American Society of Mechanical Engineers (ASME). The B31 piping codes consist of several sections or books that describe the requirements for systems of a specific type. These can readily be broken into the two basic types—a piping system and a pipeline system.

The differences between the two are that a piping system can be generically defined as being inside a localized area to connect various vessels that are for reaction and/or storage. A pipeline system is more like a pure transport medium between two geographical positions. Within both are elements of the other. There are many pipelines within a plant or localized area, and along the pipelines between distant points are stations that have piping systems necessary for some pipeline element such as a compressor station.

For these reasons, the various sections or books of the B31 codes allow piping system owners to determine which code would apply to their particular project. In making this decision the owners are also advised to take into account which code the jurisdiction(s) for their projects might consider applicable.

All system requirements basically set standards of calculation to establish a safe end result. Those qualification standards are outlined with specific calculation procedures in the codes. Some things are required to be taken into account without details of how to consider them. Some calculations require base calculations to arrive at the point where the code calculation can be used. In this book, we address many of the grayer areas.

As one goes through the steps of meeting the requirements of particular codes, he or she will also find many other standards included by reference. This is a practical way for the codes to cover many common elements in the design and construction of a system. Any calculations required for the component that are covered by the referenced standard need not be outlined in the code. The use of that component needs no further proof of compliance with the code than its compliance with the standard. Since different standards provide different methods of providing the calculations, those differences are also addressed.

The B31 piping codes are primarily construction codes for new facilities. They can be used successfully in replacing or extending a piping facility. With few exceptions, notably the pipeline sections, there are no maintenance and ongoing requirements. The pipeline sections have relatively extensive detailed requirements for continuous maintenance. There is a growing set of postconstruction requirements, some of which are published, that give methodologies for repairing and assessing the need for repairs. Some small offerings detail the methodology for certain more complex areas of analysis, and these are discussed in this book.

It should be noted that some of the calculations provided are not necessarily required by the codes. However, one must really understand those calculations to have the depth of understanding needed to do a good job when performing the calculations required.

Part I of this book provides an overview of the codes and standards, including what they are and what they aren't. It provides a detailed discussion of the "metric problem." Chapter 3 discusses piping materials, as well as other materials, that might be required to complete a system.

Part II covers some specific calculations and their formulas and has examples of how to do such calculations.

The Appendix contains a set of charts, graphs, and other helpful tables and guides that should make doing some of the calculations easier or faster. In this computer/calculator age, some tables and graphs are still a good way to look at alternative solutions to a problem before going into an in-depth mathematical analysis.

It has been a few years since the first edition of this book was written. One might point out that the amount of time between the draft of a book and the actual publishing of same may take some time. As I read the preface to that first edition I find no major differences in what was written there.

Pipe is still round with a hole in the middle. It still has all the codes and safety requirements of those codes. Stress is still force divided by area. With minor exceptions what was in the first preface still holds as true.If anything some piping systems have become more complex. There are even new editions of the codes that existed then. There are some newer codes.

The technology has changed, particularly in the sense of digital analysis and complex programs to solve the old problems in newer and easier ways. That does not change many of the essential variables. This book is not about the newer digital techniques as much as it is about the base knowledge that the reader must have an understanding to make the engineering judgment required to determine the adequacy of their analysis. The book discusses and recommends those digital approaches as deeper analysis is recognized as required.

The materials may have evolved and some techniques may have changed. These have been reflected in this edition. In some cases we have attempted to clarify that which may have been obscure in the earlier discussions. Where it was found necessary we have added some approaches and explanations to make ones analysis more complete.

As in all efforts the author was not perfect in his description and where errors were detected they have been corrected. The copy editors have been helpful in the need for clarification. The cloud mentioned in the previous acknowledgement has continued to be helpful and supportive. Unfortunately like a true cloud there has been change and movement of that cloud. We do have some new, miss some who have passed on and still thank them all for that is the beauty of sharing ideas and knowledge. It lives on.

Major Codes and Standards

Contents

OVERVIEW

The world of standards may seem to many to be something like the Tower of Babel—there are so many different standards, some of which are called codes, that the problem seems daunting. This book is meant to help remove some of that difficulty.

One concern for any reader would be his or her geographical area. Or, to put it another way, which code does the jurisdiction for my area recognize, if any, as the one to use for my project? This is a question that can only be answered in that particular area.

One can say in general that there are three main codes in the piping and pipelines realm: the ASME codes in the United States and many parts of the world; the DIN codes in Europe and other European-leaning parts of the world; and the Japanese codes, which have a great deal of significance in Asia.

The standards of the International Organization for Standards (ISO) are an emerging attempt to simplify the codification process by cutting down on the multiplicity of codes worldwide. As users of these codes and standards become more global in their reach, the need becomes more prevalent. However, there is a long way to go before we become a world where a single set of codes applies.

Piping and Pipeline Calculations Manual
ISBN 978-0-12-416747-6
http://dx.doi.org/10.1016/B978-0-12-416747-6.00001-2

The dominant themes here will come from the American codes and standards such as those from the American Society of Mechanical Engineers (ASME). Where appropriate, we will point to other sources, some of which are specifically mentioned in the following text. The main allowance for worldwide use will be the translation to metric from the U.S. customary units of measure. The ASME codes and other U.S. code-writing bodies are in various stages of conversion within their written standards. Particularly in those parts of this book where calculation procedures are given, we will show them in both methods of measure.

It should also be pointed out that we will cite other standards-writing bodies and use their techniques as we explore piping and pipelines. They include, but are not necessarily limited to, the following: Manufacturers Standardization Society (MSS), American Petroleum Institute (API), American Society of Testing Materials (ASTM), Pipe Fabrication Institute (PFI), and American Welding Society (ASM).

In mentioning codes and standards one should also mention that in many nations there is a national standards organization. In the United States it is the American National Standards Institute (ANSI). Again, each jurisdiction may have a different format, but the main emphasis is that a code with the national standards imprimatur is the de facto national standard.

In the United States once a standard has met the requirements and can call itself a national standard, no other standard on that specific subject can claim the imprimatur of a national standard for that subject. One of the relevant requirements of becoming an ANSI standard is *balance*. To obtain this balance as the standard is being written it must be reviewed and agreed on by people representing the major aspects of the subject, including producers, users, and the public. Before it can be published it must go through an additional public review and comment phase. During this process all comments and objections must be addressed and resolved. In short, a national standard gives an assurance that all relevant aspects of that subject have been addressed.

With the exception that a jurisdiction may set a requirement that a particular standard must be utilized as a matter of law in that jurisdiction, a standard is only a basis or a guideline as to good practice. As previously mentioned, it might be the law in certain jurisdictions, and it certainly can be a requirement in any contract between parties, but as a code it is not needed until one of those requirements is met.

This may lead one to question what the difference is between a code and a standard. The simple answer is nothing of significance. When one reads the title of a B31 section, he or she will find that a code is a national standard. *Code* is a descriptive word that usually designates that the standard has some legal status somewhere. The major practical difference is that a code will have several aspects while a standard is primarily about one thing.

It is also true that many standards establish dimensions for the product that the standard represents. This establishes a consistent set of dimensions that a designer or specifer, who is often early in the development stage, can use for planning. A standard also often establishes a pressure-temperature rating that eliminates the requirement for further calculation on the designer's part when it becomes part of the code. One definition of a standard is that it is a pre-engineered solution to a common problem. The net result is a cost savings to the parties involved.

Some standards-writing bodies call their offerings something slightly different. For example, the MSS calls their offerings standard practices (SPs). The MSS has recently started converting some of their SPs to national standards. Because their membership is limited to manufacturers of flanges, valves, and fittings, they have to follow a different methodology to obtain the balance required by ANSI. This is called the *canvass method*[1], which is a part of the overall protocol of ANSI's requirements. It is designed for just such a situation as noted with the MSS, where their standards focus on a single category—that is, manufacturers—and therefore do not meet the balance requirement.

STRUCTURE OF CODES

The basic structure of the ASME piping codes is fairly standard across all of the books. By following this nominal standard order a rough cross-reference between various books is achieved. Each book's paragraphs are numbered with the number of the book section as the first set of digits.

For example, for a paragraph in B31.1, the first digit is 1, while a paragraph in B31.3 has a first digit of 3, and in B31.11 it would be 11. As much as possible sequential numbering is common. This cannot be adhered to exactly because all books do not have the same concerns and therefore the same number of paragraphs. It does, however, guide a searcher to what another book says about the same paragraph or subject by leading him or her to the proper vicinity within the book.

---

[1] The canvass method requires that the standard writing body recruit independent reviewers who agree to read and comment on the standard. These recruits must be from categories that will fulfill the balance required as mentioned in the previous description. Any comments must be appropriately addressed befor moving the poposed standard through the ANSI protocol.

The major exceptions come from B31.8, which has a different basic order of elements. Even though this order is different, the elements that are required to build a safe system are included, albeit in a different section of the book.

It is also true that there are significant differences in detail. For instance, B31.3 basically repeats certain paragraphs and numbers for different risk media. It has complementary numbering systems with a letter prefix for the number. For example, where B31.3 sets requirements for nonmetallic piping, the prefix is A3xx and the numbering again is as close to the same sequence as possible. Where applicable, in each paragraph something like the paragraph in the base code (nonprefix number) applies in its entirety or "except for," and then the exceptions are listed. When something has no applicable paragraph in that base code the requirements are spelled out completely.

Some sections of the codes are not in all codes. These are usually standalone portions of that particular book. Some have been previously mentioned. Not every code has a reference to operation and maintenance. The pipelines, in particular, have extensive sections that are not in the piping codes. These include things like corrosion protection for buried piping, offshore piping, and sour gas piping.

CODE CATEGORIES

The eight major categories that the code covers are scope; design conditions; pressure design; flexibility and stress intensification; materials; standards; fabrication and assembly; and inspection, examination, and testing. Each is described in the following sections.

Scope

This is where the primary intent of the piping requirements is defined in a particular book. Scope will also include any exclusion that the book does not cover and will offer definitions of terms considered unique enough to require defining in that particular book. I repeat here that the final decision as to which code to specify for their project is up to the owners, considering the requirements of the jurisdiction(s).

Design Conditions

In this section the requirements for setting the design parameters used in making the calculations are established. These will generally include the design pressure and temperature and on what basis they may be determined. As applicable to the system considered in the scope, there will be discussion of many loads that must be considered. Many of these are addressed in later parts of the code, some in specific detail and some left to appendices or the designer. All must be considered in some appropriate manner. There is also a section that defines how the allowable stresses listed within the code are established. If allowed, a procedure for unlisted material can be computed. It will also establish limits and allowances.

Pressure Design

This section gives the calculation and methodology to establish that the design meets the basic criteria. It is probably the most calculation-intensive portion of the code. There are additional parts as required by the intended scope to define requirements for service in piping components and piping joints.

Flexibility and Stress Intensification

These sections set the requirements for the designer to be sure that the piping is not overstressed from loads generated from aspects other than the pressure. There may be loads generated from the thermal expansion of the piping system and they may also come from other sources such as wind and earthquake. In this section most codes give only a partial methodology after some critical moments and loads have been generated by some other means such as computer programs or similar methods. The codes also address piping support requirements for both above ground and, where applicable, below ground. (Part II addresses concerns that these codes may create for readers.)

Materials

This section addresses those materials that are listed and those that may not be allowed and, if allowed, how to establish them. Often, it is in this section that the low temperature toughness tests are established. This is generally known as Charpy testing, but there may be other methods allowed.

Standards

This is the section where the other standards that the code has reviewed and consider applicable to that book are listed. The listing also includes the particular issue that is recognized by that book.

Fabrication and Assembly

It should be noted that aboveground piping systems are most times fabricated in a shop in spools, and then taken to the field where they are assembled by various means such as final welding, or if the spools are flanged, bolted together. On the other hand, the majority of the time pipelines are constructed in the field with field welding. This is not to say that in both cases other methods will not be used. It does describe why some books call it construction and some call it fabrication. Needless to say, there are differences in the requirements.

Inspection, Examination, and Testing

These three elements are grouped together because they essentially define the "proof of the pudding" requirements of the codes. In some manner all systems need to be tested for integrity before being put to use. These requirements vary from book to book. The codes in general create a dual responsibility in the area of inspection and examination. The examination and documentation is the responsibility of the builder, fabricator, or contractor performing the work. The inspection is the responsibility of the owner's representative and he or she may perform an examination of the product and check the documentation in order to give the final approval.

With the exception of portions of B31.1 piping, namely boiler external piping, there are no requirements for third-party inspection and code stamping such as is required for some boilers and pressure vessels. This type of requirement may be imposed contractually as a certified quality-control system check, but in general is not mandatory.

As previously mentioned, each book may have special requirements areas for specific kinds of media or system locations. They are addressed individually in the book within that special area.

Let's set the field for the different B31 sections. In the process we can give a small background for each book. The original ASME B31 Code for Pressure Piping was first introduced in 1935 as the single document for piping design. In 1955, ASME began to separate the code into sections to address requirements of specific piping systems, as follows.

B31.1: Power Piping is for piping associated with power plants and district heating systems as well as geothermal heating systems. Its main concern is the steam–water loop in conventionally powered plants. More recently, it has added a chapter to require maintenance plans for the plants that produce the power.

B31.2: The Fuel Gas Piping code was withdrawn in 1988, and responsibility for that piping was assumed by ANSI Z223.1. It was a good design document, and although it has been withdrawn, ASME makes it available as a reference.

B31.3: Process Piping (previously called the Chemical Plant and Petroleum Refinery Piping Code) is the code that covers a larger variety of piping systems than any other. To cover this variety it has sections for different types of fluids. These fluids are basically rated as to the inherent risks in using that fluid in a piping system; the code has more restrictive requirements for the more difficult fluids.

B31.4: Liquid Transportation Systems for Hydrocarbons and Other Liquids basically is a buried pipeline transportation code for liquid products. It is one of the three B31 sections that are primarily for transportation systems. As such, they also have to work with many of the transportation regulatory agencies to be sure that they are not in conflict with those regulations.

B31.5: Refrigeration Piping and Heat Transfer Components is rather self-explanatory. It is primarily for building refrigeration or larger heat transfer systems.

B31.7: Nuclear Piping was withdrawn after two editions and the responsibility was assumed by ASME B&PV Code, Section III, Subsections NA, NB, NC, and ND. This code had some very good explanations of the requirements of piping design. This book may refer to those explanations, but will not specifically address the complex nuclear requirements.

B31.8: Gas Transmission and Distribution Piping Systems addresses the transportation of gases, and it too is primarily for buried piping. It is another pipeline code. The Code of Federal Regulations (49 CFR) is the law for these types of piping systems. As such, that code must present complementary requirements. Also, a gas pipeline would generally cover a fair amount of distance, and this may have several different degrees of safety requirements over the pipeline as it progressively proceeds through various population densities. Also, since natural gas has so much inherent risk, it is quite detailed in its safety and maintenance requirements.

B31.8S: Managing System Integrity of Gas Pipelines is a recently published book. This is a book defining how to establish a plan to handle the problems those inherent risks present.

B31.9: Building Services Piping addresses typical pressure piping systems that are designed to serve commercial and institutional buildings. Because these systems are often of less risk in regard to pressures, toxicity, and temperature, they have restrictions on these parameters. When the limits are exceeded the user is often referred to B31.1.

B31.11: Slurry Piping Systems is another transportation pipeline code that mostly applies to buried piping systems that transport slurries. It has increasingly limited usefulness as a standalone document, and may someday be included as a subset of B31.4. The expected use of slurries to transport such things as pulverized coal has not materialized.

B31.12: Hydrogen Piping and Pipelines was first published in 2008. The code is composed of three major sections. First is the GR section, General Requirements, which includes common requirements that are referenced in the other sections. The other two sections are IP, Industrial Piping, and PL, Pipelines, including distribution systems. This code incorporates information specific to hydrogen service. Either by reference or inclusion this will have many similar sections to B31.3 and B31.8. One major difference is the materials section, where the detrimental effect of hydrogen gas on some materials requires the elimination of some materials and different factors. Hydrogen is an odorless, highly flammable gas. Generally it is operated at a higher pressure and as such requires unique safety precautions.

B31 also has been developing a series of shorter books. These are not numbered but designated by a letter after the B31. Their use is universal and the development allows the incorporation of the special material by reference. The oldest one is B31-G, entitled "A Manual for Determining the Remaining Strength of Corroded Pipelines"; it is still used extensively in pipeline work.

B31-E is a standard for the Seismic Design and Retrofit of Above-Ground Piping Systems. It is, among other things, an effort to bring continuity to piping design. It is hoped that this standard will be included by reference in various B31 books.

B31-J is the Standard Test Method for Determining Stress Intensification Factors (i-Factors) for Metallic Piping Components. Most of the major books have an appendix stating the i-factors to use for certain geometries. These are based on tests on standard components. As the technology has changed, a need has developed to determine factors for other geometries that are not in these existing appendices. To provide more objective evidence, as allowed, this standard was developed to reflect how the original intensification factors will be developed. It includes a testing methodology to develop a stress intensification factor (SIF) for a geometry. Research is underway to develop more comprehensive SIF formulas that are intended to be included in this book.

B31-T details the Standard Toughness Requirements for Piping. This is a compilation of the requirements and methods of performing tests for low temperature toughness, including acceptance requirements. The intent is to have a standalone reference on the variation of toughness requirements that can be utilized by codes and standards.

B31-Q discusses Pipeline Personnel Qualifications. This is a development of a consensus standard in response to federal laws requiring pipelines to have programs that delineate the training of personnel who have the responsibility to maintain the pipelines and periodically examine the condition of those pipelines.

It is anticipated that other such books will be developed over time. Some are in development at the time of the writing of this text.

PCC-1: While not a B31 book, this has great applicability to piping. It is called Guidelines for Pressure Boundary Bolted Flange Joint Assembly. This is a very commonplace event in any piping system and the Guidelines go through how to properly "bolt up" such a set of flanges. It is specifically for "raised face" type flanges that are in standards such as B16.5 and are often specially designed by methods described by BPVC Section VIII, Appendix 2.

The ASME B16 standards committee basically covers flanges, fittings, and valves. The most familiar of these standards would be the following:

B16.5 Flanges and Flanged Fittings

B16.47 Large Diameter Steel Flanges

B16.9 Factory Made Wrought Butt-Welding Fittings

B16.11 Forged Fittings Socket Welded and Threaded

Details of these standards and any of the others that apply will be discussed as we go through the calculations in Part II and the Appendix.

A similar result occurs with the standard practices that are written by the MSS. They have several, and not all are recognized by the piping codes. If the standard practice is recognized by that code it may be used in that code. If it is not recoganized there may be additional proof of compliance required to use the standard in that code. Some of the most familiar ones are the following:

SP-97 Integrally Reinforced Branch Connection Fittings

SP-58 Pipe Hangers and Supports

SP-75 Specification for High-Test Wrought Butt-Welding Fittings

SP-44 Steel Pipeline Flanges

SP-24 Standard Marking System for Valves, Fittings, Flanges and Unions

MSS has several other SPs that are quite useful but often do not require any calculation or subsequent work for the user. These are standard practices that offer guidance for such things as a standard for Positive Material Identification and others.

API standards and specifications will be addressed in a similar manner as the need arises in the specific calculation methods. Some API standards, such as the flange standard, have been incorporated in B16.47 and as such are no longer supported by API. API has several standards for valves and high-pressure flanges and are best known for their 5L line of higher-yield piping.

The calculation requirements for elements like pipe sizing and flow will be introduced in Part II and the Appendix to give readers some insight into how to perform those calculations. The process of getting started in any piping project is not specifically covered by a specific standard in its entirety. Often there is interplay between the process engineer and the system or pipe designer as well as the equipment designer.

In all engineering situations, economics come into play regarding project initiation. One must determine, somehow, the most economical throughput, balancing any economies of scale from increased throughput and budget limitations. Then the problem becomes one of larger pipe size versus equipment size to produce the throughput. These issues are based on equivalent lengths of pipe, pipe size, fluid friction within the pipe, and so forth. Other than a rudimentary discussion and demonstration of the basics of those decisions, much of the detailed analysis lies outside the scope of this book. It is quite well covered by other disciplines and their literature.

It is important for readers to note that while the various codes and standards offer what appear to be different approaches and calculation procedures to arrive at a specific solution, that difference may not be as great as it first appears. A question I have repeatedly asked myself as I complete a particular set of calculations—How does the pipe or component know which code it was built in accordance with?—has been quite helpful in making the final decision as to whether it is proper for the situation. Mother Nature does not read codes; she just follows her laws.

Of course, you must use the code's required calculation or its equal or more rigorous requirement. More complete listings of codes relevant to piping and pipelines can be found in the Appendix of this book. The mathematics must be correct, but then the question forces the technical reviewer to face the inherent margin that the particular code he or she is working with has established. This comes from the inherent risk the fluid, temperature, and pressure offer within the area that would be affected by a failure, as well as the damage to people, property, and systems that a failure due to an incorrect calculation might incur. When you can answer that question in the affirmative, you are willing to stand behind the result of your work.

Having met that challenge, we must address the contentious question of the metric system of measurement versus the U.S. customary system of measurement. For that, we move to Chapter 2.

CHAPTER 2

Metric versus U.S. Customary Measurement

Contents

OVERVIEW

Whenever one writes anything that includes a measurement system in the United States, he or she is confronted with the problem of what system to use in presenting the data and calculations. This is especially true when writing about codes and standards. Most U.S. codes and standards were originally written some time back when the metric system was not necessarily the dominant world system.

The metric system itself has several minor variations that relate to the base units of measure. This will be discussed more thoroughly in the remainder of the chapter. The metric system has evolved in dominance to the point that only three countries do not use it as their primary measurement system: Myanmar, Liberia, and the United States. It is now known as the International System of Units (SI).

The United States has played with converting to the SI system for as long as I have been working in this field, which is a long time. Americans have not made the leap to make it our primary system. This lack of tenacity in converting to this system is difficult to understand completely. The most plausible argument revolves around the installed base of measurement and a modicum of inertial thought regarding the seemingly inevitable conversion. This argument is belied by the fact that England has converted, and it only took a few years.

To those who have worked with the SI system it is immensely preferred due to its decimal conversion from larger to smaller units. What could be

Piping and Pipeline Calculations Manual
ISBN 978-0-12-416747-6
http://dx.doi.org/10.1016/B978-0-12-416747-6.00002-4

11

simpler than converting a length measurement from something like 1.72 meters to 1720 millimeters? Compare that to converting 1 yard, 2 feet, and 6 inches to 66 inches or 5.5 feet.

On the other side, there is the problem of what you grew up with. It is rather like translating a language that is not your native language. You first have to perform some mental conversion of the words into some semblance of your native tongue. As one becomes fluent in another language, he or she can begin to think in that language.

HARD VERSUS SOFT METRIC CONVERSION

All of this is a descriptive example of some of the difficulties of converting an ASME code into a metric code. The generic classification of this problem is hard versus soft conversions. The terms *hard conversion* and *soft conversion* refer to approaches you might take when converting an existing dimension from nonmetric units to SI. "Hard" doesn't refer to difficulty, but (essentially) to whether hardware changes during the metrication process. However, the terms can be confusing because they're not always consistently defined and their meanings can be nonintuitive.

It's simplest to consider two cases: "converting" a physical object and conversions that don't involve an object.

When converting a physical object, such as a product, part, or component, from inch-pound to metric measurements, there are two general approaches. First, one can replace the part with one that has an appropriate metric size. This is sometimes called a hard conversion because the part is actually replaced by one of a different size—the actual hardware changes. Alternatively, one can keep the same part, but express its size in metric units. This is sometimes called a soft conversion because the part isn't replaced—it is merely renamed.

Another, and possibly simpler way, to explain it is that in a soft conversion a dimension of one foot would be converted to 305 millimeters or 304.8 depending on the accuracy required. In a hard conversion one might convert to 300 millimeters, as that would probably be the size one who was doing the design would choose if he were designing in a metric system.

If the latter sounds odd, note that many items' dimensions are actually nominal sizes—round numbers that aren't their exact measurements—such as lumber, where a 2 × 4 isn't really 2 by 4 inches, and pipe, where a 0.5-in. (NPS ½) pipe has neither an inside nor an outside diameter of 0.5 in. With pipe, the international community has come to a working solution to this anomaly because comparable SI pipe has different dimensions than U.S. schedule pipe. However, when working with pipe above NPS 4, the metric or DN number is 25 × NPS; for smaller or fraction pipe it is made an even number like DN 15 for a NPS ½ pipe.

An even more difficult problem comes about when one is making nonproduct-type decisions while determinign pipe calculations. For instance, how does 1720 mm compare to 5.5 ft in your sense of the two distances? That is to ask, which is longer?

The answer is 1720 mm converts mathematically to 5.643045 ft. However, for some of us, even those who have worked with but are not fluent in metrics, the answer is not obvious—until we do the conversion. We may sense that they are close. In some calculations 5.643045 may not make a significant difference. In others, it may make the difference between meeting or not meeting a certain requirement.

This points to another problem in working with things developed in one system as opposed to other systems. As it relates to conversion, there are often many decisions to be made. If for some reason we were developing a U.S. customary design and arrived at an answer that came to 5.643045 ft, we might call it any of several dimensions in our final decision. It woul depend on the criticality of the dimension in the system.

This would bring us face to face with the oddness of our fractional notation. Normally we think of fractions of an inch. However, we could be dealing with fractions of a foot. Where we are concerned with a dimension that only needs to be within the nearest ⅛ in. to be effective, we might choose 5 ⅝ (5 ft, 7.5 in.) or 5 ¾ (5 ft, 9 in.). The original 5.643 can be converted to something within 1/32 of an inch as 5 ft, 7 ²³/₃₂ inch, and so on. Mind you, all this is for converting 1720 mm into U.S. customary dimension. A similar exercise could be presented for converting something like 5 ¾ (5 ft, 9 in.) into millimeters, which would be 1752.6 mm. One would then have to make comparable decisions about the criticality of the dimension.

SI SYSTEM OF MEASUREMENT

As mentioned earlier, there are several variations of metric systems. Fortunately, they are not as complex as the U.S. customary system (USC). For instance, in distance measurement the name and unit of measure changes with the size of the distance. We have miles, furlongs, chains, yards, feet, inches, and fractions of an inch, all of which can be converted to the other, but not in a linearly logical base 10 fashion as the SI system does.

One system for length, weight, and time in metric is the centimeter, gram, and second system. Another is the kilometer, kilogram, and second system. Note that the change here is just the prefixes for length and weight; the decimal relationship is constant and thus the conversions between the two systems are simple.

The International System of Units (SI) includes some other base units for use in other disciplines:

1. Meter, the distance unit. USC usually uses the inch in pipe work.
2. Kilogram, the weight and force unit. In USC it is the pound.
3. Second, the time unit. Interestingly, a second in France is the same as a second in New York.
4. Ampere, the electrical unit.
5. Kelvin, the temperature unit. Since most of us live and work in the atmosphere, the Celsius measure is more commonly used. But a degree in either is the same; the difference is the 0 reference point. Absolute zero in Kelvin and the zero temperature in Celsius (freezing water) is a difference of some 273.15 (often the .15 is ignored).
6. Candela, the measurement of light, is similar to the U.S. term candle-light; the luminous intensity of one common candle is roughly one candela.
7. Mole, basically the measure of atomic weight. The exact definition is different but the use is similar.

Converting back and forth between the two systems is at the least time consuming. In the Appendix there is a conversion chart as well as a chart that focuses on the conversions that apply to the type of calculations commonly used in piping. Some standard charts don't give those calculations and the dimensional analysis to make them can be quite time consuming, if not nerve-racking. There is also a chart that lists the common prefixes as one goes up and down in quantity. Many need to be used only rarely, but it is often maddening not to find them at the moment you need them.

It is also good to have a calculator with some of the fundamental conversions built in. Barring that, there are some common conversions that should be committed to memory so one can quickly move from one to the other. For example, there are 25.4 mm in an inch and 2.2 lbs in a kilogram. A degree in Celsius is equal to 1.8 degrees Fahrenheit, and there is a base difference at the freezing point of water of 0°C to 32°F (as a side note, −40°C is equal to −40°F). None of these are accurate beyond the inherent accuracy of the conversion numbers, but they are good rules of thumb or ballpark conversions.

METHODS OF CONVERSION FROM ONE SYSTEM TO THE OTHER

It is also a good idea to get a conversion program for your computer. There are several good ones that are free on the Internet. It is quite handy as one works calculations at the computer to just pop up the conversion program and put in the data and check. From the previous discussion, the conversion of 1720 mm to a six-place decimal was made in less than a second on such a computer program.

Several documents give detailed information regarding how to convert to metric from U.S. customary units. The most general one, which includes guidance and conversion charts, is ASTM SI-10. SP-86 is somewhat simpler and was developed by the MSS to guide the committees that chose to add metric to their U.S. customary dimensions. It has a very good discussion of conversion and the implied precision in conversions, and is written in plain language for users who are somewhat at a loss regarding conversion other than the strictly mathematical multiply-this-by-that chart or calculator. It also has some good discussion of rounding and the conversion of fractions to two-place decimals for computer and calculator use.

The ASME B31 piping codes and standards are in various stages of converting their codes to metric. Not all codes lend themselves to metric conversion urgency, so the pace in the various book sections varies according to international usage. Some are quite local to the United States and therefore lag in conversion. Many of the B16 fittings and flange standards have been converted.

In most cases the B16 conversions have made the determination that the metric version is a separate standard. This is a direct result of the problems just described. When making a practical conversion some of the dimensions are not directly converted or are rounded, which means that a component made from one set of dimensions might not be within tolerance of the other set of dimensions. Where that is the case, the standard or code has a paragraph establishing this fact. The paragraph points out that these are two separate sets of dimensions—they are not exact equivalents. Therefore, they must be used independently of the other.

In the flange standards this created a much more mixed set of dimensions. For tolerance and relevant availability the metric version of the flange standards kept U.S. bolt and bolt hole sizes. The standard metric bolting not only did not offer equivalent heavy hex nuts, but also presented other significant issues, as bolting is important in pressure rating calculations, and metric bolts that are not necessarily the same exact area create difficulties in establishing ratings and margins. More is given on this subject later in Part II and the Appendix.

In the piping codes themselves B31.3 is probably the most international of the codes. Since many process industries like chemical and petroleum plants have international operations, B31.3 has broad worldwide usage. It is even mentioned as the normative reference code in the ISO 15649 standard. For this reason, it is probably the most advanced in its establishment of a metric version.

The main remaining pieces of the puzzle in the conversion of B31.3 are the stress tables, which are not yet completely established. In the 2012 edition of B31.3 some of the allowable metric tables for some of the materials are in the publication.

Stress tables create an almost double problem for the codes. The tables are presented material by material in what is a regular temperature range. In U.S. customary units that range is 100° in the lower temperature ranges and 50° steps as the temperatures get higher. These are in Fahrenheit, and the fact that they do not directly translate to Celsius causes a problem. Also, the stresses are in thousands of psi (pressure per square inch) and again not evenly translated into MPa, creating another problem. These two problems create a requirement for a very large amount of interpolation, which in turn has to be checked for accuracy by an independent interpolation. This, coupled with the 16 temperatures and hundreds if not thousands of those interpolations, means a slow process.

The notes in the stress tables indicate the methodology that can be used in getting an equivalent stress from the current U.S. customary tables. Where a metric stress is required those notes will be used to establish an allowable stress for the example problems in this book. The code books themselves already establish any changes in metric constants that may be required to complete calculations.

The partial addition of metric tables for stresses in the B31.3 2012 edition adds another complication. General note (c) to the stress tables states that the values given in tables A1M and A2M (the two metric tables in that edition) are for information only. The values in the USC tables are the ones required. This adds some additional conversion requirements.

The procedure is relatively lengthy. Assume you have a temperature in Celsius. You must first convert that temperature to °F. Then you have to interpolate that temperature to a value in Table A1 or A2 and determine the allowable stress in psi. Finally, you must multiply that stress by 6.895 to obtain the basic stress in MPa.

It is not quite clear why the code asserts that the USC table is the standard. When one checks the tables for one grade of pipe, A285 GrC A55, the metric table has a temperature of 600°C and an allowable stress of 6.89 MPa, whereas the USC table has an allowable stress of 1 psi at 1100°F, which seems comparable. It is probably just caution.

The intention is to convert the codes to metric completely. This of course cannot realistically happen until the United States takes that step. As previously noted, for reasons that can only be surmised, it hasn't happened yet, but it will happen. When one buys a beverage container, the metric equivalent is often noted. Those who work with automotive equipment might need a new set of metric wrenches to work on newer devices. Likewise, if one is into antique cars, he or she might need an older set of U.S. customary wrenches.

CHALLENGES FOR CONVERTING FROM ONE SYSTEM TO THE OTHER

One of the vexing problems is when one is doing calculations that include standard elements such as the modulus of elasticity, moment of inertia, section modulus, universal gas constant, and other similar standard elements. When one is accustomed to working in one system, he or she may not know all of the standard units that are used in the other. This causes some concern when working a particular formula to get the correct answer in a working order of magnitude. Inevitably, the question is: What unit do I use in the other system?

One example could be the section modulus, Z in most B31 codes. It is often used in concert with moments and stresses and other calculated parameters. Not infrequently there is a power or a square root involved. Which values should be used in such calculations? The best advice is to use a consistent unit of measure such as meters or Pascals, which are defined in Newtons/m^2, when converting from USC or something like inches. The metric system is helpful with its decimal conversions from one size to the other. For instance most of us working in the USC system use something like 29 x 10^6 for Young's modulus. The proper number in the SI system is probably something like 200 GPa, but it could be 200,000 MPa. One must be careful because some disciplines develop the formulas in foot measurements when converting from SI to USC. Fortunately, the way the world is going, most conversions are from the USC system to the metric system.

The saving grace in all this is that whichever system you are working in, you can calculate the result in it and then compare what you get to the result you get in the system to which you are converting. This will essentially allow you to develop your own conversion factor for the combination of units to which you had converted the components. Here again, Mother Nature has been kind to us even if the measurement gurus have not. The stress, for instance, is the same order of magnitude no matter which set of units you calculate in.

When I was first learning how to do beam calculation, one of the problems given as an exercise was to calculate the size of a ladder rung that would hold a man of a certain weight on a ladder a certain distance wide. I had to calculate it in both the USC system and what was then the metric system. After I converted the weight from pounds to kilograms, converted the width from inches to millimeters, calculated the moment of inertia, and so forth, the size of the rung came out to 1 inch (or very close) in USC. To my surprise, the rung in millimeters was 25 (or very close), because in the calculation we used integer numbers in the weights, widths, stresses, and so forth, so the answers came out in whatever accuracy that the slide rules allowed. Nowadays, the same exercise would most likely give an answer for the rung diameter in several decimal places. The wise engineer would decide to make the rung 1 in. in diameter, and in metric make it somewhere near the standard size of round wood in his or her geographical region.

Two lessons were learned. One, Mother Nature doesn't really care what system of measurement you use. If your math is right you will get the same special diameter and you can call it what you want. Second, unless you are in some high-precision situation, you can pick the nearest standard size that is safe.

It is hoped that someday there will only be one set of unit-sized equipment. However, it is unrealistic to think that all of the older equipment will disappear overnight should that conversion occur.

The calculations will be done in both U.S. customary and metric units in any sample problems that are presented in this book, of course, when it is necessary to walk through the calculations. There are some that are self-evident and need not be done in detail.

CHAPTER 3

Selection and Use of Pipeline Materials

Contents

OVERVIEW

When one thinks of materials for use in the piping codes the usual thought is about the materials that make the pipe, fittings, and supporting equipment—the materials that the codes address. However, there are more materials than that to be considered.

The material that the piping will be immersed in is important. In aboveground piping, this is usually just air, and is not always significant. Even then one has to consider the environment—for example, the humidity levels and whether the location has extreme temperatures or winds. If the location is earthquake prone, that has a bearing on the design calculations and the construction.

Buried piping has another set of concerns. One has to know the topography and soil conditions that the pipeline is routed through. Usually there is need for some kind of corrosion protection. Does the route cross rivers, highways, canyons, or other things that can cause special problems?

All these questions must be considered, and they are not usually spelled out in the piping codes. They may be mentioned as things that must be considered; however, there is often little guidance. There is a whole new set of code requirements for offshore and underwater pipelines. The pipeline codes explain those requirements in detail.

One also needs to consider the fluid or material that the pipe system will be transporting. Often, the code's title is the only indicator of the fluid.

Piping and Pipeline Calculations Manual
ISBN 978-0-12-416747-6
http://dx.doi.org/10.1016/B978-0-12-416747-6.00003-6

19

B31.8 is specifically for gas transmission. Most people relate it to natural gas, however it is for gas transmission and distribution. This code does have specific requirements in it for certain gases.

As mentioned before, B31.1, Power Piping, is primarily involved with steam-water loops. It does, however, involve several portions of the power plant piping that do not carry steam, so there are considerations for those fluids.

The newest code, B31.12, Hydrogen Piping and Pipelines, is similar to B31.8 in that it is for a gas. However, it only involves one gas, hydrogen. In that sense it is like B31.8. On the other hand it is also somewhat like B31.3 in that hydrogen is also a process gas. The code is a hybrid in that it covers both the technical things that are unique to pipelines and also those aspects that are unique to process piping. Because hydrogen has effects like embrittlement on materials it comes in contact with and other disabling properties, it has different requirements than the two grandfather codes, B31.8 and B31.3. These will be discussed in more detail as we move through the book.

B31.3, because of its broad range of application to a variety of process industries, has the most information about transport fluids. It defines four types of fluid:

1. *Category D service.* These fluids must meet certain requirements and are basically low pressure, not flammable, and not damaging to human tissue.
2. *Category M service.* These fluids are the opposite of Category D fluids and therefore must be treated by separate requirements.
3. *High-pressure fluids.* These are fluids that have extremely high pressures as designated by the owner and have independent requirements.
4. *Normal fluid service.* This is not your everyday normal Category D fluid service, but it does not meet the requirements in 1, 2, or 3, and is generally called the "base code." One can use this base code for Category D fluids, as it is sometimes simpler when Category D service covers the entire project.

SELECTION OF MATERIALS

By and large what type of fluid is used for a project comes as a given. The specifier or designer then chooses an appropriate material to handle that fluid under those conditions. In general, codes do not have within their scopes which material should be used in which fluid service. Each code will have a listed set of materials and their allowed stresses regarding the temperature of the proposed system. However, they may limit which materials can be used in certain system operation conditions, like severe cyclic conditions or other effects that must be considered. Many of these do not give specific methods to make those considerations. Some methods are discussed later in this chapter.

At this point, given a fluid and the need to calculate which piping material should be used, there comes a little bit of interaction with regard to sizing the pipe. This is especially true when there is the opportunity to have more than one operating condition in the life of the system. In these multiple-operation situations, a series of calculations must be made to find the condition that will require the thickest pipe and highest component pressure rating. For instance, it is possible that a lower temperature and a higher coincident pressure may result in use of heavier pipe than a higher temperature and a lower pressure. This combination may not be intuitively obvious. Such considerations will be discussed and demonstrated in much more detail in Part II and the Appendix.

The sizes required may have an effect on what materials are available. All components may not be available in materials compatible with the chosen pipe materials. This conundrum was common when higher-strength, high-temperature piping was developed in the late 1990s for high-temperature service. Material to make components compatible with the new piping was not readily available for several years.

It is also true that when newer materials are developed the fabrication skills and design concerns take a little time to develop. New techniques are often required to achieve the same net margins one is used to with the older materials. That and similar problems explain why the adoption of new materials proceeds at a less-than-steady pace.

Having explained generically some of the potential issues with materials, we can turn our attention to the materials of construction for a pipe system. Each code has what is generally called *listed materials*. These are materials that the various committees have examined and found to be suitable for use in systems for the type of service that that book section is concerned with. It stands to reason that those books that work with a wider variety of materials have more types on their "preferred" list.

ASTM AND OTHER MATERIAL SPECIFICATIONS

In piping these are usually ASTM grades for materials. For ferritic steels, they usually are ones from ASTM Book 1.01. In many instances, there is also a list of API 5L piping materials. One major exception is boiler external piping, listed in B31.1, which requires SA materials rather than ASTM.

It is basically true that one can substitute SA for ASTM materials of a similar grade. The SA materials are often the same as ASTM materials of the same grade, as in SA-515 or A-515. Section II of ASME's Boiler and Pressure Vessel Code (BPVC) is the materials section; it reviews the ASTM materials as they are developed for applicability to the boiler code.

There is a little hitch that always occurs when one standards-writing body adapts or references another's standard for their purposes—a time-lapse problem. If standard group A issued a change to their standard, the adopting group B cannot really study it for adoption until after the publication date. And then they can't necessarily adopt it in time for their next publication date, which is most likely to be out of sync by some amount of months or possibly years with the change. So the lag exists quite naturally.

In addition, sometimes the change made by group A is not necessarily totally acceptable to group B. Specifically for the SA/ASTM problem, there are some SAs that say they are the same as the ASTM of a specific edition with an exception. Or they might just keep the earlier edition that they had adopted.

Because of this inherent lag, standards groups spend a fair amount of effort letting you know which edition of a standard they have accepted for use in that code. Typically, B31 and other standards will list the standard without an edition in the body of their code. Then they will offer an appendix to the code that lists the editions that are currently approved. Every attempt is made to keep the inherent lag in timing to a minimum.

In recent years, ASME has determined that they can trust other ASME standards. In the appendix the standards that are trusted include an instruction to use the latest edition rather than listing the approved edition. This saves a lot of time for those in the committees whose responsibility it is to review the editions.

In addition to these listed materials, sometimes unlisted materials are accepted with certain limitations. Also, some standards discuss unknown materials and used or reclaimed materials. Table 3.1 shows what each B31 book section says in general about this subject.

Other standards have materials requirements that often point back to ASTM or an acceptable listing in another standard. This helps to eliminate duplication of effort and the lag problem is again minimized. Some standards develop their own materials. The most notable of these is MSS SP-75,

Table 3.1 Unlisted Materials

Book	Listed	Unlisted	Unknown	Reclaimed or Used
B31.1	Yes, including SA	Yes, with (non–SA) limitations	No	Not allowed
B31.3	Yes	Yes, with limits	No	Yes, with limits
B31.4	Yes	Yes, with limits	Yes, with limits on fluids	Yes, with limits
B31.5	Yes	Not addressed	Not addressed	Yes, with limits
B31.8	Yes, with specific types	Addressed in types	Addressed in types	Yes, with limits
B31.9	Yes	Yes, with limits	Structural only	Yes, with limits
B31.11	Yes	No	Not addressed	Yes, with limits
B31.12	Yes	Yes, with limits	No	Yes, with limits

which has a material called WHPY that has a defined chemistry and other mechanical properties.

LISTED AND UNLISTED MATERIALS

The listed materials are those in the B31 books, which include the allowable stresses at various temperatures for the materials that they have listed. These tables are necessary because there is a wide range of temperatures utilized in the systems based on these books. Over a wide range of temperatures the yield and ultimate strengths will be lower than at the ambient temperatures. In addition, at some temperatures, time-dependent properties, such as creep and creep rupture, become the controlling factor.

To establish the allowable stresses at a specific high temperature could require expensive and time-consuming tests. ASME determined a method that, while it doesn't completely eliminate the tests, reduces them to an acceptable level. It uses this method to establish the allowable stress tables.

In cases where the material one wants to use in a project is not listed in the particular code, the first step is to determine whether that code allows the use of such a material. Some guidelines of where to look are in Table 3.1.

The careful reader will note that Table 3.1 only lists B31 piping sections. There are of course several other piping codes in the world. As noted some refer to B31 codes, however, many are standalone for that country or region. It is not a difficult reach to believe that the separate codes have their own "listed materials" that would be unlisted in another code. Those listed materials would be from a materials standard from that region. There are even several other material standards in the United States.

A discussion of the process in the B31.3 code to establish the applicability of using an unlisted material in that code follows. This book cannot describe the process in a myriad of other codes. Suffice it to say that the process described below covers the essential elements that one should be concerned about when not using a listed material for the code for which he or she is working.

For those who lack expertise, the choosing of a different material or even knowing details of the multitude of materials is a daunting process. ASTM has a handbook that compares world steel standards (ASTM DS67, which includes A, B, C, etc., for the various editions). Its editor is a well-known materials expert. In this book he points out that for any two steels there are rarely equivalent materials. He uses the terms comparable and closest match as the means to choose between materials.

This strongly implies that there are different reasons for choosing certain materials. In some instances the first choice criteria might be the mechanical properties, such as tensile or yield strength. They, along with time-dependent properties such as creep, are the basis in most design codes for establishing allowable stress. The comparison might go so far as to detail the many different testing methods used to establish those values. At some point in time the attention to such details are in the province of the metallurgist rather than the designer.

In other instances, the chemical composition may be the first criteria for choosing one material over the other. These would be situations where corrosion caused by the fluid being transported or the atmosphere in which the piping is situated is more important. For instance, in general the paper industry has low-pressure applications with highly corrosive fluids. To be safe with standard sizes it would be advantageous to have a more corrosive-resistant chemistry than a higher tensile material.

The various material standards in the world are listed below with a brief notation as to their country or region. This listing gives an idea of the complexity of the "unlisted choices" even before one establishes the proper values to use when choosing a material for design purposes.

- ASTM: This is basically an international standard.
- API: This is also international in use.
- AFNOR: This is a French material standard.
- BSI: This is a British standard.
- CEN: This is a European standard.
- CSA: This is a Canadian standard.
- DIN: This is a German standard.
- ISO: This is the world or international standard.
- JIS: This is the Japanese standard.
- SAE:T his an automotive (fasteners, etc.) standard.
  And in fact there are many more.

B31.3 is the most adaptable to unlisted materials, so a brief discussion of that procedure is given. It is important to note that the code does not give one license to use it in compliance with other codes; however, it is a rational method to determine acceptable stresses for temperatures where there isn't a published table of allowable stresses.

The nonmathematical part is to select a material that is in a published specification. The odds of a material being in a published specification are quite good because of the proliferation of national or regional specifications that for one reason or another have not been recognized by the codes in either direction. That is to say, the code from one country does not specifically recognize another country's or region's material specification. There is progress in the direction of unifying these different specifications, however it has been slow.

To be useful, they must specify the chemical, physical, and mechanical properties. They should specify the method of manufacture, heat treat, and quality control. Of course, they also must meet in all other respects the requirements of the code. Once the material is established as acceptable, the next priority is to establish the allowable stress at the conditions, particularly temperatures in which the material is intended to be used.

This discussion assumes one is intending to use that material at a temperature that is above the "room" temperature or temperature where normal mechanical properties are measured. Measuring mechanical properties at higher temperatures is expensive and can be very time consuming if one is measuring properties such as creep or creep rupture. The ASME code, recognizing that this process is difficult, developed a trend line concept to avoid requiring such elevated-temperature mechanical tests for each batch of material made, as is required for the room temperature properties. This is called the *trend curve ratio method.*

The method is relatively straightforward. Some of the difficult extended temperature tests have to be conducted. While as far as is known there is no set number of tests, it stands to reason that there should be more than two data points to ensure that any trend line that is not a straight line will be discovered from the data points. It also stands to reason that the temperature range of the tests should extend to the highest temperature for which the material is used. This eliminates extrapolating any curve from the data and limits any analysis to interpolation between the extreme data points, which is just good practice.

Obviously, if the intended range extends into the creep or creep rupture range, those tests should be run also. This decision becomes a bit of a judgment call. As a rule of thumb the creep range starts at around 700°F or 371°C. However, depending on the material, that may not be where those temperature-dependent calls control the decision.

So now one has a set of data that includes the property in question at several different temperatures. For purposes of illustration, we make an example of a set of yield stresses. This is not an actual material but an example. The data for listed materials can be found in ASME Section II, Part D, and these are already in tables so there is no need to repeat that data here. We will call this material Z and the necessary data to establish the trend curve ratio are listed in Table 3.2.

Table 3.2 Material Z Test Data for Trend Curve Ratio

Room Temperature, °F	Tested Yield, kpsi	Ratio to Room Temperature Yield
70	32	1
100	32	1
300	29	0.906
500	24	0.750
600	20	0.625
800	10	0.3125

Given these tables, a regression on the temperature versus the computed ratios can then be established. It should be noted that the original data might be in the same degree intervals that the table is intended to be set up in, but in general this is not the case. Therefore, a set of data that ranges from the room or normal temperature to the highest intended temperature can then allow a regression that is basically interpolative rather than extrapolative. It is unlikely that the material supplier has test data at the exact temperature at which one is going to use the material.

One might note when delving into Section II of the boiler code, which is the basic material and stress section, that these yield temperature charts rarely go above 1000°F. This is due to the fact that this is a temperature that is usually within the creep range and that yield is the less-dominant mechanical property. Yield above that temperature is not as critically needed.

Regardless, the regression supplies formulas that allow one to predict the yield at any intermediate temperature. For the previously presented data, one regression is a third-degree polynomial that has a very high correlation coefficient. That formula is

$$\text{Ratio at temperature}\left(R_y\right) = 1.00361 - (2.08E - 0.06)T$$
$$- (9.5E - 0.07)T^2 - (1.58E - 10)T^3$$

One might think that the latter terms might be ignored, but if one thinks of, say, a temperature of 500, that 500 is cubed; therefore, that small constant changes the yield by over 500 psi in the current example, and that is a significant change in stress.

This explanation applies to the method ASME has developed to avoid the requirement for each batch of material to go through extensive high-temperature testing.

A test of tensile strengh and yield at room temperature (generally defined as 70°F or 20°C) satisfies the requirement. The temperature values are that room temperature value multiplied by the appropriate temperature, R_y or R_t. The same general technique is used for both yield and tensile properties.

ALLOWED STRESS CRITERIA FOR TIME-DEPENDENT STRESSES

The other criteria for establishing allowable stresses are that of creep and creep rupture. The criteria involve a percentage of creep over a length of time. These have been standardized in ASME as the following values:

1. 100% of the average stress for a creep rate of 0.01% per 1000 hours. This can be described as causing a length of material to lengthen by 0.01% in 1000 hours when a steady stress of a certain amount is applied at a certain temperature. Obviously this requires many long tests at many temperatures and many stresses.

2. 67% of the average stress for a rupture at the end of 100,000 hours. Once again, many stresses at many temperatures are tried until the part breaks or ruptures.

3. 80% of the minimum stress for that same rupture. Again, many stresses at many temperatures are tried.

These criteria are basically the same over all the ASME codes. The double shot at the rupture criteria (2 and 3) comes about to eliminate any possibility of having a test that gives a wide variability of highs and lows. It is essentially an analogy for having a rather tight standard deviation in the data. One can also assume that there are expedited testing methods for the creep-type tests. A full-length test of 100,000 hours would last over 11 years and several different stresses would have to be tested. Even a full 1000–hour test would take over 41 days.

Having assembled all that data, the decision for any given temperature is then made to allow the lowest stress. The *tensile stress* has a percentage applied to it that is set, as much as possible, to ensure that the material has some degree of ductility. The main stress factor is *yield stress*. The percentage of yield that is allowed is dependent on the code section. Generally, the two most often used criteria are 67% of yield and a divisor of 3.5 on the ultimate tensile stress, all at the desired design temperature. The creep criteria are included in this survey, and the one that yields the lowest stress is established as the *allowable stress* at that temperature.

This is not true in the books where the applications have a limited range of operating temperatures, mostly in the pipeline systems. In those, they simply set the specified minimum yield of the material as the base allowable stress. Their calculation formulas then have a few variable constants based on the pipeline's location class and the temperature and any deviation for the type of joint that is employed in making the pipe. It is noted that the temperature range for pipe containing natural gas or hydrogen, for instance, would be quite small. On the other hand, that pipeline can go through a wide variety of locations. These locations might vary from, say, a desert, to the downtown of a large community. It is easy to determine that the margin of safety to avoid calamity at those two extremes would be quite different.

STRESS CRITERIA FOR NONMETALS

When one considers nonmetals, the presentation of stresses is considerably different. Nonmetals have a much wider set of mechanical properties with which to contend. There are several types of nonmetallics. Those recognized by the various codes are thermoplastic, laminated reinforced thermosetting resin, filament-wound and centrifugally cast reinforced thermosetting resin and reinforced plastic mortar, concrete pipe, and borosilicate glass. One doesn't need to be an expert to recognize that they represent a wide range of reactions to stress or pressure. The allowable stresses are set this way as well. For instance, B31.3 refers to five different stress tables for the above-mentioned materials. A brief listing of how those tables vary is as follows:

1. The thermoplastic pipe table lists several ASTM designations and allowable stresses over a limited temperature range for each ASTM designation. It is the most like the metal tables.
2. The laminated reinforced thermosetting pipe table lists an ASTM specification with a note stating the intent is to include all of the possible pipes in that specification. That specification gives allowable usage information.
3. The filament-wound materials (e.g., fiberglass piping) table lists several ASTM and one American Water Works Association (AWWA) specification with the same note as that in item 2.
4. The concrete pipe table lists several AWWA specifications and one ASTM, and it states the allowable pressure for each pipe in the specification. The specification itself defines the controlling pressure-resisting dimensions and attributes, eliminating the need for any wall thickness calculation.

5. The borosilicate glass table lists one ASTM specification and an allowable pressure by size of pipe.

B31.3, which for now is the only high-pressure design for pipe code (B31.12 is for high pressure but for a single fluid), has a separate allowable stress table for the limited number of metals that are recognized for use at those high pressures. Those tables do have an unpredictable difference in allowable stress values for common temperature. Like everything in the chapter, it is mandatory to comply with the code once the owner has defined a piping system as a high-pressure system. Many times it is asked: What is high pressure? The general requirements are that it can be anything, with no specific lower or upper limit. It is high pressure only if the owner specifies it as so. For the purposes of writing the chapter, the committee used the definition of any pressure at a certain temperature in excess of the pressure at that temperature for the material that is defined as Class 2500 in the ASME B16.5 pressure-temperature charts.

CORROSION AND OTHER FACTORS

A main remaining consideration in material selection is what is called the material deterioration over time, commonly referred to as corrosion allowance. This corrosion can occur on the outside of the pipe due to the environment the pipe is in, and can come from the inside due to the fluid and the velocity and temperature of that fluid.

The acceptable amount of corrosion allowance is dependent on the rate the corrosion will occur over time and the expected lifetime of the particular system. The calculation after the corrosion allowance is set is addressed in Chapter 5 when we calculate pressure thickness. Setting that allowance is outside the scope of the codes. There is a suggestion in B31.3, Appendix F, Precautionary Considerations, that points the reader to publications such as "The Corrosion Data Survey" from the National Association of Corrosion Engineers. This would help guide the setting of corrosion allowance.

The Appendix contains a list of common materials from U.S. ASTM Book 1.01, which lists the vast majority of the materials used in piping. As was mentioned, ASME has a Division 2 listing of materials, which have an SA or SB designation. By and large, these have been adopted as ASTM materials. Some have restrictions on elements like the chemistry, or some other portion of the current ASTM material may be invoked when adopting them. Those restrictions are noted in the listing. The primary purpose of these materials is for use in the boiler code sections; therefore, they are not treated in this piping-related book more than they have been already.

There are materials standards from other parts of the world. Some of the considerations regarding their use as unlisted materials were previously discussed. Many of them are similar to ASTM materials, but some are quite different. It appears on cursory examination that often these standards have a greater number of micro-alloyed materials. The mélange of materials has not been resolved into some simple "these are the materials of the world" standard. There is considerable work going on in that area, but it might take a long time to get to the finish line in that effort. For those who feel the need, there are books that attempt to be conversion sources to compare world materials—for example, Stahlschussel's *Key to Steel*. It is quite expensive and very detailed, and works primarily with European steels, but lists many regional steels. I have used it with success in untangling the web of various steels.

There are a few standards that discuss hydrogen embrittlement. The industry is aware of the phenomena and it has been researched. There is some ASME material available relative to use with fracture mechanics. The publishing of B31.12 has encouraged some more recent research on the rate of deterioration of piping materials exposed to hydrogen. The hope is to find a reliable predictive method or test to include in the code.

There is a little more to consider in preparing to do the calculations required by or suggested by the codes: the business of sizing the pipe for a particular system. This includes the flow in the system and the attendant pressure drops, which, as mentioned, are not really a code-prescribed concern. However, a basic understanding of the methods employed in this process is background for the user of the codes and as such is addressed in Chapter 4. A description of the calculations and examples with certain parameters are given rather than an explanation of the development of those parameters.

The reader will note that the metals listed as acceptable are often ASTM standards. One of the interesting things about ASTM steels is that they are segregated into different forms. The steel might have almost exactly the same chemistry, and therefore in the casual reader's eye be the same material. This could be considered true. Certainly, it is true if the various elements in the steel are within the chemical tolerance of the specification for the particular form being reported. However, the chemistry is not the only thing that ASTM and other standards would specify. The major forms of the same material would most likely have different mechanical properties and minimum stresses. Those things depend to an extent on things like the method of manufacture and post-manufacture treatment, as well as the chemistry. It is true that chemistry is the main ingredient; however, the other factors will make a difference and that is why the same chemical material would have a different number depending on the form the material takes—pipe, plate, or forging or casting.

CHAPTER 4

Piping and Pipeline Sizing, Friction Losses, and Flow Calculations

Contents

OVERVIEW

After reading this chapter, you should be acquainted with the complicated field of fluid flow or, as it is known, *fluid mechanics*. You will be aware of the basics and have an understanding of the important issues in this discipline. If you choose to delve deeper into the subject, Elsevier has many titles to choose from that can give you more understanding.

For the most part the following issues will be treated as givens in the final design and erection of a system of pipes: fluid, pressure, and temperature, and how they will vary during the life of the process that is involved. The givens may also include which material is appropriate for this system.

Necessarily, there is often some interaction in the early stages of establishing these givens. As the project is in its formative stage certain trade-offs are made, including considerations from an economic point of view to establish the cost/revenue returns the project might require. Often these trade-offs involve fluid mechanics considerations.

It is the intent of this book to provide a level of understanding of those fluid mechanics considerations to the subsequent systems designer. Understanding how they may have arrived at a certain set of givens makes

Piping and Pipeline Calculations Manual
ISBN 978-0-12-416747-6
http://dx.doi.org/10.1016/B978-0-12-416747-6.00004-8

the business of moving forward somewhat easier. At the least, one can move forward with more confidence.

FLUID MECHANICS CLASSES

There are two major classes of fluids. The first is *incompressible fluids*, which are generally liquids. The second is *compressible fluids*, which are generally gases. We discuss the incompressible fluids class first, as many of the techniques are transferable from that type to the compressible fluids class. In fact, we find that in some instances that some compressible fluids can be treated as incompressible. There are other differences that we will discuss as well.

There are differences within each of the classes, which we will point out as the chapter progresses. For instance, in incompressible fluids there are Newtonian fluids and non-Newtonian fluids. In compressible fluids there are the perfect gas laws and the degree that the fluid differs from a perfect gas.

In all cases some calculation procedures are given and explained. Many of these procedures are complex. In some cases a simpler, less accurate or precise procedure is pointed to for simple rule-of-thumb calculations or ballpark estimates. When appropriate, charts and graphs are provided in the Appendix for many of the issues.

Since this is basically a manual, readers who already feel familiar enough with the fluid mechanics field may skip this chapter. There is little in the other chapters that will require the calculations given here. In most cases these givens are brought to the table when performing the other calculations. If necessary, the reader is referred back to this chapter or the appropriate chart or graph in the Appendix.

Now we must familiarize ourselves with the fluid mechanics terms. Following is a discussion of the less common terms along with a short description of that characteristic of the fluid. Those discussed are important to successful calculations. Where appropriate, there are some supporting calculations. At the end of the list there are examples that put it all together for a small piping system.

VISCOSITY

The short definition of *viscosity* is the resistance of a fluid to flow. Many of us are familiar with the expression "as slow as molasses in January." This of course has more meaning to those who live in northern climates, where

January is often very cold. Its deeper meaning is that the resistance to flow is dependent to a great degree on temperature. It has, for the most part, very little dependence on pressure.

A more scientific definition of viscosity involves the concept of fluid shear. Many readers who have worked with metals or other solids understand shear as the force that causes a material to break along a transverse axis. Fluid, being fluid, doesn't really break—it moves or flows. Naturally, being a fluid, it has to be contained, say in a pipe, and when the force is along the free axis of the containment, flow occurs.

The containment material has some roughness on its surface that causes the fluid to "drag" or move more slowly on that surface and more rapidly as it moves away from that surface. The net result is that for any small section of the fluid, the velocity pattern is a parabola.

There are two basic measures of viscosity. The first is *kinematic viscosity*, which is a measure of the rate at which momentum is transferred through the fluid. The second, *dynamic viscosity*, is a measure of the ratio of the stress on a region of a fluid to the rate of change of strain it undergoes. That is, it is the kinematic viscosity times the density of the fluid. Most methods of measurement result in dynamic viscosity, which is then converted by dividing by the density when that is required.

We use the following symbols in this book:

- Dynamic viscosity, μ
- Kinematic viscosity, ν
- Density, ρ

Therefore, the basic viscosity relationship is

$$v = \frac{\mu}{\rho}$$

Example Calculations

The dynamic viscosity of water at 60°F is 2.344; the units are lbm.s/ft^2 (pounds mass per second/ft^2) $\times$ 10^{-5}. You will notice the lb has an m, which means those units are in slugs, or what we normally think of as weight divided by the acceleration due to gravity (which for engineering purposes can be 32.2 ft/sec^2).

The density of water in slugs at 60°F is 1.938, which means that the specific weight of water at that temperature is calculated as 62.4.

Therefore, the kinematic viscosity of water at 60°F is 2.344/1.938, which comes out to 1.20949 on a calculator. Those units are ft^2/sec $\times$ 10^{-5}. It should be noted here that a table of viscosities would most likely note 1.20949 as 1.210.

The same procedure in the metric system would most likely give you the following numbers at 20°C, which is the nearest even degree for the Celsius scale. One could do some interpolation between, say, 10 and 15, but the changes are not necessarily linear, so the calculation is more complex and there is some concern about the necessity for increased accuracy in a rough calculation.

- Dynamic viscosity = 1.002 N.s/m^2 × 10^{-3}
- Kinematic viscosity = 1.004 m^2/s × 10^{-6}
- Water density = 998.2

With the preceding we begin to see some of the differences between the U.S. customary system (USC) and the metric system. Numerically, the metric system is all about shifting the decimal point. The major difference between dynamic and kinematic viscosity is the -3 and -6 exponents of the numbers. The density doesn't change much with the design of the system.

To say that the U.S. customary system was designed is to stretch one's credibility. The units tend to stay the same size, but there is little or no numerical significance. It is interesting to convert from one to the other system after calculating. However, in converting final calculations from charts one must be sure that the temperatures are the same.

On many charts for water the only temperature that is the same is the boiling point, or 100°C and 212°F. At those temperatures the kinematic viscosities are 0.294×10^6 for metric and 0.317×10^5 for USC. The conversion factor from ft^2 to m^2 is 0.093, and in the other direction it is 10.752. The respective kinematic viscosities for metric are 0.294×10^{-6}, which converts to 0.316×10^{-5} against a 0.317 on the comparison chart. For USC, it is 0.317, which converts to 0.295×10^{-6}. The error is very small.

This gives readers an idea of why the business of fluid mechanics, as well as moving between metric and USC units, is computationally complex. And we have not even discussed the many different forms of viscosity units that exist. The Appendix contains a discussion and conversion details for many of those units. Computer-oriented conversion programs usually work better for things like viscosity. Between dynamic and kinematic viscosity, the author found at least 25 different ways to express the measure (poise, centipois, stokes, and several different measures like square inches or square millimetres per hour or second and so forth.

It also begins to explain why techniques such as CFD (computational fluid dynamics) programs and their skillful users are in demand. The programs are essentially finite analysis programs and beyond the scope of this book. Suffice it to say, this is not where the non–fluid mechanic wants to spend much time in turning the crank, which explains many if not all the charts, graphs, and other assists that are available. However, we have other fish to fry before we leave our discussion of fluid mechanics.

REYNOLDS NUMBER

The Reynolds number gets its name from Osborne Reynolds, who proposed it in 1883 when he was 41 years old. It is a dimensionless number that expresses the ratio between inertial and viscous forces. This set of dimensions often occurs when one is performing a dimensional analysis of fluid flow as well as in heat transfer calculations.

The number in relation to flow defines the type of flow. There are several types of flow with a low Reynolds number (R_e) when the viscous forces are dominant. This is characterized by smooth, more or less constant fluid flow. As the Reynolds number gets higher, the inertial forces begin to dominate and the flow then becomes turbulent. This flow is characterized by flow fluctuations such as eddies and vortices.

The transition from laminar to turbulent does not occur at a specific number. It is gradual over a range where the types of flow are mixed up and in general the Reynolds number becomes indeterminate as far as being a reliable indicator as to what happens in the pipe or conduit. This range is not specific, but in general is $R_e > 2000 < 5000$.

In its simplest form for flow in pipes the Reynolds number is

$$R_t = \frac{VD}{v} \tag{4.1}$$

where

V is the velocity, ft/sec or m/sec
D is the internal diameter of pipe, ft or m
v is the appropriate kinematic viscosity, SI or USC
Since we know the relationship of dynamic viscosity to kinematic viscosity, Eq. (4.1) can be rewritten in terms of the dynamic viscosity as

$$R_t = \frac{\rho VD}{\mu} \tag{4.2}$$

where one just substitutes the density and dynamic viscosity. Since you need to know the density to use this equation it is simpler to compute the kinematic viscosity and use Eq. (4.1). As always, it is prudent to keep consistent measurement systems.

Example Calculations

Using the kinematic viscosities of water found previously in the "Viscosity" section, and adding the information that we are using an 8 NPS schedule 40 (S40) pipe with water flowing at 7 ft/sec, we make the following calculations:

$$8 \text{ NPS S40 pipe ID} = 7.981 \text{ in. or } 0.665 \text{ ft or } 0.203 \text{ m}$$

$$7 \text{ ft/sec} = 2.13 \text{ m/sec}$$

$$\text{Kinematic viscosity at } 60°\text{F} = 1.210 \text{ USC}$$

Kinematic viscosity at $20°\text{C} = 1.004$; at $10°\text{C} = 1.307$, both at 10^{-6}

The USC Reynolds number is

$$7\left(\frac{0.665}{1.2105 \times 10^{-5}}\right) = 384,711$$

The SI Reynolds number is

$$\text{At } 20°\text{F} : 2.13 \times 0.203/1.004 \times 10^{-6} = 430,667$$

$$\text{At } 10°\text{F} : 2.13 \times 0.203/1.307 \times 10^{-6} = 326,167$$

Interpolating up as $60°\text{F} = 15.55°\text{C}$, one calculates R_e as 384,269.

This basically shows that by using the appropriate units in either system one will get the same Reynolds number (i.e, it is dimensionless). It is important to be sure to convert the temperature exactly. One would get a slightly different number if the interpolation were made on the kinematic viscosity.

As one might expect about something that has been around since 1883 there are many forms of the Reynolds number, but they all eventually boil down to these results, and the other forms are left to your exploratory inclinations.

FRICTION FACTOR

The drag of a fluid at the contact between the fluid and the container (mostly pipe in this discussion) is caused by what is called a *friction factor*. In fluid mechanics there are two major friction factors: the Fanning friction factor and the Darcy–Weisbach factor, which is sometimes called the Moody friction factor.

The two factors have a relationship where the Darcy factor is four times larger than the Fanning factor. This can cause confusion when using the factor. It is important to be certain which factor one is using, or the answer one achieves will not be correct.

In laminar flow the factor doesn't change over the range of laminar flow, so when one is using a chart or graphical solution it is fairly easy to determine which factor is presented. The Fanning factor in laminar flow is

$$\frac{16}{R_e}$$

while the equation for the Darcy factor is

$$\frac{64}{R_e}$$

So it is easy to determine which factor one is using. If one is using a chart, simply read the factor for an R_e of 1000, and then you will read either the decimal number 0.064 or 0.016, which will give you the factor being used. The factor used changes the form of the head loss equation that one uses to calculate the pressure drop in a pipe section or line.

It is common for chemists to use the Fanning factor, while civil and mechanical engineers use the Darcy factor. So if you are a civil engineer and get a Fanning factor chart, multiply the factor by 4 and you will have the factor you need, or use the Fanning formula for head loss.

The two equation forms used with the proper form of the head loss equation will give the same loss for that line segment of pipe:

Darcy–Weisbach:

$$h_f = f \frac{L}{D} \frac{V^2}{2g} \tag{4.3}$$

Fanning:

$$h_f = \frac{2fV^2L}{gD} \tag{4.4}$$

where

    L is the length of straight pipe, ft or m
    D is the pipe interior diameter (ID), ft or m
    V is the average velocity of the fluid, ft/sec or m/sec

G is the acceleration of gravity, in the appropriate units

F is the appropriate dimensionless factor for the form being used

h_f is the head loss, ft or m

In both cases all symbols are the same except for the f factor, which changes.

Example Calculations

Use the pipe and velocity in the Reynolds number example (i.e., pipe 8 NPS S40 and 7 ft/sec velocity) and the appropriate SI dimensions.

- The acceleration of gravity is 32.2 ft/sec² in USC and 9.81 m/sec² in SI.
- The length of pipe is 100 ft or 30.5 m.
- $R_e = 1000$.

Darcy-Weisbach calculations:

$$\text{USC}: h_f = 0.064 \times (100/0.665) \times \left(7^2/2 \times 32.2\right) = 7.32 \text{ ft}$$

$$\text{SI}: h_f = 0.064 \times (30.5/0.202)\left(213^2/2 \times 9.81\right) = 2.23 \text{ m}$$

Fanning calculations:

$$\text{USC}: h_f = 2 \times 0.016 \times 7^2 \times 100/(32.2 \times 0.665) = 7.32 \text{ ft}$$

$$\text{SI}: h_f = 2 \times 0.016 \times 2.13^2 \times 30.5/(9.81 \times 0.202) = 2.23 \text{ m}$$

From the example, the formulas give the same answers in both unit systems and either equation form if the appropriate factor is used with the form. Chemists and civil engineers will get the same answer whichever method they choose.

This example was for a laminar flow regime; most regimes are not in laminar flow. In the case of turbulent flow the calculation of the factor is not so simple, which was one reason that Moody, for whom the Darcy factor is sometimes named, developed his graph. This was for many years the preferred way to establish the factors. The graph is developed for both the Darcy form and the Fanning form. In the remaining chapters, we will work with the Darcy factors and forms. The graph in the Appendix is presented mainly for reference.

The advent of computers and calculators has reduced by a significant amount the work involved in calculating this factor. This is because the calculating equations involve what used to be tedious work, like computing logarithms or making an iterative calculation. Both are done much more simply by today's electronic wizardry.

The base equation is known as the Colebrook equation, which was developed in 1939. It is a generic equation and is based on experiments and other studies, but it can be used for many if not all fluids in the turbulent region. It is not useful for laminar flow, and as discussed, for it to be effective one must first calculate the Reynolds number.

One drawback of the equation is that it has the unknown factor f on both sides, so it must be solved iteratively. For those who have an Excel-type spreadsheet with a goal seek tool, this is not as difficult as it used to be. The equation is

$$\frac{1}{\sqrt{f}} = -2 \log_{10}\left[\left(\frac{1}{3.7}\right)\left(\frac{\varepsilon}{D}\right) + \frac{2.51}{R_e\sqrt{f}}\right] \tag{4.5}$$

where the symbols have the previous meanings given with the exception of ε, which is the roughness factor for the pipe material.

For reference, a roughness factor for new steel pipe is 0.00015. As might be expected, this is not a precise factor. It is a reasonable estimate for a particular material. Several materials have different factors and some sources give different estimates. A table of reasonable factors used in this book and by several sources is given in the Appendix.

One way to calculate the factor in spreadsheet form is to make a column for all the variables in the formula. Set up three different cells. In one cell set the formula for

$$\frac{1}{\sqrt{f}}$$

In the other cell set the formula for the right side of the equation. Then in the third cell set the difference between the two cells. Then use the goal seek function to make that third cell zero by changing the input cell for f. This will let the computer do the iteration. If your spreadsheet doesn't have the goal seek function, you can perform the iteration manually by changing the cell for the variable f. A sample spreadsheet layout is given in the Appendix.

Example Calculations

Using a roughness factor of 0.00015 and the diameters and Reynolds numbers calculated previously for a speed in the turbulent regime, the Darcy factor is calculated as follows (for USC): The friction factor using the spreadsheet method described calculates to 0.016032 with the difference between the two sides at 2.9×10^{-5}.

Before spreadsheets were developed there was a need to find a direct solution to the Colebrook equation. This is the sort of thing that mathematicians do—fiddle with expressions to make them either simpler or more difficult. In this case, at the price of some accuracy, another equation was developed. When a statement at the price of some accuracy is utilized one must recall that this may not be a major problem given such things as the uncertainty of the roughness factor that was used in the original calculation. In fact, the natural deviation between the two is quite small and for all practical (engineering) purposes, zero. This equation is known as the Swamee-Jain equation:

$$f = \frac{0.25}{\log_{10}\left[\left(\frac{e}{3.7D}\right) + \frac{5.74}{R_e^{19}}\right]} \tag{4.6}$$

When one computes this in USC units the factor calculates to 0.016108, which is a minuscule 0.000076 difference and far inside the probable uncertainty of the roughness. This uncertainty is expected to be in the +10 to −5 percent range.

There is another relationship that can be used: the rough-and-ready relationship. Chemists deem it sufficient for plant construction and calculations. It can be found in *Perry's Handbook* so one must recall that it is in Fanning factor form. For purposes of this book it has been converted to the Darcy format; the formula is

$$f = \frac{0.04}{R_e^{0.16}}(4) \tag{4.7}$$

As such, it calculates to 0.0204 as opposed to the more exact calculations presented before with Colebrook and Swamee-Jain. It is conservative in that it is approximately 25 percent high. This higher factor would present a need for either higher pumping energy or larger pipe. However, it can be a very quick field-type estimate that would rarely if ever be low.

It must be pointed out that all of the previous equations and discussions relate to the line flowing full. That is, it is assumed that there is an incompressible fluid touching all of the inside surfaces of a round pipe. This is not always the case in the real world. The problem is handled by introducing the concept of equivalent diameter, or as it is technically known, hydraulic radius. This will be discussed later in this chapter.

This then is the process for straight pipe. But how does one handle the pipe for situations where valves, elbows, tees, and other elements are added to that pipe? This is covered in the next section.

EQUIVALENT PIPE LENGTHS

The previous discussion covered calculating the friction and head loss for straight pipe. However, any pipe system has elements in it that also add friction, such as valves, fittings, entrance changes in direction, and so forth. So a method is needed to work with those sets of frictions as well. Basically the solution is to compute an equivalent length of pipe for each of these elements and then add them to the length of straight pipe.

The question then becomes, how does one do that? Recall Eq. (4.3) in the last section that calculated the friction loss in a section of straight pipe. It looked like this in the Darcy-Weisbach form:

$$h_f = f\frac{L}{D}\frac{V^2}{2g}$$

It is relatively simple to break the formula into two parts. The last part of the right side is

$$\frac{V^2}{2g}$$

which is known as the velocity head. The rest of the right side is basically the friction component per length of pipe. The method is to simply replace that with a new factor, often called K or the resistance coefficient.

Manufacturers and others have run tests and developed the K factor for their product, or one can use common K factors (see the Appendix). Multiply the appropriate K factor by the velocity head and you have an expression for the head loss for that element.

If the run is horizontal, all the elements and their respective K factors can be added and then multiplied by the velocity head to get the total head loss for that horizontal run. Elevation losses need to be added separately. If there is a need to calculate the equivalent length, one can just substitute the head loss achieved by the K factor method and solve for L in Eq. (4.3).

Example Calculations

Assume a fully open globe valve is in the line we have been working with (i.e., 8 NPS S40 with a velocity of 7 ft/sec). The common K factor for such a globe would be 10. One might get a different number from a specific manufacturer.

$$\text{USC: Head loss} = 10 \times (7^2/2 \times 32.2) = 7.608 \text{ ft}$$

$$\text{SI: Head loss} = 10 \times \left(2.13^2/2 \times 9.81\right) = 2.312\,\text{m}$$

$$\text{USC: Equivalent length} = 7.608 \times 0.665 \times 2 \times 32.2/(0.016034 \times 49)$$
$$= 414.7\,\text{ft}$$

$$\text{SI: Equivalent length} = 2.312 \times 0.202 \times 2 \times 9.81/(0.016034 \times 4.537)$$
$$= 125.95\,\text{m}$$

A globe valve was picked for demonstration because it has a high and therefore dramatic effect that shows how important it is to include these "minor losses." The losses are called "minor" mainly because they are independent of the Reynolds number for calculation purposes. As one can see from the example, they may not be minor in terms of actual size.

Saying they are not dependent on the Reynolds number applies only if you do not convert to equivalent length. When one converts to equivalent length, the Reynolds number and the kinematic viscosity come into play in the computations.

HYDRAULIC RADIUS

The discussion so far has been in regard to round pipe that is flowing full. This is not always the case when doing fluid flow problems with liquids. Sometimes the pipe is not full and the geometry is not a circle. There is a method to use these formulas and techniques for flow in noncircular devices, which is what the hydraulic radius is all about.

The basic definition of a *hydraulic radius* is the ratio of the flow area to the wetted perimeter of the conduit in which it is flowing. For starters, consider the hydraulic radius of the round pipe flowing full. For illustration purposes assume an inside diameter of 0.75 m and calculate the flow area to be 0.442 m^2. The circumference of that same diameter would be 2.36 m. The ratio of area to wetted perimeter is then 0.1872, which then is the hydraulic radius numerically. How does that relate to the diameter that we started with and used in the previous calculations?

This is one of those anomalies of language. Geometrically the diameter of a circle is twice the radius of the circle. Twice 0.1872 is clearly not the 0.75 starting diameter. Four times 0.1872 is 0.7488, which rounds to 0.75.

This indicates that the hydraulic diameter is four times the hydraulic radius. It also points out the vagaries of numerical calculations. If one had used 3.141592654 for pi in the calculation procedure, the ratio would have come out to 0.1875, which when multiplied by 4 would have been 0.75 exactly.

For this reason it is somewhat more customary now to speak of the hydraulic diameter and define it as four times the area of the wetted perimeter ratio. This eliminates the language confusion of the different radius meanings. However, old habits die hard, so one must remember that hydraulic radius is different than the geometric radius by a factor of two. It is fortunate that for full flowing pipe the two diameters are the same.

The same fortunate relationship works out when one considers a full flowing square tube. The flowing area is the side (S) squared and the wetted perimeter would be $4S$. That ratio would then be S over 4, and using the definition of four times the ratio, the hydraulic diameter becomes S, the length of the side.

This becomes only slightly more difficult mathematically when the conduit is not full. It also makes it fairly easy to calculate the hydraulic diameter of a channel that is not fully enclosed as a pipe or tube. Consider a rectangular device that is flowing partly full (Figure 4.1).

The flowing area would be $W \times H$ and the wetted area would be $W + 2H$. So the hydraulic diameter would be

$$\frac{WH}{W + 2H}$$

If the rectangle were a square of dimension 5 and the height were 4, then the hydraulic diameter would be 6.15, whereas it would be 5 if it were flowing full. Observation shows the flowing area denominator is smaller and the wetted perimeter is even smaller, so the ratio of those smaller diameters is more than 1, which predicts that the hydraulic diameter would be larger by that ratio.

The fundamental expression for hydraulic diameter (D_h) is

$$\frac{4 flowarea}{wettedperimeter}$$

and this works in all situations regardless of the geometric shape and amount of flow.

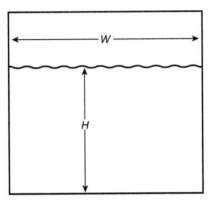

FIGURE 4.1 Rectangle flowing partly full

COMPRESSIBLE FLOW

The information provided so far in this chapter is all about incompressible flow that changes to compressible flow when some of the factors change. In general, compressible flow means a gas, and as such it means that it is primarily subject to the ideal or, for old-fashioned folks, the perfect gas law.

Most readers are aware that for the perfect gas there is a relationship among the pressure (P), the volume (V), and the absolute temperature (T). That relationship has two proportionality constants: the first is mass (m) and the gas constant (R_g), and the second is the number of moles (n) and the universal gas constant (R_u). As might be expected, the two proportionality constants are strongly related. And given the proper use of units, they are the same in both measuring systems.

The relationship is as follows. The gas constant R_g is the universal gas constant divided by the molecular weight, and 1 mole is the molecular weight in mass. This means that if you work in a unit of 1 mole with the law, it is not necessary to know the molecular weight until you start to work with the actual flow rates. The perfect gas law can be stated as

$$PV = R_g T \tag{4.8}$$

It is important to remember that the absolute temperature is either in degrees Kelvin or Rankine depending on the unit system being used. This relationship can be utilized to tell the temperature, volume, or pressure at another place in an adiabatic system by writing the equation in the form

$$\frac{P_1 V_1}{T_1} = \frac{P_2 V_2}{T_2}$$

where 1 is considered the upstream point and 2 is the downstream point. If you know the upstream point you can calculate a downstream point characteristic when any of the other two are known.

This can be helpful in calculating pressure drop. It must be pointed out that most gases only approach being a perfect gas, and therefore a modifying factor called the compressibility factor has been added for most accurate calculations. This factor is highly developed in the gas pipeline industry and is called the Z factor.

As an example, air at 1 bar pressure from $-10°F$ to $140°F$ has an average Z of 0.9999, and at 100 bars the average across those same temperatures is 1.0103. So for a very wide range of temperatures and a wider range of pressure the average is 1.0051. This is not to say that other gases don't have a wider range, but to point out that unless one is striving for high accuracy like those who are measuring thousands if not millions of cubic feet or meters of a substance flowing through their pipeline, it is reasonably safe to ignore the Z factor for common engineering calculations.

The tables that compute these factors, including a factor called super-compressibility, run six volumes long. To simplify these tables, the Pacific Energy Association developed an empirical formula that estimates the Z factor. This equation also requires some additional adjustment for the highest degree of accuracy. It is given without the subsequent adjustments for things like the inclusion of CO_2 and other nonvolatiles (see Appendix). The degree of accuracy that is required for the commercial selling of things like natural gas may require such detail. It is not normally required for the designers.

Before we begin to discuss seriously the fluid calculations for friction loss in compressible flow it is important to point out that it may require no change in calculation technique. Many authorities assert that if the pressure drop from pipe flow is less than 10 percent, it is reasonable to treat that fluid as incompressible for that pipeline. Further, it is generally acceptable if the pressure loss is more than 10 percent but less than 40 percent based on an average of the upstream and downstream conditions. Recall that the specific volume changes with the change in pressure by the relationship previously discussed. Having given that caveat, it stands to reason that only large pressure drops are left, which implies very long pipe. This of course means pipelines where the length of the pipe is often in miles.

Therefore, we must talk more specifically about what is important in the design and sizing of such longer pipes. Probably the most important thing after, or maybe even before, the topography and the selection of the exact route is how many cubic feet of gas need to be available and/or delivered. All pipe systems are designed for the long term, but in plants and such, that pipe is just a portion of the project; in the pipelines, pipe and the pumping or compressor stations are the project.

Determining the pipeline route is the job of surveyors and real estate professionals. As such, it will not be discussed here. For those with a long memory, the Alaskan pipeline stands as evidence of the time it takes and the struggles that intervening terrain causes in that process. The existing pipeline is for crude oil, not gas. Along with politics and other such problems surrounding natural gases, a pipeline for this hasn't ever been started, even though they were thinking about it at the same time as the construction of the oil pipeline.

There are miles of existing and planned gas pipelines to reference for these compressible flow problems. Suffice it to say that the design elements used are not as simple as those of incompressible flow. For one thing they would fall into the category of a pressure drop of more than 40 percent, where the two simplifying uses of the Darcy-Weisbach formula and its friction factor, along with velocity head, are not common.

We discussed earlier how the compressibility factor was not particularly important. The average compressibility factor of air was used as an example of how little error would be introduced when taking the factor as 1 (and it thus not playing a part in such a calculation). This is not quite the same when dealing with millions of cubic feet of gas, which is measurably more compressible than air, delivered over several miles at a higher pressure. The compressibility factor is most often a measured factor that is then published in tables. One of those sets of tables is six volumes of tables plus a seventh volume of correction factors for nitrogen and carbon dioxide content. Even then, they often require extensive manual correction factors. Several formulas have been developed that are helpful in computing the factor. The Pacific Energy Association developed one of the simplest for natural gas. In this method a supercompressibility factor is first calculated and then the compressibility factor is calculated from that. The formula is

$$F_{pv} = \sqrt{1 + \frac{k_1 P(10^{5-k_2 G})}{T_f^{3.825}}} \qquad (4.9)$$

where

F_{pv} is the supercompressability factor itself

k_1 and k_2 are factors dependent on the specific gravity of the gas

T_f is the temperature, degrees Rankine

G is the specific gravity of the natural gas

Since natural gas can have a large range of specific gravities depending on what else is found in the well, there is a table of k_1 and k_2 values based on specific gravity. Natural gas has a very wide range depending other contents from the well.

k_1 and k_2 Factors for Pacific Energy Formula 4.9

Range of Specific Gravity G	k_1	k_2
$0.600 \leq G$	2.48	2.020
$0.601 \leq G \leq 0.650$	3.32	1.810
$0.651 \leq G \leq 0.750$	4.66	1.60
$0.751 \leq G \leq 0.900$	7.91	1.260
$0.901 \leq G \leq 1.100$	11.63	1.070
$1.101 \leq G \leq 1.500$	17.48	0.900

Example Calculations

As an example consider a natural gas that has a measured specific gravity of 0.73. The flowing pressure is 400 psig and the flowing temperature is 40°F. Use Equation (4.8) to calculate the supercompressability factor.

$$F_{pv} = \sqrt{1 + \frac{(4.66)(400)\left(10^{5+(1.6)(0.73)}\right)}{(460+40)^{3.825}}} = 1.063$$

Now there may be additional computations for CO_2 or air in the gas but they are found in even more adjustment tables. They are not shown is this example. The actual compressibility factor Z_f is defined as

$$\frac{1}{F_{pv}^2} = \frac{1}{1.063^2} = 0.885$$

This is a far cry from the lack of effect that was asserted earlier for air. It shows how important compressibility is to the industry when dealing in large amounts of compressible fluid like pipelines

Similar types of highly complex ways to calculate other properties of gases are available either in chart form or, in some cases, empirical formulas. We will not go into specifics of these as they are beyond the scope of this book, which is not to say they are not important. Natural gas is the most common gaseous medium that we work with, so there is more discussion addressing it.

There are several formal methods to calculate what pipeline owners and operators usually desire: the pipeline's capacity to flow in millions of standard cubic feet (or meters) per day. Those formulas are the Weymouth, Panhandle A, and Panhandle Band, but there are several others. These equations can be and have been modified to eliminate the friction factor. In fact, there are several proposed friction factor equations, but the Darcy-Weisbach equation is applicable to any fluid. It has some inherent conservatism that may be best for the estimating uses most readers will be involved in.

Before approaching the ways to calculate these millions of standard cubic feet or meters of gas, there is another element of gaseous flow that must be presented. Gas has a limit—the speed of sound in that gas—to the velocity at which it can travel. This can most simply be described by saying that the pressure waves can only travel at that speed of sound. Therefore, as the pressure drops further none of the fluid upstream can receive the pressure wave signal of a further change in pressure. This is a little like Einstein's thought experiment about moving away from a clock at the speed of light. As he surmised, he would never see the clock's hand move, so for the ride time it wouldn't change.

One of the many ways that speed of sound in gas can be calculated is by the following formula:

$$V_s = \sqrt{kgR_gT}$$

where k is the ratio of specific heats, and for methane (close to natural gas) it is 1.26.

The molecular weight of methane is 16, so R_g in USC is 96.5 (1544/16) and in SI it is 518.3 (8314.5/16). The universal gas constant can have many different units; in USC units it is customarily taken as 1544 (1545.349 more precisely). Then, in some formula where mass is involved rather than pound force, for the acceleration of gravity (32.2 ft/sec^2), as in the speed of sound formula above, one must multiply or divide depending on the exact formula to get the value in mass units, or slugs.

As noted, one of the advantages of the SI system is that somewhat awkward conversion is not required because of the definitions. In that case the g is dropped out of the velocity formula. T is assumed to be 40°F or 500°R for absolute temperature and 277.5°K for SI. The velocity then is

$$\sqrt{1.26 \times 32.2 \times 96.5 \times 500} = 1400 \text{ ft/sec in USC}$$

$$\sqrt{1.26 \times 518.3 \times 277.5} = 427 \text{ m/sec in SI}$$

This might seem quite high and not likely inside a pipe, and that is reasonable. But one must remember that as the pressure drops, for the flow to continue absent any dramatic change in temperature, the volume of gas must expand and that can only happen with an increase in flow velocity.

The previously mentioned flow equations are in use in the United States and may be in use worldwide, but rather than discuss them here, we will talk about the fundamental equation of flow in compressible gas. The equations mentioned are all in some way a variation of the fundamental equation through algebraic manipulation or a change of factors (like the friction factor). For instance, the fundamental equation has a correction factor for converting to "standard conditions." However, these vary. For instance, some data have a standard temperature of 0°C, others 20°C, and in United States it might be 60°F or 68°F. All have to be converted to absolute values. Goodness only knows how many different units are recorded in some of the other properties. The fundamental equation is

$$Q = C\frac{T_b}{P_b}D^{25}e\left(\frac{P_1{}^2 - P_2{}^2}{LGT_aZ_af}\right)^{0.5} \tag{4.10}$$

where
  C is the constant, 77.54 (USC units) or 0.0011493 (metric units)
  D is the pipe diameter, in. or mm
  e is the pipe efficiency, dimensionless
  f is the Darcy-Weisbach friction factor, dimensionless
  G is the gas specific gravity, dimensionless
  L is the pipe length, miles or km
  P_b is the pressure base, psia or kilopascals
  P_1 is the inlet pressure, psia or kilopascals
  P_2 is the outlet pressure, psia or kilopascals
  Q is the flow rate, standard cubic ft/day or standard cubic m/day
  T_a is the average temperature, (°R) or (°K)
  T_b is the temperature base, (°R) or (°K)
  Z_a is the compressibility factor, dimensionless
It should be noted that in these equations it is customary to use the arithmetic average temperature across the length of the pipeline. There is a

generally agreed method of calculating the average pressure. These two averages are used in calculating the compressibility factor. The average pressure equation is

$$P_{av} = \frac{2}{3}\left(P_1 + P_2 - \frac{P_1 P_2}{P_1 + P_2}\right)$$

Comparison was made between several formulas given the same conditions, which were

- 10-in. pipe ID
- 100-mile pipeline
- P_1 of 550 psia and P_2 of 250 psia
- Temperature of 95°F
- Standard condition of 60°F and 14.7 psia
- Gas-specific gravity of 0.65

For purposes of comparison, an efficiency of 1 was used.

For the two calculations, a calculated friction factor of 0.01344 was used. For the Weymouth and Panhandle A calculations, the form of equation that had eliminated the friction factor by including it in the constant employed was utilized. The results of this comparison are shown in Table 4.1.

Since everything is at an efficiency of 1 it is obvious that the only difference is in the accuracy of the constants used or the friction factor. The efficiency factor is usually based on some value between 0.9 and 1.0. It comes from experience, and a designer could use some value based on his or her experience.

PIPE SIZING

As a quick means to size pipe for the fluid flow one can use a simple relationship between flow in cubic feet per second as a starting point. As an example, take the flow of 2×10^6 standard cubic feet per day, which the

Table 4.1 Comparison of Various Gas Pipeline Calculations for Millions of Standard Cubic Feet Per Day

Weymouth formula	18.96×10^6
Panhandle A	23.63×10^6
Fundamental equation with f as 0.01344	19.9×10^6

Note: The Weymouth and Panhandle formulas are adjusted empirical formulas that eliminate the need to develop a friction factor. They are implied as some factor divided by the Reynolds number to some power. As such, they can be shown as higher or lower than a flow by the fundamental equation, which has a more rigorously calculated friction factor. All are estimates.

table shows as the general equation for standard conditions. This translates to 231 ft³/sec at those conditions. We know from the parameter of the problem that the gas never sees standard conditions of 14.7 psia in the pipe, as the lowest pressure is 250 psia and the average pressure shown by the last formula is 418.7 psia. Since that was the pressure used to calculate properties such as viscosity it is a good one to use. It is also one that would be available when the starting and ending pressures were established. So converting the 231 to that pressure and assuming no change in temperature from the averages used, the flow would fall to 8.11 cfs (0.230 m³/sec).

The next issue is, what is the target velocity? For discussion let the assumption be that the target velocity is 15 fps (4.6 m/sec). The size of the flow can now be estimated using the following formula:

$$ID = C\sqrt{\frac{F}{V}} \tag{4.11}$$

where

ID is the calculated internal diameter, in. or mm
C is a constant that is 13.54 in USC and 1133 in SI
F is the flow, ft³/sec in USC or m³/sec in SI
V is the velocity, ft/sec in USC or m/sec in SI

Then calculate

$$ID = C\sqrt{\frac{F}{V}} = 13.54\sqrt{\frac{8.11}{15}} = 9.95 \text{ in. in USC}$$

$$ID = C\sqrt{\frac{F}{V}} = 1133\sqrt{\frac{0.230}{4.75}} = 254 \text{ mm in SI}$$

Both of these equations would lead one to pick a pipe close to the NPS 10 or DN 250 pipe given in the sample problems.

When the fluid is a liquid, one is not concerned so much with the conversion to or from standard conditions as one is with gases. The volume in gas is highly dependent on pressure and, for that matter, temperature. It is not so dependent in liquids.

Once the trial size is chosen based on the desired amount of flow, the friction losses and amount of horsepower for pumping or driving the fluid over the length of the pipe can be estimated and the economic calculations made. As the pipe size goes down, the friction and therefore energy requirements grow higher. As the pipe size and its fixtures grow, the energy required goes down but the capital costs increase. At some point an economic decision can then be made. Of course, there are many more ways to calculate these hydraulic mechanics concerns.

One of the most difficult aspects is being sure that one is using the correct set of units. In charts and tables from other sources, they are using different approaches. The universal gas constant includes energy, time, temperature, and space or distance units. For such situations, a constant should be reasonably standard and there should be one constant for SI and one for USC, and usually this simply is not the case. One source listed 24 different values for the universal gas constant. This is of course the same constant expressed in different units.

Readers are cautioned to read very carefully which data units, and on what basis, different sources are using to make their calculations. Confusing the data unit will give an incorrect numerical answer. Any data found in the Appendix of this book will specify this as completely as possible. That, along with the conversion chart included and a good dimensional analysis when one is not sure, will give one the best opportunity to obtain the correct numeric calculation. Simply let it be said that when going through calculations like this, the changes in the numbers can be drastically affected by the units used in the calculations.

In my experience, using the wrong constant has caused more problems than almost any other consideration. Often it is relatively easy to avoid and sometimes quite hard to find. For instance, the universal gas constant is known as 1545.35 when one is using pressure in lbs/ft^2 and volume in cubic feet as the units of measure. But change to pressure in $lbs/in.^2$, and the universal gas constant becomes 10.73. Close examination reveals that the 144 conversion from a square foot to square inches is the difference. That is, divide 1544.35 by 144 and you get 10.73. Finding that when it is buried in the calculations may be difficult. By the way, that comparison was assuming the temperature involved was in Rankine, not Kelvin, and that gas computations were in degrees absolute. Change your temperature to Kelvin, keep the pressure in psi and cubic feet, and the R is now 19.3169. In the SI system it is a little simpler. Quite often it is just a matter of shifting the decimal point correctly as one moves between measures such as cubic meters and cubic millimeters. However, it still requires a great deal of attention.

Some of the discussion here will be repeated in Chapter 14, on valves. Since this one chapter can be skipped, and some of the calculations for valves use some of the calculations just presented, the background necessary to understand valves is repeated there as needed.

CHAPTER 5

Piping and Pipeline Pressure Thickness Integrity Calculations

Contents

OVERVIEW

One of the primary issues in pipe design is the minimum wall thickness for pipe sizes when exposed to given temperatures and pressures. To establish that wall thickness, the material and its allowable stress at those conditions are the first consideration.

In the B31 code, establishing the allowable stress is different for different books, as was discussed in Chapter 3. There are two ways to choose the basic allowable stress with variations, which are also discussed. If the same material is used and the same service conditions apply, that basic allowable stress may still be different.

This comes about because of the different levels of concern for the pipe to be in a safe condition at that service state. There is also some allowance for the level of analysis of the pipe as it is being designed.

It is common to discuss the margin between a design, say right at the yield point and at a lower point, by calling that the safety factor, which of course it is, but the level of safety is dependent on the knowledge of the condition one is designing for. That knowledge comes from the certainty that the loads used in the equations are accurate, the allowable stresses are correct, and the method of analysis utilized all of the possible variations in computing the results. So the size of the safety factor can correctly be called a measure of what you don't know.

Piping and Pipeline Calculations Manual
ISBN 978-0-12-416747-6
http://dx.doi.org/10.1016/B978-0-12-416747-6.00005-X

The base codes are usually simplifications, meaning that in their analytical approach they strive for conservatism. When setting the allowable stresses, they use (as does ASTM or ASME) the minimum values to ensure that the real property is equal to or above that value. The amount of analysis is dependent on what is perceived as the need for more consideration.

There are two basic approaches to setting the stresses. One is to give a table of allowable stresses for a given material form over a relevant and wide range of temperatures. This is because the major properties change with temperature changes. As the temperature goes up, the strength goes down; however, at some temperatures the strength may not be the controlling factor.

As temperature changes, the material begins to creep with no increase in load and thus distorts. Sometimes that distortion even involves what is called creep rupture, where for instance a pipe will just burst. Those codes that give you tables over a temperature range indicate what the controlling factor, be it strength or temperature-dependent properties, determines the allowable stress to be.

Some of the newer chemistries of piping materials actually have no perceptibly stronger strength properties but excel in creep. When they are used in higher temperatures, they have higher allowable stresses and can require less actual material to make the same high-temperature pipe. It must be pointed out that this is not a free lunch—the base material is more costly and often it requires a more costly fabrication, but when the total cost is less the material will be chosen.

The other major set of book sections operate over a very small temperature range, so they basically work from specified minimum yield strength (SMYS) and control any variation by factors against that SMYS. In the case of B31.4 and B31.11, they only have a temperature range up to 250°F (121°C), whereas B31.8 and B31.12 (pipeline section) allow up to 450°F (232°C), so both have a temperature correction factor.

The new B31.12 does have tables that set allowable stresses at a range of temperatures. Because that code is essentially for hydrogen and this gas is not compatible with temperatures where in metals the time-dependent or creep properties begin to control the stresses, they are in the process of eliminating those higher temperature stresses. The pipeline section of B31.12 utilizes SMYS in the same way other pipeline codes use it.

Each of the codes establishes a limit on the amount of shear and bearing or compressive stress that may be used. This is some percentage of one of the allowable stresses that was already established. And in some manner each code tells you how to use materials that are not on its preferred list. That manner varies from "thou shall not" to "here is how you compute the stress for this material." A sample of some of those calculations is shown in the Appendix.

BASIC WALL THICKNESS CALCULATIONS

In calculating the wall thickness for pipe the basic formulas for the primary (hoop) stress have been around for ages. There are many variations. At last count there were more than 20. Each of these addresses the basic problem somewhat differently to account for the variations in failure modes that can occur. But there are two fundamental differences: the thin–wall approach, which we call the Barlow equation, and the thick-wall approach, which we call the Lame equation. This then raises the question: When does a thin wall become thick?

When the problem is thought about, it is not too hard to figure out that the pressure is higher on the inside of the pipe than on the outside. That may not be true if the pipe is buried in a very deep underwater trench. There, the outside pressure can be higher than the inside or at least the same order of magnitude.

From that logic, for the more general case a man named Barlow surmised that if the pipe is thin one can assume that the thinness of that wall allows one to average the stress across the thickness (see Figure 5.1). So he devised a simple formula by splitting a unit length of pipe through the diameter. He then said the pressure across that diameter creates a force equal to the pressure times the diameter, and the two unit thicknesses create the area that resists that force.

Thus, the stress equation becomes

$$S = \frac{PD}{2t}$$

This is the basic equation that the code presents. Since the goal is to find the unknown thickness, the formula is rearranged to solve for t given the other three parameters: pressure, outside diameter (OD), and allowable stress. The formula then becomes

$$t = \frac{PD}{2S}$$

One can note many things from this simple formula. For instance, for a given pressure the stress is proportional to the ratio

$$\frac{D}{t}$$

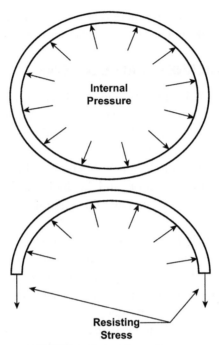

FIGURE 5.1 Barlow force diagram

This is sometimes called the standard dimension ratio (SDR). It can be manipulated to represent

$$\frac{outsidediameter}{insidediameter}$$

and offers many interesting ways to think about the stresses and pressures in a pipe.

Some B31 code equations have added a factor Y to adjust and mathematically move the actual average toward the middle of that thickness. This movement depends on the material and the temperature. They also have an E factor to correct or allow for the efficiency of the way the pipe is made.

Recently, a W factor was also added to those codes that operate at high temperatures. This is to make a correction on certain welds when they will be in high-temperature service. This W factor was the result of some unpleasant experiences from not taking into account the fact that most often the weld and its attendant heat-affected zones do not have the same strength as the parent material.

In spite of the adjustments they are the same basic equation with frills for things that have become known over the nearly hundred years the code has been in effect.

Mr. Lame developed the thick-wall theory of pipe. Knowing that the pressure on the inside is different than the outside, he deduced that as the wall gets thicker, that difference becomes important enough to consider. His formula is somewhat more complicated. It is built mathematically around radii rather than diameters. A simple form of that equation is

$$S = \frac{Pb^2\left(a^2 + r^2\right)}{r^2\left(a^2 - b^2\right)} \tag{5.1}$$

where

S is the equal stress at intermediate radius r
a is the outside radius of pipe
b is the inside radius of pipe
P is the pressure

Note two things: when calculating the stress at the point where r equals the inside radius of the pipe, the stress is higher than the pressure only, and if r is set at the outside radius, the stress continues to have a component from the inside pressure, and the difference of the two components is the pressure. This issue applies no matter how thick or thin the pipe wall might be. So we now have a tool to begin to answer the question of when a thin wall becomes thick.

A general answer is when it becomes more than 10% of the inside radius. We have a tool to check that. Simply use the Lame equation on any inside radius, make the thickness 10% of that, and find that maximum stress when on the inside.

Say the pressure is 150, the inside radius is 3, and the wall thickness is 0.3, making the outside radius 3.3. You will note that leaving the units off the measurement system only requires that you keep all the units compatible with the system you use. The maximum stress on the outside wall is

$$S = \frac{150 \times \left(3^2 + 3.3^2\right)}{\left(3.3^2 - 3^2\right)} = 1578$$

Using the Barlow equation, we obtain

$$S = \frac{150 \times 6.6}{2 \times 0.3} = 1650$$

which comes to an approximate 5% difference. In the conservative directions, assuming an E of 1, seamless pipe, and low temperature, Y would be 0.4 and W would be 1, and the code equation would give you a somewhat lower stress of 1594. By changing the Y factor and keeping the same thickness the stress drops to 1544, and by making the Y factor 0.5 the stress is 1575. It appears the Y factor adjustments do a pretty good job of calculating the maximum stress per Lame. As in all comparisons like this, the scale factor, higher stress, and so on, may change the relative values, but the adjustments to the simple formula seem to work.

Again the careful reader will note that we were comparing stress results from the formulas. In conventional practice we are given a pressure and temperature along with the material. The temperature allows us to determine the allowable stress by one of the methods described. So stress is not a regular calculation made in the code; it is thickness. Recall that the ratio

$$\frac{D}{t}$$

can be related to stress, and with some algebraic manipulation that can be related to the thick/thin puzzle. A relationship between the thickness and the internal radius can be derived, and then this expression can be established:

$$1 + \frac{t}{internalradius}$$

From this one can establish an index of the maximum stress to the internal stress and get an index of how much that maximum stress exceeds the simple Barlow equation (not the code-adjusted Barlow). Then, keeping in mind that the allowable stresses are established at a margin below yield, one can determine the severity of using the simpler equation.

In Table 5.1 you can see that the K factor representing a thickness of 10% or less of the internal radius represents a maximum Lame pressure of 5% or less than the average pressure.

Table 5.1 Ratio of Maximum Stress to Barlow (Average)

K^1	1.1	1.2	1.4	1.6	1.8	2.0
Ratio[2]	1.05	1.10	1.23	1.37	1.51	1.67

[1] K equals the expression $1 + \frac{t}{internalradius}$.

[2] The ratio is the maximum stress calculated by Lame divided by the stress calculated by Barlow (average).

B31.3 has an enigmatic note in its equation that says that $t < D/6$ and does not require any further consideration, but if it changes to $t > D/6$ one must consider other things such as theory of failure, effects of fatigue, and thermal stresses. This is related to the thick/thin problem. Interestingly, the standard pipe dimensions (i.e., schedule pipe in the United States, which more or less adopted by ISO) do not have t thicknesses that exceed $D/6$ above the three double-extra strong schedule. A chart showing the SDR (D/t) of common pipe is in the Appendix. It is as we get into nonstandard pipes that the problem can occur.

This then is the general discussion of calculating which pipe thickness to use under given conditions. There are other equations than the code and even a few within the codes. We will discuss these in the following section.

BASIC CODE EQUATIONS

As discussed, the codes offer several variations of the equations. The equations presented in each book section are listed in the code equations table and are discussed individually as well as in general. Within certain parameters the different results can narrow considerably when one sets up the conditions properly. The differences are the code books' responses to the particular problems in the type of service that the specific book was written for.

There are some special code equations for high-pressure design in B31.3 (Table 5.2). There are basically three different equations. The third equation is specific materials, such as solution heat-treated austenitic stainless steel and others at specific temperatures, which utilize the von Mises stress criteria and strive to initiate yielding on the outside surface. As such, readers are referred to the codes and other sources before using that formula. They all take a form similar to the Lame equation. The equation for the thickness using the outer diameter is

$$t = \frac{D - 2c_a}{2}\left[1 - \exp\left(\frac{-P}{S}\right)\right]$$

The high pressure B31.3 equation eliminates the need for the Y factor adjustment. It does not give the same thickness as the other equations because it has an algebraically manipulated form for ID calculations. In standard pipe the constant dimension is the OD, and as the schedule or thickness changes, so does the ID. The ID forms of the equations are more for convenience when for internal reasons one purchases the pipe to a specified ID.

Table 5.2 Code Equations

Code Designation	OD Formula	ID Formula
B31.1	$t_m = \frac{PD_o}{2(SE+Py)} + A$	$t_m = \frac{Pd+SEA+2yPA}{2(SE+Py-P)}$
B31.3[1]	$t = \frac{PD}{2(SEW+PY)}$	$t = \frac{P(d+2c)}{2[SEW-P(1-Y)]}$
B31.4[1,2]	$t = \frac{PD}{2S}$	N/A
B31.5[1]	$t = \frac{PD}{2(S+Py)}$	$t = \frac{Pd}{2(S+Py-P)}$
B31.8[3]	$P = \frac{2St}{D}(FET)$	N/A
B31.9	$t_m = \frac{PD}{2SE} + A$	Option B31.1
B31.11[1,2]	$t = \frac{PD}{2S}$	N/A
B31.12(IP)[4]	$t = \frac{PD}{2(SEM_f+PY)}$	$t = \frac{P(d+2c)}{2[SEM_f-P(1-Y)]}$
B31.12(PL)[5]	$P = \frac{2St}{D}(FETH_f)$	NA

Note: The symbols are the same across the various book sections: P is the pressure; D and D_o are the outside diameter of the pipe (not the nominal diameter); d is the inside diameter; y and Y are the adjustment factors as discussed in the general equation section; A is basically the same as c, the sum of mechanical tolerances; E is a weld joint efficiency factor for some welded pipe and is given in the books (seamless pipe has an E of 1); and W is the weld joint efficiency factor for longitudinal welds when the temperature is in the creep range as defined in the code. The current editions of B31.1 and B31.3 have tables for the factor for certain materials and certain conditions. The factor is required for longitudinal and spiral welded pipe. It is left to the designer to determine the needs for other welds, such as girth welds. When the temperature is below the creep range, that factor is 1.

[1] These have a separate formula to calculate the minimum acceptable t, which is $t_m = t + c$, where c is the sum of mechanical tolerances like thread depth, corrosion, or erosion allowance.

[2] These equations adjust the stress by multiplication of specified design factors and, if applicable, an efficiency factor for some pipe that is not seamless. The proper stress to use in the formula is the adjusted SMYS.

[3] The B31.8 formula is given in pressure terms for various reasons. It can and is rearranged to solve for t. As the pipeline is monitored over its use the t may vary, and an allowable operation pressure is recalculated using this formula among other lifetime calculations. The F factor is a location factor that recognizes a difference between highly and sparsely populated areas. The T factor is a temperature derating factor to recognize that the material loses strength as temperature rises.

[4] Much the same as B31.3 with the addition of an M_f factor that addresses loss of material properties in hydrogen service.

[5] Much the same as B31.8 with the addition of an H_f factor that addresses loss of material properties in hydrogen service.

There is an emerging equation that is a modern calculator substitute for the Y factor. It essentially automatically sets the Y factor to 0.5. As such it is quite accurate with the B31.3 formula when the Y factor is 0.5. This is because B31.3 makes the adjustment for the corrosion and mechanical allowances outside the wall thickness calculation as was described in the notes of Table 5.2. This equation loses accuracy to a Y factor equation when that factor varies (as it does due to temperature). However, like the Barlow equation, it is a good first approximation if you have it built into your calculator or laptop. The equation is

$$t = D\left(\frac{1 - e^{\left(\frac{-P}{S}\right)}}{2}\right)$$

As might be expected the equations for nonmetallic piping are different. Some differences are obvious. The E factor for those metal pipes that are not cast or seamless is not required. This is because all code recognized for nonmetal pipe is seamless, so the factor would be 1 and is not necessary. The W factor for welded metal in the creep range is not necessary. This is because, while nonmetals might creep, they are not welded in the same sense as metals, nor are they used at the temperatures where the effects that W is intended to correct for occur. Finally, there is no use of a Y factor to correct for the stresses moving through the thickness of the wall. The basic nonmetal ASME formula is

$$t = \frac{PD}{2S^* + P}$$

The asterisk ($\star$) indicates that for some materials, such as reinforced thermosetting resins and reinforced plastic mortar, a service factor needs to be included. This multiplier in the code sense is established by procedures established in ASTM D 2992. The designer is to set that service factor after fully evaluating the service conditions. The code limits the maximum service factor depending on whether the service is cyclic or static, but not otherwise.

For comparison purposes we calculate the thickness by the different formulas and comment on the rationale for any difference discovered. To be fair, the calculations will be done for two different conditions: one in the lower temperature ranges that are compatible with pipeline service and the other in the non-high-temperature service. The second set will address the higher-temperature and higher-pressure service that mainly only affects the first two codes. These comparisons are found as charts or tables in the Appendix.

Naturally, when one is working with a particular code it is important to use the code equation from that code to establish any values, such as thickness for a given pipe at a given pressure and temperature for a given allowable stress, from that code's rules. This is especially true when converting from one system to another. A particular code as shown by the different formulas for B31 codes will give slightly different answers based on the formula and the requirements of that book section. The same is true for any other code.

However, it is a truth of nature that the material does not know which code was used to calculate the thickness. Within the accuracy of our knowledge the stresses at the same conditions are the same regardless of the code. As an aphorism on this it has often been asked: How does the pipe know which code it was calculated for?

The careful reader is aware that with a major portion of pipe being purchased in standard wall thicknesses it is a rare occasion that the maximum pressure that a particular pipe would bear is the same as the maximum pressure for the system that the pipe was chosen. As mentioned before, in multicase piping systems the design conditions are set such that the case that has the maximum wall would be the design case. For all other cases the wall thickness required would most likely be less.

It is often the case that one might want to know the maximum pressure for reasons other than a code calculation. The differences might not be significant in the decision that the question is intended for. There is a table in the Appendix titled Pressure/Stress Ratio and Various Pipe Properties that makes the calculation of maximum pressure for a given pipe relatively simple.

For example, if in that table you choose NPS 6 (DN 150) pipe with a standard wall you would find a pressure/stress (P/S) ratio of 0.07684. Then determining the allowable stress for that pipe size and multiplying it by the ratio gives you the maximum pressure. Assume you have a material and temperature for that pipe that has an allowable stress of 20,000 psi (138 MPa). The maximum pressure will be

$$S * (P + Sratio) = MaxP = 20000(0.07684) = 1536.8\,psi\ (10.6MPa)$$

It may not be the same exact pressure one would calculate in a code. It is an independent measurement system. The table uses standard NPS and DN dimensions as a start. This calculates the wall thickness reduced by standard manufacturing tolerance of 12.5% and it calculates it at the midpoint of the wall (a Y factor of 0.5) However, it is useful in back of the envelope calculations and as a check against the code calculation made or reported from a computer. The only thing one must do is multiply the P/S ratio by an allowable stress.

It is time to remind readers that the discussions so far have been about hoop stress only. One will note that most of the codes either require a specific identification of mechanical allowances, including manufacturing tolerance, to determine the minimum required thickness, or advise you that through their modification factors they have taken into account such things and that the nominal pipe size is the result calculated.

Manufacturing tolerance in standard pipe is usually 12.5% of the thickness and is therefore an important inclusion. If the pipe is made from plate, which usually has a much smaller tolerance, it is still important but is not as significant. It is important to remember that the word *minimum* means minimum, and in using things like manufacturing tolerance, it means one has to be sure that what they are using is within that tolerance. This is also the basis of some codes allowing measured thickness. They define in some manner how the measurement must be made.

Other mechanical allowances include corrosion allowance and erosion allowance. Both of these are usually beyond the scope of a particular code. They both have little or no influence when the pipe is new and the system is just starting up. They do, however, have a significant influence on the life of the pipe. As a pipe corrodes or erodes it loses strength and material. If no material is added to allow for this loss when the service causes it, the pipe soon will not have the required strength to withstand the service. Sometimes the amount of corrosion or erosion is learned from experience in that service.

The National Association of Corrosive Engineers (NACE) offers publications that give guidance on the corrosion that may occur. Erosion can be quite heavy in flow that has entrained solids like sand. I am aware of conditions of highly erosive flow that caused failure through a high-pressure drop device such as a valve in hours rather than years. There are basically too many variables in erosive quantities to predict a rate. The best one can do is to increase the bend radii as much as possible, add protective coatings or linings in the pipe, and/or work on the hard coating.

In the rapid erosion mentioned, a very hard weld coating was added to the device that increased the life to a matter of days rather than hours. Even so, the process was never deemed economical.

Other mechanical allowances would be the depth of any threads in the pipe or grooves and other incursions on the integrity of the pipe wall. The formulas of these are simple—one just adds the material to the calculations. This means that you have extra material for the stretches of pipe that do not have threads, grooves, or the like. This may not be economical for the entire length of the pipe. The designer should then consider other means or components to achieve what those threads and grooves provide. One solution might be to insert a pup piece of the thicker material for a short distance.

This discussion has to this point been concerned with hoop stress at a steady state or temperature. This is not always the case. In some cases the system is designed to go up or down in temperature or pressure. In daily operation these changes may occur in an unplanned way. Aspects such as changes in flow may cause some severe pressure shocks. They will be discussed in detail in Chapter 12, which covers fluid transients. Many of the other loads are considered in the flexibility analysis, including longitudinal stress calculations. Flexibility and fatigue analysis is a subject by itself and as such has a chapter devoted to it (see Chapter 7). Longitudinal stresses do have a component coming from the pressure equal to half of the hoop stresses in the simple calculations. They have other components that create moments and other stresses, so they will be discussed in Chapter 7, which covers flexibility.

What we have examined so far is straight pipe. In a piping system there is rarely only straight pipe. Piping has elbows, bends, tees, and other branch connections. The methods of calculating stresses in straight pipe are not sufficient to establish the thickness or stresses in these components. In some cases there are standard fixtures that tests or another form of proof have shown provide adequate mechanical strength for the situation. These are often covered in a separate standard that is then considered by the various code committees. They determine that when the particular component is in accordance with the code requirements it is acceptable without further proof.

This still leaves certain components that require some design input to determine their compliance with the particular section of the code. The next few calculations discussed will deal with this issue.

PIPE TURNS OR BENDS

One can ask about the difference between elbows and pipe bends. The answer is relatively simple: elbows are by definition covered by some standard. As such, they have limitations as to size, bend radius, and resulting angle, usually 90° or 45°. Some of the standards that define elbows do define elbows at intermediate angles. These are called segment or segmentable elbows. It is possible to order a segment from the elbow provider. Some elbows are segmented in the field. This field segmenting may cause other problems related to tolerances and void the compliance with the standard. Any other similar product is a pipe bend. In any system there are usually some bends and some elbows resulting from a need for one of the characteristics to be different from something that is covered by one of the standards.

The basics of pipe bends are relatively simple. First, if it is a bend, it is not a sharp corner. Subject to material and thickness constraints there is a limit to how small a radius can be bent in a given pipe. Depending on the method of bending there are further constraints. These constraints will be discussed in Chapter 13 on fabrication and examination. The discussion here is about the design and considerations of the designer in his or her calculations for compliance with the stress constraints. The nomenclature of a bend is shown in Figure 5.2; it is the same whether the component is a bend or an elbow.

There are two basic criteria to determine an allowable pipe thickness. These criteria can be utilized to determine if the resulting bend is compliant with the code. They are based on the fact that as the pipe is bent two things happen:

1. On the extrados the wall of the pipe thins by some amount dependent on the bend radius.
2. On the intrados the opposite occurs; this is also dependent on the bend radius.

These two events are predictable to a certain degree and are the natural result of the neutral radius of the bend, which is the place along the bend that the thickness remains constant during the bend. As indicated by the nomenclature, this is usually considered the centerline of the pipe.

It becomes evident that neither the extrados nor the intrados is this neutral radius. The amount of thinning is as important as the amount of thickening.

Finite-element analysis of such bends varies of course as the bend radius varies with the pipe diameter, but some published studies have shown that the actual hoop stress on the intrados may be as much as 75% higher than the stress on the extrados.

The changes in geometry as the pressure or fluid moves around the bend cause the changes in stress. The hoop stresses on the outside (extrados)

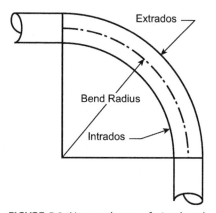

FIGURE 5.2 Nomenclature of pipe bend

become lower, and the inside (intrados) stresses intensify. It seems that what happens naturally is what Mother Nature knew would be required because the change in wall thickness is in concert with the change in hoop stress. This concert of phenomena if done properly in the bending process allows the bend to maintain the full pressure capacity of the straight pipe for which it is matched.

All of the codes put restrictions on the bends. Some, like pipelines, specify the minimum bend radius for the field bend. These minimum radii are of such length that the changes in thicknesses are minimized. In essence, the pipe behaves as if it were straight or nearly so. The older method of controlling the bend required that the minimum thickness after the bend match the minimum thickness of straight pipe. This required that the bender start with pipe that is thicker than the straight pipe to the extent that when it is bent, the thinning results in a wall that is still above the minimum wall computed for straight pipe. This can create some problems of matching up the bent and the straight pipe unless sufficient straight tangent pieces are included in the bend. It also leaves open the question of the need for the intrados to be thicker for the increased stress there. Some thickening will occur, but there is no definition of what thickening is required to keep the bent pipe compliant with the stresses. If an elbow is compliant with a standard, this represents some proof that it meets the requirements of its matching pipe.

Some codes, B31.1 for example, offer suggested minimum thicknesses before bending depending on the pipe diameter. This only covers a range from long radius to 3D elbows. One should note that when the straight pipe is thicker, there is some natural increase in the thickness at the intrados, so it is usually conservative.

A newer method is one that has been adopted by the two major codes, and is under consideration by others. It gives a method of determining what those minimum thicknesses should be. This is accomplished by introducing a factor, called I in the B31 codes that have adopted it. This factor is a divisor to the allowable stress used in the system. It is based on the amount of increased stress or decreased stress depending on where one is checking. Naturally, when one increases the allowed stress the required thickness is reduced and vice versa. The two formulas are

$$I = \frac{4\left(\frac{R}{D_o}\right) - 1}{4\left(\frac{R}{D_o}\right) - 2} \text{ for intrados} \tag{5.2}$$

$$I = \frac{4\left(\frac{R}{D_o}\right) + 1}{4\left(\frac{R}{D_o}\right) + 2} \quad \text{for extrados} \tag{5.3}$$

In both cases,

R is the bend radius, usually a multiplier of D_o

D_o is the nominal outside diameter, which is considered the same for standard pipe in DN and NPS

If one stays with those rules the factors are the same. In whatever way one calculates the thickness one divides the allowable stress by the appropriate I factor as calculated to find the stress, and from that the thickness.

Example Calculations

Use the simplest equation (Barlow) to calculate the thickness for a 6 NPS pipe (6.625 D_o) at 875 psi pressure and allowable stress of 23,000 psi. Then the thickness required is

$$t = \frac{168.275 \times 6032.913}{2 \times 158,579} = 3.200 \text{ mm in SI}$$

$$t = \frac{6.625 \times 875}{2 \times 23,000} = 0.126 \text{ in. in USC}$$

Assume you want to bend the pipe with a bend radius of three times the nominal size, or 18 in. in diameter. Using Eqs. (5.2) and (5.3) the intrados I factor is 1.1 and the extrados factor is 0.928. Note that in SI, DN can be considered to be the same as NPS or 6, so the factors in standard pipe would be the same. Thus, the new thickness calculation required would be to divide the stress by the I factor in whatever form of thickness equation you use. It should be pointed out that the thicknesses along the opposing centerlines would not be changed, as there would have been no thinning or thickening. Thus, the new thicknesses required would be

$$t = \frac{6.625 \times 875}{2 \times \frac{23,000}{1.1}} = 0.139 \text{ in. in USC}$$

and 3.531 mm in SI for the intrados.

The same procedure would be used to calculate the thinner wall on the extrados; substituting the 0.938 in the previous equation, the results would be 0.117 in. in the USC system and 2.972 mm in SI. Naturally if you were using one of the code equations that includes a Y factor, the actual numbers would be some amount smaller because you would have increased the denominator.

Note that you will be reminded from time to time that a wall thickness carried out to three decimal places is borderline silly. It is done because the U.S. system in spite of all efforts is still basically a U.S. system, and we carry the walls to three decimals in inches and do other things that make little sense in the SI system. It is obvious that a true soft conversion (starting with the SI units one would probably round the wall as calculated from 2.972 to 3 mm) would make a difference in the wall of 0.001 of an inch, which is not a significant change from an engineering point of view. This is not to say that one can go below the minimum of the specifying code. It is just to remind the reader that mathematical conversions from one measurement system are not necessarily law, but the code written is law. This is especially true when the code, as many U.S. codes are, is mixed.

Any comments on the different manufacturing methods are reserved for Chapter 13, which covers fabrication. There are some differences that have a real impact on the operation, from both flow and safety perspectives.

MITER BENDS

In spite of the increasing ability to bend pipe, there are just some situations where a bend can't be made. This may be because of bending equipment size or tooling availability. This is especially true as the size of the pipe gets larger. The larger sizes usually don't have enough demand to justify the huge tooling expenses involved in machine bends. It may be because of the size and wall thickness that the pipe involved is not strong enough to withstand the bending forces without creating ovality or flat spots, which are not acceptable in the finished bend.

These do not mitigate the need for changing the flow direction in the piping system. A frequent solution to this directional change is the choice of a miter bend. A miter is succinctly defined in B31.3 as "two or more straight sections of pipe matched and joined in a plane bisecting the angle of junction so as to produce a change in direction." Normally, when that angle is less than 3° no special consideration is needed as to the discontinuity stresses that might be involved in the joining weld.

Care should be taken here to be sure that when one is speaking of an angle in a miter there are actually two angles involved. The first angle is called θ, which is the angle of the cut on the pipe. Naturally, for the pieces to mate properly for welding the same angle cut needs to be made on the mating piece of pipe. When joined, this creates the second angle, α, which is sometimes called the total deflection angle of the pipe. See Figure 5.3 for an example.

Usually if θ is more than 22.5°, problems are going to occur (that is, unless it is a single miter for a total deflection of more than 45° but less than 90°). B31.8 does state that if the operating pressure creates a hoop stress of 10% or less of the SMYS, you can have a miter where the total angle is not more than 90°.

There are two different kinds of miter bends. The first is a single miter, as just mentioned. The second is a multiple miter, where the direction change needs to be of a higher degree.

Multiple miter bends come in two varieties, closely spaced and widely spaced. As we examine the way to determine the minimum thickness of the miters, we will begin to understand the difference. It comes about as the length of the individual sections becomes longer. The difference comes when the centerline of the section in question changes from less to more than the mean radius times the factor $(1 + \tan \theta)$. This is the definition of closely spaced. The reasons for this change can be dictated by constraints on the narrowness of the crotch or shorter length of the section and surrounding requirements, or by the resulting equivalent bend radius.

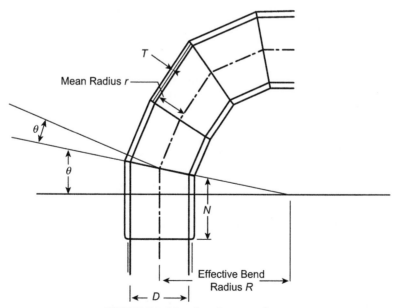

FIGURE 5.3 Miter bend nomenclature

Only two of the B31 code books give a specific formula for calculating wall thickness for the miter section. The rest of the codes give you limits to the pressure that may be utilized along with the percent of allowable stress that may be used with that pressure. These are fairly consistent, and are limited to the pipeline or low-temperature codes where one doesn't have to make significant adjustments for the mechanical property changes that occur as the temperature rises.

Basically these constraints limit the pressure that can be utilized or the amount of hoop stress that pressure can develop in the pipe. The theory is that the increased stresses that may occur due to the discontinuities from the changes in direction will not raise the stress in a miter so much that it will make it inappropriate to use the same wall thicknesses that were calculated using the lower stress for the pipe thickness sizing.

Cursory mathematical examination of possible situations indicates that the increased stresses expected in the miter will not be over the limit. Some judgment has to be made in performing such checks, as the radius and other factors can change the resulting climb in the stresses. Suffice it to say, the prudent engineer would also perform some analysis and/or increase the thickness by some percent, and thus the need exists to determine that percent by analysis.

B31.1 and B31.3 take different approaches, and both will be explained and discussed in the formulas and calculations. As suggested, the prudent engineer might perform some calculation to be sure that the restriction as applied to the actual system under consideration does not violate stress limits.

One might ask why we don't use the same technique of arbitrary restrictions with B31.1 or B31.3. After reflection, the answer would seem evident: both of those codes are for systems that, unlike pipelines, do not expect to have continuous operation at one state. Their systems might not work by lowering the pressure, whereas in a pipeline lowering the pressure may be economically undesirable. However, the net result is primarily only less flow, but there are not many intentional state changes as in the process industries. In power plants, lowering the pressure may change the quality as well as the pressure of the steam, which might have a very serious effect on the turbines. And lastly, in many of their temperature regimes lowering the allowable stress further might make the system one that can't be constructed economically, since as the temperature climbs the allowable stresses fall, sometimes rather steeply. The result is they must calculate the thicknesses and pressures for the given conditions.

Before we examine the two methods of calculating this thickness we need to look at the nomenclature of the miter bend so the various new symbols in the resulting analysis are understood.

The equations have the look and feel of empirical equations that revolve around the somewhat arbitrary function of a θ, which is a saw cut and equals ½ pipe deflection.

There are two equations for the sections under the θ. The first check covers a single miter for deflections from 3° to 45°. The second check uses that equation plus a second equation that is dependent on the equivalent bend radius, which might be the controlling factor depending on that radius. There is a minimum equivalent bend based on an empirical constant and the pipe diameter.

The third equation applies again to single miters for a θ over 22.5°. This situation could make economic sense if one needed a deflection larger than 45° but less than 90° (probably several degrees less), where the extra cutting and welding would increase the cost of construction.

This might sound quite complicated, but it is a relatively simple decision-making process and will become clear as we work through the example. The checks mentioned are basically a check on the maximum pressure the miter can take at the allowed pressure given the θ and some new thicker component. You will recall that the other methodology is to limit the pressure to some amount less than what the calculated straight pipe thickness would allow. This method turns the logic around and asks: What new thickness is needed for this particular miter to be stressed at the allowable amount?

The calculations then become iterative and are presented in much the same way that they were when calculating the friction factors in fluid mechanics that we discussed earlier in Chapter 4. Thank goodness for modern spreadsheets and calculators, which can be set up to perform the iterations in a painless way. Following are the relevant equations that are used in B31.3. They are somewhat more comprehensive. This is especially true as the pressures get higher, since B31.1 places limitations on the pressures that are allowed before allowing an increase in thickness to compensate.

$$P_m = \frac{SEW(T-c)}{r_2}\left[\frac{T-c}{(T-c)+0.643\tan\theta\sqrt{r_2(T-c)}}\right] \qquad (5.4)$$

$$P_m = \frac{SEW(T - c)}{r_2} \frac{R_1 - r_2}{R_1 - 0.5r_2} \tag{5.5}$$

$$P_m = \frac{SEW(T - c)}{r_2} \left[\frac{T - c}{(T - c) + 1.25 \tan \theta \sqrt{r_2(T - c)}} \right] \tag{5.6}$$

where

c is the total mechanical allowance

E is the pipe efficiency rating

P_m is the maximum allowable internal pressure for miter bends

R_1 is the effective radius of the miter bend

r_2 is the mean radius of the pipe

S is the allowable stress

T is the miter pipe minimum wall thickness

W is the weld strength reduction factor

θ is the angle of miter cut

The minimum value of R_1 is established by a formula where R_1 is dependent on the final thickness calculated, and its smallest amount is the equivalent of 1 in. larger than the pipe radius at less than 0.5 in. (13 mm), and goes up from there. This adder varies according to a specified formula that causes one to add 1 in. (25 mm) at the thin thickness to 2 in. (50 mm) at 1.25 in. (32 mm) thickness. If one is doing an actual code calculation, a specific formula check is recommended. For demonstration purposes we treat the minimum R_1 as 2 in. (50 mm) over the pipe radius.

A little explanation of the usage of the formulas is required. Before setting up an example, Eqs. (5.4) and (5.5) are both used in multiple miter calculations where θ is 22.5° or less. One calculates the maximum pressure by both methods and then uses the lesser pressure as the appropriate pressure. If one is only intending to use a single miter where θ is less than 22.5°, then Eq. (5.4) is the only calculation required. Finally, if one is intending to utilize a miter cut of over 22.5°, only a single miter is allowed and the minimum pressure is calculated by Eq. (5.6). It is a known fact that the thickness of the miter pipe is required to be more than the thickness of the straight pipe to which it will be attached. Often this requires an educated guess or repeated calculations. Once again, setting up a spreadsheet and using a goal seek function will save a lot of time calculating several different miter bends. For now, let us explore a set of sample calculations.

Example Miter Calculations

First, establish the data for the problem. Assume a multiple miter with θ of 22.5° and the following:

- Design pressure (for straight pipe) is 400 psi (2750 Kpa)
- Pipe OD is 48 in. (1220 mm)
- W and E are 1
- Wall thickness is nominal, 0.5 in. (13 mm)
- Corrosion allowance is 0.06 in. (1.5 mm)
- Allowable stress is 23,000 psi (158.5 MPa)

Step 1: Calculate the mean radius of the pipe at 23.75 in. (603.5 mm). Note that the layout geometry requires an R_1 of 30 in. (762 mm), which is well above the required minimum. For calculation purposes, make a guess at the required thickness to meet the pressure and then check for the radius that will work. For this example, guess that the required thickness is 0.8 in. (21 mm). First, use Eq. (5.4) to calculate the minimum pressure it will allow.

$$\frac{23{,}000 \times 1 \times 1 \times (0.8 - 0.06)}{23.75}$$

$$\times \left[\frac{(0.8 - 0.06)}{(0.8 - 0.06) + 0.643 \times 0.414\sqrt{23.75 \times (0.8 - 0.06)}} \right]$$

$$= 415.5 \text{ psi } (2864 \text{ Kpa})$$

This will certainly handle the proper pressure for the bend. We must make a check using the 30-inch radius and Eq. (5.5) to find which formula yields the minimum allowed pressure. Unfortunately, that check shows a much smaller allowable pressure. A careful reading of the formula shows that increasing the minimum radius will increase the allowable pressure by that calculation. So we will estimate a new R_1 and run that formula. For this run, we will do it in SI units to be fair to our metric readers and then convert back to USC. Our guess is 1020 mm R_1.

Using Eq. (5.5), the calculation is

$$\frac{148.5 \times 1 \times 1 \times 1000(21 - 1.5)}{603.5} \left[\frac{1020 - 603.5}{1020 - 0.5 \times 603.5} \right]$$

$$= 2782.4 \text{ Kpa } (403.5 \text{ psi})$$

Note that megapascals were converted to kilopascals and that there are subtle differences in converting back and forth between the systems due to rounding, and so forth.

Now the designer has to determine if the space in the layout can fit the larger R_1 required. If not, he or she must determine what to do. There are options, such as using a different material with a higher allowable stress. That would entail other considerations. The designer could possibly use Eq. (5.6) and create a single miter for a larger θ. This might cause pressure drop problems as the fluid flows through the miter, which causes higher erosion or other considerations. From a pressure-only view, let's assume that the final required change in direction is 80°, which in a single miter means a miter cut of θ is 40°. Apply Eq. (5.6); a quick check shows that the thickness of the miter would have to be 2 in. (51 mm). This thickness in itself might cause problems. From this basis it seems that the designer might have to take a closer look at the layout and possibly make the corrections there. It stands to reason that a different layout might be the answer.

As noted previously, B31.1 puts restrictions on pressure and many other things. However, if the pressure is over 100 psi (690 Kpa), some calculation is allowed. First, the code refers to Paragraph 104.7, which among other things leads one to things like finite element analysis (FEA), testing, or calculations. Those calculations might lead one to go to the formulas in B31.3. However, formulas are provided for the minimum wall thickness that is acceptable regardless of what the calculations show would work. Those formulas are dependent on what are called closely spaced miters and widely spaced miters. These spacing definitions are based on the centerline cord of the miter section. If that cord is less than the quantity $(1 + \tan \theta)$ times the mean radius of the pipe, it is considered closely spaced. Conversely, if it is larger than that figure, it is considered widely spaced. The definition is the same in both B31.3 and B31.1. It is also used to differentiate the stress intensification factor. The use and calculation of this factor is discussed in Chapter 7 on flexibility analysis.

B31.1 also defines the effective radius of the miter bend differently than B31.3. In a closely spaced miter, R is defined as the centerline cord times the cotangent θ divided by 2, and the radius of the widely spaced miter as the quantity $(1 + \cot \theta)$ times the mean radius divided by 2. This means, of course, that to change the effective radius of the miter bend, one merely changes the length of the centerline cord, which is true.

The formula for the minimum thickness of the pipe for the two types of miter bends is similar to Eqs. (5.4) and (5.5) without the portion that converts the result of the calculation into a minimum allowed pressure. For closely spaced miters, the formula is

$$t_s = t_m \frac{2 - \frac{t}{R}}{2\left(1 - \frac{t}{R}\right)}$$

For widely spaced miters, the formula is

$$t_s = t_m\left(1 + 0.64\sqrt{\frac{r}{t_s}}\right)\tan\theta$$

Again, note that the widely spaced formula has the same factor t_s on both sides of the equation, which requires an iterative solution or, with manipulation, a quadratic solution. The spreadsheet solution is quite simple with the goal seek function.

If one experiments with the formulas for both, the inescapable conclusion is that there is a vast difference in the results to make a given miter design's geometries have a required thickness. One can only assume this is because of the difference in the design philosophies of the two committees regarding the safe margin in a given design. The differences lessen as the sizes of the pipe change. It is also true that the different fluids, and so on, that are used are the basis for those differences. It is hoped that the mechanical design committee, whose mission is to establish a standard, can find a way to minimize those differences.

There are of course other differences depending on the other analysis that one does in calculating the acceptability of the miter for that code. It seems that the simple thing to do is use the B31.3 approach and check against the minimum thickness formulas given in B31.3. It goes without saying that building an FEA model is acceptable, but that is an advanced methodology. Nevertheless, as the technology advances it may be the simple way.

Setting up B31.3 formulas in a spreadsheet gives one both speed and flexibility in calculations. The foregoing is a discussion of the methodologies to determine the pressure that a given pipe will sustain at a given temperature. Or, what size pipe is required for that temperature and pressure given the pipe and its material? However, one might recall that in every case the discussion revolves around pressure that is internal to the pipe. What happens when the pressure is external to the pipe? The first thing to note is that the tensile formulas will work, but there is another failure mechanism. Called by many names, basically it is buckling or instability based where failure doesn't occur in the same manner. There is a second check that does not necessarily have to occur in every situation but should occur, and the designer or engineer needs to be aware of the situations where that happens. Therefore, we move to the case of solutions for external pressure.

EXTERNAL PRESSURE

When we are dealing with internal pressure we are dealing with the tensile properties of the metal as the pressure is trying to expand the metal or, as one person put it, tear the pipe in two. With external pressure just the opposite occurs. It compresses the material and tries to squeeze it together.

As you may know, steels in particular have very similar properties in tension and compression. In that case the questions become: What is the big deal if the compressive strength is similar in size to the tensile strength? What difference does that make? While it is true that the yield points and thus general distortion and ultimately failure can occur, another phenomenon can occur in compression where the failure is well below the yield point. We are so used to thinking of the failures being proportional to the load applied that we tend to neglect the failure from buckling instability.

Buckling instability is well known in columns. It is described in many strength-of-materials books by using the Euler formula. A column loaded in compression has a critical load, where the column can fail by buckling before it fails due to the compressive load. This is based on the cross-section of the column, the modulus of elasticity, a constant based on the end supports of the column, and, most important, a factor called the slenderness ratio. The slenderness ratio of the column is the length divided by the radius. There are many variations of the computational ways to determine what the critical load is, but for the details of a column, readers are referred to a strength-of-materials book such as Roark's *Formulas for Stress and Strain*.

The buckling of pipes and tubes has a very similar buckling phenomenon based on the OD, wall thickness, external pressure (net), and length of pipe between adequate supports. One might ask, when is this a problem? Well, just consider the OD of a pipe or tube with a very thin wall. Surely you have handled an aluminum soda can; you can shake up and free the entrapped CO_2 while the can is sealed. Therefore, it can withstand a fairly large internal pressure. However, one can squeeze the can (especially when empty) and it will collapse. The question is at what pressure (squeeze) does that occur and with how large a can. Now, consider a pipe with a vacuum where the external pressure could be 15 psi ($\approx$ 100 Kpa), double-walled piping with pressure in the annulus, piping underwater, pipe inside a pressurized vessel, fire tube, and so on. In other words, it can happen to pipe with some regularity.

ASME Division 2 has an extensive set of graphs and charts for several materials that allow one to calculate an allowed external pressure. A excerpt from these charts is in the Appendix, but not the entire set. Here we work through the analytical aspects that were basically used to develop these charts. It should be pointed out that the calculation method here is only appropriate for one material at 300°F (150°C) or lower. Regressing the other temperatures and materials would be an arduous task and the charts for those are available through ASME. The piping codes reference these ASME charts for their requirement to check compliance with their codes. It is suggested that if the condition one is checking for this material is sufficiently resolved by these calculations, one can safely assume they have met the intent. However, unambiguous compliance with the code might require use of the graphs and tables.

A two-lobe form of a tube collapse is the scenario calculated. This is the lowest pressure that would cause this type of failure. The next higher pressure would be 2.66 times higher, and if the situation was not sufficient to cause this two-lobe failure, the higher pressure is of no consequence as the pipe has already failed.

It is important to point out that the development of the formulas introduced here is fraught with considerable highly technical mathematics, which are not shown. What we are working with is the resulting derivations from that math, substantiated by experiment.

The length of the pipe is important; like the slenderness ratio, it does come into consideration. It is defined as the distance between two end supports of a pipe. For instance, consider a length of straight pipe with a flange on the end of the spool and a valve some distance away. That distance between the flange and the valve would be the length under consideration. In making that comparison it is assumed that the flange and valve had enough moment of inertia to hold the pipe circular. In the absence of any stiffener similar to the valve or flange, one could add a stiffening ring around the pipe. The determination of what constitutes an appropriate distance is based on the results of the investigation, which in effect is a trial-and-error situation in the ASME methodology.

The chart/graph methodology is to establish a factor A with a graph or chart using as the independent variables the ratio length L previously described and the OD of the pipe as the first variable. The second variable is the OD of the pipe divided by the wall thickness. Using those two variables one can read factor A.

The factor, A, is a form of a stress strain curve related to the geometry of the tube including length. Therefore it is essentially independent of the material being used.

Then one checks for the appropriate material chart and uses factor A as a variable. Read the chart, which has a different line for that material at different temperatures, and from those two variables you get a factor B. Factor B assumes a critical elastic buckling stress modified by experiment and a reduction for safety margin to establish the allowed external pressure. If that pressure is higher than the design external pressure your estimated length and wall thickness for that size pipe is adequate. Since the length is usually established by the geometric layout, if the pressure is not high enough one starts with a new thicker pipe and repeats until there is a sufficient solution.

Subsequently, there is a need to determine the size of the stiffening rings, which will be discussed after going through this abbreviated (due to one material temperature) calculation procedure. The reader is cautioned that this process does not allow for creep-type deterioration of the material. That is a separate consideration and beyond the scope of this discussion.

The first step is to calculate the critical length. As one examines the graphical presentations, you note that factor A tends to flatten out at a certain length/diameter ratio. This is important, because above this critical length there is a different calculation procedure for factor A.

It should be noted that the following analytical procedure, as with other procedures, does not get results that will be absolutely the same as real-world values. The real world does not exactly follow the precision of the math world. As is noted there is a " butterfly effect" that occurs in complex systems. The formula is as follows; note that it is the same in inches or millimeters:

$$L_c = 1.11 D_o \sqrt{\frac{D_o}{t}} \tag{5.7}$$

where
  L_c is the critical length
  D_o is the outside diameter
  t is the wall thickness
An example with 6 NPS S40 pipe is

$$1.11 \times 6.625 \times \sqrt{\frac{6.625}{0.280}} = 35.77 \text{ in.}$$

$$1.11 \times 168.275 \sqrt{\frac{168.275}{7.112}} = 908.558 \text{ mm}$$

This then is used to determine what factor A to use for a length that is more than the L_c:

$$\text{Factor } A = 1.1\left(\frac{t}{D_o}\right)^2 \tag{5.8}$$

$$1.1 \times \left(\frac{0.280}{6.625}\right)^2 = 0.001965 \text{ in.}$$

For factor A in millimeters multiply the answer in inches (0.001965) by $25.4 = 0.04991$. *Note:* If one is using Eq. (5.8) in millimeters, the factor of 1.1 changes to 27.94:

$$A = 1.1(25.4)\frac{7.112^2}{168.275} = 27.54(.00179) = .04919$$

Once again we get minor errors when converting from US customary to Metric. This can come from the empirical nature of the solution.

If the length is less than the critical one, the formula for factor A changes:

$$\text{Factor } A = \frac{1.30\left(\frac{t}{D_o}\right)^{1.5}}{\frac{L}{D_o}} \tag{5.9}$$

A specific L is needed to calculate this. To dramatize the difference in the two factor A's, the following example uses an L of 35 in. (889 mm). Using Eq. (5.9) at the specified length, factor A becomes

$$\frac{1.30 \times \left(\frac{0.280}{6.625}\right)^{1.5}}{\frac{35}{6.625}} = 0.002138 \text{ in.}$$

As before, if one uses the millimeter units in the calculations, the factor of 1.3 changes and becomes 33.02, which gives a factor A of 0.05431 mm. This merely states mathematically what is intuitive: as the tube gets longer it takes less external pressure to buckle. Of course, it is also true that OD and the thickness of the wall play important roles in determining the buckling pressure.

Once factor A is determined, it is used to determine what ASME calls factor B, which is really the allowable stress for the material. This stress is not like the allowable stresses that are in the stress tables in the B31 books. It is based on what is called the critical pressure—the pressure at which the pipe or tube actually begins to collapse or buckle with appropriate safety factors for things such as out of roundness of the tube and other imperfections that are not known at the time of the design, as well as conservatism. The ASME Section VIII formula to calculate the critical pressure is as follows:

$$P_a = \frac{4B}{3(D_o/t)}$$

For our analytical calculations above B would be 13,800 and D_o/t is 23.66. So, the allowable external pressure would be 778 psi. Using the graph method, the author got 13,000 for B so that approach would obtain 733 psi as allowable pressure. It is another incidence of how different mathematical methods do not give precisely the same answer.

There have been many analytical ways to marry the fact that the actual pipes are not perfectly circular, thick, or smooth. These have involved empirical work done over the years. The analytical work has been done for over 100 years. There is a book, *Textbook on Strength of Materials,* written by S. E. Slocum and E. L. Hancock in 1906, that has a lengthy discussion of the theoretical calculation of this critical pressure, and then a subsequent chapter that develops the theoretical formula into what Mr. Slocum calls a more practical method. Roark's sixth edition of *Formulas for Stress and Strain (Table 35, Case 19b)* has the same formula, lists it as approximate, and attributes it to a 1937 book. The formula does not include any of the margins established by the ASME for safety. However, it is reasonably close to the critical pressure in the table below, if you assume a margin between 3 and 4, which is what is used it is within reason for the allowable pressure. These various theoretical and practical calculations vary somewhat wildly with the ASME charts.

The ASME charts are based in part on work done at the University of Illinois in "Paper 329." Those charts were done without any allowances for experimental data. The knockdown factors and allowances to account for experimental results and shape factors were included in the development. The base formula is

$$e_c = \frac{E_R}{E}\alpha\alpha_e$$

where
e_c is the critical stress
E_R is the tangent modulus
E is the modulus
α is a knockdown factor to accommodate the differences between tests or experiments
α_e is the theoretical stress

In short, a lot of work has gone into the development of the charts. There are 21 pages of graphs, most of which have four different materials, each with a variety of temperature lines to read factor B. Then ASME added some charts that in effect digitize the answers to factor B. One should use them to determine the exact knockdown factors and which exact theoretical form they used. To calculate the theoretical stress is at best difficult. In addition, the graphs and charts are made with a margin of 4. And even when one gets a factor B, it needs to be converted to a margin of 3 by multiplying by 4/3.

If one repeatedly uses the same materials at the same temperature it is possible to make a regression on those charts and then use that regression to calculate B from the calculated A with the formulas we used before. For this book and personal use, that regression was made. With a good graphing or regression program the digital charts make that fairly easy. Note it was not said that it was a quick process; it takes time. Since the charts are not in SI units, further discussion is only in USC with apologies to the SI purists. It is emphasized that any code calculation should be checked with the actual graphs and charts. The work is not official. This would be especially true if the calculation created concern by being close to failure. There may be commercial programs that have codified the work required.

That being said, readers are informed that more information is in the Appendix. Using that data and the factor A calculated earlier, one gets the results shown in Table 5.3.

The allowed pressure is calculated using the formula from ASME Boiler and Pressure Vessels, Section VIII, Paragraph UG 28 for cylinders that was discussed above.

Assuming that the pressure you are testing for is only a vacuum where the external pressure is 15 psi, you are done with the calculation. If the external pressure exceeds the allowable pressure, the first thing is to try a thicker wall. Depending on many factors, that may be expensive or difficult. And again, depending on the situation, another way to accomplish that is to shorten the L distance. As noted earlier, shorter is inherently stiffer.

The last concern would be the moment of inertia. Whatever holds the ends of the section of pipe being considered in the external pressure would have to be sufficient to handle the loads imposed by that set of external forces. Those stiffeners can be flanges, valves, structural shapes, or just plain rectangular rings. The moment of inertia calculations are standard calculations, so they are not further discussed here. The major

Table 5.3 Regressed Factor B for Selected A Factors Using Regression on Chart CS2

Calculated A*	Factor B	Allowed P (psi)	Critical P (psi)
0.001965	14,989	845	2534
0.002138	15,261	860	2580
0.002993	16,184	912	2736

*The first two A factors are either side of the critical length and close; the third is at two-thirds the critical length.

question is: What moment of inertia is required? The following formula allows one to calculate the appropriate moment of inertia:

$$I_{required} = \frac{D_o^2 L}{10.9}\left(t + \frac{A_s}{L}\right) \text{Factor } A \qquad (5.10)$$

The familiar symbols have the same meaning as before. The new symbol A_s is the area of steel that is involved in the calculation. One might question the 10.9 factor, as it is somewhat different than a typical moment of inertia factor. It is an allowance for using the pipe between the stiffener as part of the calculation. If that pipe is not used, the reader is directed to ASME UG 28 and further discussion of the other alternative.

For purposes of simplification, Figure 5.4 shows the arrangement and relationship. Note that this figure and the above formula make the assumption that there are a series of rings placed midway between a much longer pipe. Therefore the L is split halfway to either side of the ring. This configuration would be more compatible with a long pipeline-type situation rather than a flange and valve, which was posited in some of the

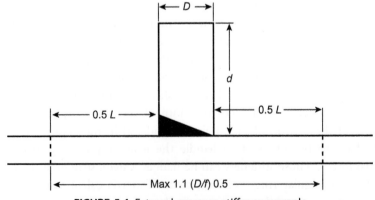

FIGURE 5.4 External pressure stiffener example

previous discussion. The figure shows a simple cross-section of a rectangular ring. If one uses some other arrangement, it will alter the actual moment calculation. But success is achieved when the calculated moment is larger than the calculated required moment.

Note that the actual calculation may require repeating if the first attempt to size the moment of inertia does not meet the requirement, since the size of the ring is also part of the requirement calculation. This is just another case where experience and judgment may be required.

Straight Pipe, Curved Pipe, and Intersection Calculations

Contents

OVERVIEW

This chapter covers the basic requirements for establishing the appropriate wall thickness for various pipe diameters and the different materials chosen depending on the type of service. In effect, that wall thickness establishes the pipe pressure-temperature rating. If the piping system were just straight pipe or, as necessary, curved pipe that was welded together, the pressure portion of the design would be complete.

However, it is the rare pipe system indeed that doesn't have intersections and for that matter places along the pipeline where something is used, most often flanges, to break the continuity of the pipe run. The intersections include such things as tees, wyes, laterals, branch connections, and, rather than bent pipe, fittings such as elbows. In fact, there are a myriad of fittings that can be used to accomplish the even larger varieties of ways that pipe can be put together.

There are, of course, many different types of piping systems. They can be broken into two broad types. One type is facility piping systems. One characteristic that helps define these is that they are generally located in one specific area such as a power plant or a process facility. The other type is transportation piping systems. These are characterized by the fact that they transport their fluids over terrain that may vary from the sea floor to a highly urbanized area. Each of these types have unique requirements and the various piping code books address these specific requirements. However they both require some or all of the fittings described above.

Piping and Pipeline Calculations Manual
ISBN 978-0-12-416747-6
http://dx.doi.org/10.1016/B978-0-12-416747-6.00006-1

Those fittings that are ubiquitous can be standardized; in fact, there are many that are codified into standards. These are often called *dimensional standards* since the most obvious thing they do is establish a set of dimensions that define how they would fit into the piping system. By pre-establishing these "take-out" dimensions they perform an important function in allowing the standard piping system design to proceed long before the actual purchase of the components. They even allow in a somewhat more limited sense the opportunity to precut the individual pieces of pipe that are to be fitted into the puzzle of the system.

One of the secondary issues in developing these standards is that these fittings are also subject to the same temperature and pressure requirements of the parent pipe to which they are attached. This gives rise to the need to establish the pressure rating of the fitting or component itself. A fundamental philosophy would be that the pressure-temperature rating of the base pipe of the system represents the minimum rating that should be installed in the system.

CODE STANDARDS

Each of the B31 code books has, as one of its responsibilities, the function of reviewing the various standards that are available and listing those that meet the requirements of that code. Naturally, many of the more common standards are recognized by all code books. Not all are for obvious reasons. One of the things that the committees look very hard at is the way any particular standard approaches the business of pressure ratings in conjunction with the requirements of the process for which they are most concerned. When a standard gets "blessed" by a code book the meaning is simple—users of a product from that standard need only be concerned about the compliance of that product with the requirements of the standard. If it meets the standard, it meets the code.

One can use a product from another standard that isn't on the approved list. However, in using that standard some compliance with the code must be established. There are several ways to do this, including using the traditional area replacement calculation method, or some other method that is deemed to give the same margin of safety that is embedded in the code. Many codes have a paragraph or refer to some other means of proving the code. Both of the general methods also apply to products that are not covered by a standard, but are special. Oftentimes, such a special product might be a variation of a standard that doesn't in some manner completely comply with the standard. This noncompliance needs to be agreed on between the manufacturer of the product and the purchaser, and that agreement should include some explicit understanding that the product meets the pressure-temperature rating as if it complied in all respects with the original dimensions.

Definitions

When a code defines what is necessary to meet its requirements, the approved methods are often spelled out, including information such as the following:

1. Extensive successful service under comparable conditions, using similar proportions and the same material or like material. This rather begs the questions: How did one get that experience, what is extensive, how close is similar, and what is like material? It is basically a "grandfather" clause for items that have historically been used. Then the question is: How did it get into codes that have been around for years?
2. Experimental stress analysis.
3. Proof test in accordance with . . . . There is a list of accepted proof tests ostensibly to cover various shapes and situations. The B31 technical committees have been working to develop a universal standard proof test.
4. Detailed stress analysis such as the growing use of finite-element analysis. One should be sure the analysts know what the special needs are. For instance, analysis in the creep range is significantly more restrictive. Plus, one asks: Does the analysis of one size cover other sizes? What are the acceptance criteria?

The preceding paraphrases some of the existing lists. The comments are not intended to point out that the list is not sufficient. When done properly, any of the methods is more than adequate to define compatibility. Codes also point out that the particular requirements do not replace good engineering judgment in the use of the procedures.

The comments basically point to the fact that the use of the method is not sufficient unto itself. This is especially true of the user/purchaser's acceptance of the use of one or more of the systems. It points to the reasons that the technical committees are working on the universal procedure. I was privileged to read a report from one of those methods. To paraphrase, "We tested it, it passed." Now if that had come from a source that one knew well and was familiar with over a broad range of situations, it might be marginally acceptable, but even then I would ask by how much and what was tested, etc. There is more discussion on this later.

Intersections

The point so far is to say that the code acceptance of a product other than a piece of straight or bent pipe is somewhat more complex than the computation of a minimum wall thickness to establish the product's pressure-temperature rating.

The fundamental philosophy of such intersections should revolve around this fact. Nothing should be added to the pipe that reduces its ability to safely perform its duty of transporting its designated fluid through the designed system at the designed pressure and temperature. This is a rather wordy way of saying that a chain is as strong as its weakest link. That weak link in the piping system should be the pipe, and we are discussing here the minimum pipe. Of course, there is nothing that says the designer can't use and, in many cases for many reasons, will use stronger pipe.

For instance, if one calculates that a minimum wall of a 6-in. pipe needs to be 0.3, the designer has a decision to make. The minimum wall of an S40 pipe is 0.245 in. (0.280 × 0.875 manufacturing tolerance), while the minimum wall of an S80 pipe is 0.378 in. The designer might petition to lower the pressure-temperature rating, which is unlikely to happen. Therefore, he or she might specify a special pipe wall size, which might be cost prohibitive. Likely, he or she will choose the S80 pipe and have a stronger-than-necessary pipe. This particular dilemma may follow the designer through the course of this pipe spool design. Many of the components, as mentioned, are standardized and for that reason it is often more cost efficient to use a standard-size component that has a pressure rating higher than the particular system for which it is being specified.

In transport piping where the size or length of the pipe may make it probable that the pipe be made from a special thickness of plate then rolled and welded to the proper diameter, it might be quite feasible to order the special pipe wall size. In some such pipelines there are miles of pipe of a given required wall.

However, when an intersection is going to be made in the base pipe other considerations are necessary. It is likely that U.S. piping codes and other countries' codes will have something like the following statement: "Whenever a branch is added to the pipe the intersection weakens the pipe.

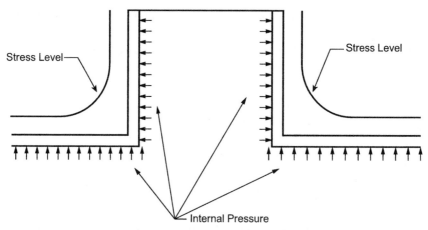

FIGURE 6.1 Intersection stress increase diagram

Unless the pipe has sufficient wall strength beyond that required without the branch, additional reinforcement is required." See Figure 6.1 for an example.

Then the codes proceed to tell you what kinds of reinforcement can be used. All piping codes also have the ability in certain circumstances to use other methods. Some of those methods discussed are already methods that require work or calculation that is not required by the approved or listed standard.

The grandfather method known by most pressure code users and many nonusers in general is what is known as *area replacement.* Area replacement, although simple in concept, is somewhat complex within the different codes. We discuss it here in the context of a generic concept that will describe the fundamental process. Readers are left to choose a particular code and follow its specific process. Many of these differences in methodology are brought about by the fact that some of the pipeline codes work with nominal walls in their calculations, while others work with minimum walls. In addition, some have restrictions on what you can and cannot use regarding sizes and methods.

Fundamentally, one knows the required wall thickness for pressure on both the run and the branch. As Figure 6.1 shows, the increased stress in the assembly extends in a decreasing manner as one moves away from the discontinuity. Those distances are not the same. There are very complex theoretical analyses that determine the distance of such discontinuities. They are important in limiting the area where any reinforcement is particularly effective. The codes' intentions were to make the assembly safe.

So they rather simplified the analysis. Looking at Figure 6.1, how far should one go for an acceptable decrease in the intensity of the stress? The expression that defines the amount of stress as one moves from the center of the hole out is rather simple:

$$\sigma = \frac{1}{4}\left[4 + 3\frac{r2}{x2} + 3\frac{r4}{x4}\right] \tag{6.1}$$

where σ is the stress at distance x from the hole and r is the radius of the hole.

ASME STANDARDS

Given those factors, the decision to be made is how far along the run should the distance x be set as acceptable. ASME decided to make that distance one diameter of the opening in either direction. Noting that one diameter equals two radii, let us check what occurs at the edge of the hole where it would be highest and then at the two-radii distance. One should note also that the use of 1 in the numerator of the fraction ¼ in Eq. (6.1) is a unit substitution. If one were doing a real calculation, that 1 would be replaced by the nominal l hoop stress that was calculated for the unbranched pipe. Assuming an opening of 6 in. at the edge (one radius from the center), the stress would be

$$\frac{1}{4}\left(4 + 3\frac{3^2}{3^2} + 3\frac{3^4}{3^4}\right) = 2.5$$

So whatever the nominal stress is, it is 2.5 times the stress at the edge of the hole. Now moving out to the diameter (2r) from the center, what is the stress?

$$\sigma = \frac{1}{4}\left[4 + 3\frac{3^2}{6^2} + 3\frac{3^4}{6^4}\right] = 1.23$$

At this point the stress is 1.23 times the nominal. One would note that if the reinforcement were there and completely integral, the new stress would be somewhat lower because of the additional material. A good question might be: How far does one have to go to make the stress absolutely nominal by this simple calculation? Well, for this size hole, it turns out that the distance to 1 is ≈ 92 in. For smaller holes it is less, and for larger holes it is more. That is one of the complexities.

But how far up should one go along the branch from the run? Here again, there is a simplification of the complex math that can be used. This factor involves the use of Poisson's ratio and, indirectly, Young's modulus (modulus of elasticity) and the moment of inertia, as well as the radius. Without going through the rigorous math, that ratio in terms of Poisson can be expressed as the allowed length up the branch equal to

$$\frac{1}{\beta}$$

where

$$\beta = \frac{\sqrt{rt}}{1.285}$$

This equation was worked with Poisson's ratio set at 0.3, which is for steel. This is sensitive to both the radius and the wall thickness. ASME decided to use the thickness as a guide and to note that the walls of standard pipe can be very roughly equated to 10 percent of the radius of that pipe, especially in the standard and extra-strong wall dimensions. This 10 percent wall thickness calculates out as follows. Setting r equal to $t/0.1$ and rewriting the preceding equation for l in terms of t, we get the following:

$$\frac{\sqrt{\frac{t^2}{0.1}}}{1.285}$$

This equation gives the same answer as setting t in the previous equation to $0.1r$. That result is 2.46. ASME chose 2.5 times the header wall thickness or a combination of branch and header walls as the standard.

Without the variations per book, and understanding that in most cases we are talking about the minimum required wall thicknesses, the reinforcement zone is defined as one diameter of the opening on either side of the center of that opening and 2.5 times the thickness above the surface of the pipe. It should be noted that all material within the established reinforcement zone is usable.

It is also true that one can extend into the internal diameter of the pipe the same amount. This is not often done in piping because that would interfere with the flow of the fluid through the pipe. It is done fairly frequently in vessels and/or tanks where fluid flow past the nozzle or break is far less common.

The basic idea is simple. When you cut an opening, you remove an amount of metal that has an area through its diameter that is equal to the required thickness of the pipe from which you have cut the hole times that diameter. Then you calculate the necessary reinforcement zone with the rules just established. The material that is originally required for the two pieces—branch and run—is within that zone; however, the amount required for pressure integrity is not available because it has been removed. So you must add an equal amount of metal to the area that is removed within the reinforcement zone. There are at least four areas of potential reinforcement within that zone:

1. Excess metal in the run from which you have just made the opening
2. Excess material on the branch pipe
3. Material in any ring pad added
4. Attachment welds

If one designates the metal cut out as A_R and the excess metal in the areas as $A_1 + A_2 + A_3 + A_4$, the area replacement becomes a simple calculation of making those two sets equal. Some of the ASME codes add an additional factor for cases where the nozzle is inclined away from the 90° intersection with the centerline of the pipe by an angle. This is called θ and is defined as the angle from the centerline of the run pipe—that is, the calculated diameter is multiplied by the factor of $(2 - \sin \theta)$. This manipulation changes the area required to an equivalent of the major axis of the ellipse that is formed when a cylinder is cut on a bias. So the designer goes through the steps, as indicated in Figure 6.2.

As a reminder one must always actually use the specific requirements of the code the designer is working with, as each has idiosyncrasies specific to the code. We will keep the generic focus in the example.

Assume that we are preparing to put a 3 NPS (80 DN) S40 branch on a 6 NPS (150 DN) S40 run. The respective minimum walls are 0.189 and 0.245 in., based on the U.S. manufacturer's tolerance being deducted. The design temperature is 350°F (175°C) and the material allowed stress is 23,000 psi (158 Mpa). The pressure of the system is 1750 psi (11.8 mpa). This basically uses all of the minimum wall of the 150 DN pipe, and therefore, only uses 0.129 in. of the 80 DN pipe. The opening cut in the run for a set on branch is a 3.068-in. diameter, so the area required is $A_R = 3.068 \times 0.245 = 0.752$ in.^{2}. There is no excess metal in the run and the height of the allowed reinforcement zone is $0.245 \times 2.5 = 0.6125$ in. The excess of the branch is $0.189 - 0.129$, or 0.06×0.6125, or the inherent excess area is 0.036 in.^{2}. Now the designer has to decide what to do.

One solution is to change the run to S160 pipe. The minimum wall of S160 pipe is 0.491 in., which is twice the 0.245 in. of the required wall of 6 NPS pipe. That could, depending on the geometry of the spool piece, be a

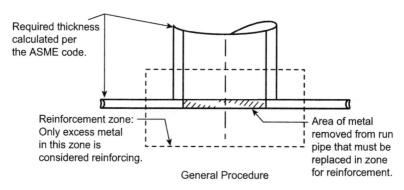

Required thickness calculated per the ASME code.

Reinforcement zone: Only excess metal in this zone is considered reinforcing.

Area of metal removed from run pipe that must be replaced in zone for reinforcement.

General Procedure

1. Calculate area of metal removed by multiplying required thickness by the diameter of hole; adjust for angle as allowed.

2. Calculate total area of excess metal in reinforcement zone; this includes attachment welds.

3. Increase pad, run, or branch wall until item 2 is equal to or greater than item 1.

FIGURE 6.2 Generic area replacement

short length of that pipe. Certainly that would be less expensive than making the entire run out of heavier pipe. A short pup piece might not be economical because of the extra welding that might be required.

The designer could of course put in a pad piece. The basic OD of the ring (outside diameter of the reinforcement zone) is 6 in. and the OD of the branch is 3.5 in., so the ring would have an excess length of 2.5 in. Such a pad at $5/16$ -in. thick would have sufficient metal to meet the required 0.751 area, and have a little excess from the branch and welds for tolerances, and the like. There is the cost of welding on the pad piece and testing the pad, as most codes required for bubble tightness would add extra expense. I haven't mentioned the possibility of increasing the branch wall, which is also an exercise of the same type, given the very short height usable frustrations.

Most likely, an experienced designer would have specified other means of establishing the branch. There are tees in those sizes. There are branch outlet fittings for that type of branch connection. These would all be covered by an accepted standard and thus eliminate the dilemma of what to do. In most cases for standard size, this is the preferred solution. The area replacement method is there and as noted can work. In some cases it is the simpler solution.

The caution to always work with the specifics of the project's code of record is even more important when one is working with a non-U.S. code.

Codes such as EN13445, BS E11286, AD Merkblatt, and others all use a different approach. The approach is often called the pressure area approach. In this approach the area of the design pressure within the compensating limits multiplied by that area must not exceed the area of the metal in the same area multiplied by the allowable stress.

These methods, while not precisely equivalent in terms of numerical equivalence, give reasonably the same margins. The major differences are due to the way the reinforcement boundaries of the compensation limits are set. The methods can be expressed as follows:

$$B = \sqrt{(D_i - t_r)t_r}$$

where

B is the distance from the outside edge of the nozzle to the end of the compensation limit

D_i is the ID of the run

t_r is the run (the proposed thickness for the final solution)

and

$$H = \sqrt{(D_{ob} - t_b)t_b}$$

where

H is the height up the branch for compensation

D_{ob} is the OD of the branch

t_b is the thickness of the branch (again, the final proposed thickness)

The procedure is basically simple and realistically the same for all configurations. Figure 6.3 is a sample configuration of a branch with a pad. The EN code and others have drawings of several different configurations. The basic formula is the same for all configurations, except one must adjust for the actual variations in the fundamental areas due to different configurations.

Looking at Figure 6.3, the basic formula is the same. The weld material is not shown in the sample figure because each code may have different rules regarding where the weld goes and how much to do. The rule is that any weld material in the zone may be included in the computation. In each case, for any material the appropriate allowable stress is the minimum of that material's stress and the allowable stress of the run that is being reinforced. Given all that, the formula is

$$(A_s)(S_s - 0.5P) + (A_{pad})(S_m - 0.5P) + (A_b)(S_m - 0.5P)$$
$$\geq P(A_{shellpressurearea} + A_{branchpressurearea} + 0.5A_{shellpressurearea})$$

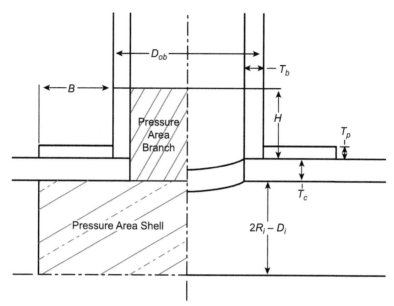

FIGURE 6.3 Sample pressure areas

It should be obvious that A_s is the quantity $(b + t_b)t_s$, the area of the pad is $(B - t_b)T_p$, and the area of the branch is HT_b. Each of those areas should include the area of weld that is in the zone. The $A_{shellpressurearea}$ area is $(B + T_b +$ internal radius of branch$)$ R_i and the $A_{branchpressurearea}$ area is $(H + T_s)$ (internal radius of branch) for this sample configuration. However, each area can be configured differently and the amount would change as the size or thickness changes.

EN13445 explicitly states that this problem will most usually require an iterative solution. As we move further into the electronic world this is not the deterrent that it once was.

Using the minimum thickness for the solution we worked before, we find that the boundaries are a little larger. The outside boundary circle would have been 7 in. (179 mm) and the height would have been 0.79 in. (20 mm). This would have required some thought as to how to solve the problem as it did in the area replacement example. The solution might have been the same or there might have been another solution.

The B31 codes don't specifically allow the pressure area method. On the other hand, they don't specifically exclude the use of such devices. Many of the B16 codes, which are fittings standards, say one can use mathematical methods approved by a recognized code. Certainly, the EN codes are recognized. We are after all one world. In the United States the procedure is somewhat different than the EN method just discussed. These methods are quite useful, short of finite-element analysis, since often they are the most difficult to use with the area replacement method (see Figure 6.4).

One immediately notices in the figure that it would be very hard to calculate an area replacement for this as there is no definite hole in the elbow. Note also that the picture is primarily for a forged elbow rather than one formed out of pipe. The examination shows that the extrados of the elbow is intended to be thicker than the nominal, or even minimum, wall of the pipe that is formed. The equations in the calculations for pipe bends discussed earlier actually call for that thickness increase. Fortunately, the bending process tends to give one that thickening as a result of the compression in the bending. The forging process that basically forges a solid elbow and then bores out the waterway passage also tends to give the thickening as a result of the process.

The question then remains as to how much thickening occurs. Experience has shown that it is not always enough, which is the reason the codes call for proof. Generally, the forged fittings can be proven more readily by the mathematical process. The bent ones were proven by the proof testing

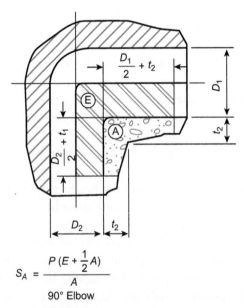

$$S_A = \frac{P\left(E + \frac{1}{2}A\right)}{A}$$

90° Elbow

FIGURE 6.4 Pressure area elbow

process. The advent of the equations given to define the minimum thickness of the intrados has greatly reduced the need for any proof but measurement. There is not quite the same universal acceptance of the allowed thinning of the intrados. This may be due more to concerns about other factors than pressure design only. For instance, if one has a fluid with entrained solids, there may be faster or enhanced erosion, and the thinner wall would not be an advantage. That should be built into the corrosion or mechanical allowance, taking that erosion into account.

The pressure area method has been used most in forged wyes and laterals. In the Appendix there are drawings similar to Figure 6.4 for elbows and laterals. There is also a drawing of a tee where that same method can be used. It is suggested that the reader take a normal tee and run it through the calculations. As a hint, use square corners rather than radii in the corners. Using the same pipe wall as the matching pipe makes it difficult with this method to prove that the tee can be fully rated. Readers are reminded to reread the portion on extruded penetration discussed earlier. The reducing tee situation is much easier. A full tee almost demands that the pipe wall be thicker.

When examining many tees one finds that often the starting material is of a heavier wall than the matching pipe wall. As one reads B16.9, the fitting standard that covers the majority of the butt-welded tees, he or she will find such words, as it is expected that some portion of the formed fittings will have to be thicker than the pipe wall for which the fitting is intended. Chapter 6 of that same standard discusses that one may have to make certain adjustments if he or she needs certain things like full bore, as the bore away from the ends is not specified. There are other cautions regarding what one gets from a standard B16.9 fitting.

The forged fittings of B16.11 do not have the same problem. In a forging, particularly one that is a socket weld end or done to fit the pipe, the OD does not have to match the mating pipe. These are details to be concerned about when one is working with the pressure design of the fitting.

As mentioned, there are other forms of the U.S. version of the pressure area method. They have been used successfully in certain cases for compliance with the B31 codes. Modified versions of this methodology are creeping into the ASME codes. This has started with the boiler code (Section VIII, Division 2).

Regardless of the method one uses to analyze the intersection, all codes make reference to the fact that if you use reinforcing material that has different allowable stress at the design temperature, there are additional things you must do. If the allowable stress of the reinforcing pad or material is less than the material one is reinforcing, the amount of reinforcement must be adjusted by a code-specific formula to allow for that lower-strength material. In the B31 codes that is the ratio of the allowable stresses. If the material has a higher allowable stress nothing extra may be taken for that stronger material. In general, this applies also to all situations where one is utilizing different materials.

This can cause some problems when one is using higher-strength run pipe to reduce the weight of the purchase of the larger amount of pipe. Also, when one is attaching smaller branches the need to make the intersection or reinforcing area larger due to the lower stress can become quite stressing (pun intended). This situation of mixed-strength material has been relatively more prevalent and newer alloys have been developed to reduce weight in the larger usage components, such as pipe, but the material hasn't yet worked its way into the lower-volume (weight) market.

These two traditional methods of computing the required reinforcement are for welded branch connection intersections. Many times an intersection can be made without welding something directly to the pipe. This is called an *extrusion*. Essentially, weld preparation on the branch pipe is done by extruding a surface out of the run pipe. This is accomplished in several ways. The traditional method is to cut a smaller hole and put on it some pulling device with a tool the size one wants for the final branch opening. Then, after heating the area, one pulls the larger tooling out and creates a radial lip on the base pipe. The weld prep is then prepared and the branch can be welded to it. This method is often used in manufacturing. Tees, especially reducing tees, are often used in large pipeline manifolds to reduce the required welding when the branch is smaller than the run. There is more discussion of this in Chapter 13 on fabrication. But how does one calculate the design itself?

The more general pressure area method only requires a change in configuration, which results in a different calculation set due to the change in geometry. The B31 codes state that the previous method can essentially be used. However, it recognizes that there may be a difference depending on certain factors. The main requirement is that the extrusion must be perpendicular to the axis of the run. In addition, there are specific minimum and maximum radii on the extruded lip. Both relate to the OD of the branch, and there are graphs in both USC and SI units in the Appendix. There is a change in the way that one calculates the height of the reinforcement zone; that new formula is

$$H = \sqrt{D_b T_x}$$

where T_x is the corroded thickness at the top of the R_x.

The other major change in extrusion that meets the required limitations is that there is a factor K that reduces the size of the required replacement material, such that when the standard area of the opening is calculated, it is multiplied by this factor K, which is 1 or less depending on the ratio of the branch to the header (D_b/D_h). It is a straight line function between 0.7 and 1.0 from a ratio of 0.15 to a ratio of 0.6. A graph of that function is in the Appendix.

As a generic illustration, Figure 6.5 shows the other difference to be that the actual area in the radius is not included in the calculations. This is not, area-wise, a significant loss. The radius, however, has significantly fewer rough spots than a weld, and so less spurious discontinuities. This accounts for some of the reduction in the required area.

Careful observation of Figure 6.5 shows a taper to the wall thickness of the branch. This is most likely to happen, but is not necessarily required. Good practice requires that there be a minimum mismatch between two butt welds. The taper can be inside or outside, but when one is thinking about minimizing the amount of reinforcing material, it would be wise to make the extruded ID as small as possible. The process flow may enter into the determination. In addition, there are maximum mismatches in wall thickness between two butting joints. This is discussed further in Chapter 13 on fabrication.

There is another analytic-type method that is working its way through the B31 code cycle. It has already been adopted as a code case for intersections in the boiler code. It does have limitations on situations it may be

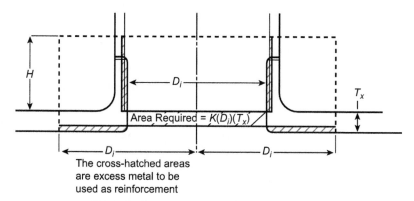

FIGURE 6.5 Generic extrusion area replacement

used for, but it is based on research work reported in the Pressure Vessel Research Counsel (PVRC) Bulletin 325 by E. C. Rodabaugh. It is currently in the project list to be adopted by the B31 codes. It has a competitor procedure that is adopted in ASME Section VIII, Division 2, which is also being considered for adoption by the piping codes.

The Bulletin 325 method is based on determining the wall thickness that will bring the burst pressures ratio of the assembled branch and the run connections into agreement. As such, it is more compatible with the current methodology, although it is also more computationally complex than either of the current methods. The benefit of the more computationally complex approach is the expected saving of materials. That calculation generally shows less reinforcing material, pointing out the conservatism inherent in the older methods. It should be noted that in the case of area replacement this method has been around for over 50 years.

It is basically written around a combination of some empirical fitting of test data curves and theory. It consists of a few equations that can be solved iteratively, and can determine the maximum wall thickness of either all reinforcing material in the branch or all reinforcing material in the run. And with more work one can determine the effectiveness of some combination. Those equations are shown here for information and a reference to them with some worksheet suggestions is in the Appendix.

The first equation is a parameter that is a function that relates to the theoretical stresses in a cylindrical shell. That parameter is

$$\lambda = \frac{d_o}{D_o}\sqrt{\frac{D_o}{T}}$$

where

d_o is the mean diameter of the branch
D_o is the mean diameter of the run
T is the thickness of the run

For code, work should be translated as the minimum thickness. Then that is used in an equation based on data from over 150 different types of tests. That equation is

$$SCF = \frac{2 + 2\left(\frac{d_o}{D_o}\right)^{1.5}\sqrt{\frac{t}{T}} + 1.25\lambda}{1 + \sqrt{\frac{d_o}{D_o}\left(\frac{t}{T}\right)^{1.5}}}$$

where the terms are as before, and t is the wall thickness of the branch.

After calculating the SCF (Stress Correlation Factor) for a particular combination of branch and minimum wall thicknesses, the procedure is to find a set where the ratio of the pressure on the branch side is equivalent to the burst in pressure on the run side—that is, where $P_b/P_{burst} = 1$. To do this, another empirical equation was developed:

$$\frac{P_b}{P_{burst}} = A(SCF)^B$$

For this equation the correlation coefficients are $A = 1.786$ and $B -0.5370$ It has been found that when the SCF as calculated with the above equation is 2.95 the desired ratio of one is received:

$$1.786(2.95)^{-0.537} = 1$$

On a calculator, one might get 0.99904, which clearly rounds to 1, again demonstrating the difficulty of digital calculation or empirical regression. Extending the decimal places in any of the three elements of the preceding formula might possibly find an exact 1, but at what effort and result?

It is obvious that an iterative solution is required, which can best be done in a spreadsheet-type situation. One can solve in one way for full reinforcement in the branch and in another way for full reinforcement in the run. This would bracket the solution. When one has some excess in the material in the branch and runs with a standard wall thickness, it can be checked to see if it approaches the desired ratio of 1.

The Division 2 procedures develop a method to calculate stresses, which is not the way the current codes tend to settle the issue. That methodology might be easier to do through the computer with finite-element analysis to determine the need for reinforcement material. There are commercial programs that provide the stresses for any sort of configuration that one can draw. It is currently in the project list to be adopted by the B31 codes. It has a competitor that tries material sizes until a resulting stress is found that satisfies the requirements of the particular code. Many of these programs are linked in some way to drafting programs. Such programs utilize much more detailed analysis and generally have more accuracy. They certainly provide more precision. Many codes were written in the "slide rule age" when detailed computation was very time consuming, so they simplified assumptions and by intention were also conservative.

The codes are aware of this and therefore always point out two things: They are not handbooks and they do not substitute for good engineering judgment. This is true for such tools as finite-element analysis or any other more rigorous method of doing the calculations. Usually the user of such methods is required to justify the method used that is not explicit in the code but is acceptable to the owner or to a certifying agency, if any. To be complete the method should be spelled out in the design in a sufficient manner that a reviewer can determine its validity. All users of computer programs are aware of the phrase "garbage in, garbage out."

I prefer to say "wrong assumptions in, wrong answers out." A particular computer program may not produce its answers with the same stresses that the codes require. Many of the commercial programs have default settings in their system that preset those assumptions. Users of such programs need to assure themselves what the assumptions are, why they are made, and whether they are the correct ones for the particular code at hand. Depending on the size of meshes and the thickness of the material, this can get into problems that require some rationalization of the answers.

Readers will recall that we discussed that the stresses actually change through the thickness. For internal pressure, that varies from high at the inside and low on the outside. The more recent finite-element programs, especially the solid-model ones with solid mesh, give those incremental stresses cell by cell. They do not make the assumption that Barlow did that it is okay to average. Nor do they make the Y factor adjustments that some codes make to set those stresses to some specific point through the wall. They require what is generally known as linearization to get from a comparable stress to a "code stress." It is true that there are FEA programs that do that sort of thing for users. However, in most cases this is left to the program operator. It is not an impossible job but it does require skill and expertise. Since this book is not about FEA, there is not much more included except to point out these areas. I suggest that one work some calculations, compare them, and get a feel for what is happening. In fact, in reality one needs to have a "feel" for what order of magnitude the answer should be.

The proof test is another area that is an important way to prove that a particular intersection meets the fundamental philosophy of not reducing the pressure retention strength of the base pipe. Readers will recall that the third point in the list of ways to prove an intersection is a proof test. This is the way that several component standards offer as a means to prove their product. As noted, the codes that offer this possibility usually list potential proof test procedures as acceptable. The procedures come from standards.

The original proof test is listed starting with BPV code, Section VIII, Division I, Paragraph UG 101. It was developed for use in pressure vessels and is quite detailed. This is not to say it is not useful for piping components, but that it is not designed with them in mind. The UG 101 test allows one to establish the maximum working pressure of the tested component. In piping one is trying to prove that the component has an equal or better pressure rating than the pipe it is intended to mate with. This establishes a target pressure rating. Many standards such as B16.9 and MSS-SP-97 offer proof tests developed specifically for piping components. It should be pointed out that proof tests are expensive to perform. It is desirable to have one test cover many instances. In many of the piping codes the paragraphs addressing it state that extrapolation is not allowed while interpolation is allowed. This makes proof testing for a family of components such as tees, which are covered among other things in B16.9, more viable as a method.

One of the reasons that the two standards tests are mentioned here is that the two standards cover two basically different types of components. B16.9 covers components like tees, which are inherently "inline" components—that is, all portions of the part that are connected to the base pipe have the same nominal size as the pipe. They are not necessarily the same exact wall thickness because of the need for reinforcement. But the working thought is that they are configured to match the inline flow in some direction. This is not true of the components covered in MSS-SP-97. Those are branch outlet fittings, which only match the flow of the pipe. In the direction of the branch, regarding the run, a given branch size can and is contoured to fit on several size runs within geometrical rules. That difference is schematically illustrated in Figure 6.6.

In most respects the requirements are the same. The major difference is which materials one uses in establishing the test to determine the target or burst pressure. The main purpose of the proof test is to take the test specimen to a pressure that is a burst pressure, and show that at that pressure, the component being tested does not burst prior to that pressure. The standards that have a burst test as an alternative do not preclude the use of mathematical proof of the pressure integrity of the design. They point out that this usually requires some portion of the fitting to be thicker than the matching pipe, and that is allowable. It is allowed at the manufacturers' option to perform a proof test. An acceptable procedure is outlined in the standard.

As mentioned, the B31 codes committee is in the process of writing a universal procedure that will include all of the concerns of the standards-writing bodies, as well as the concerns of the codes. There are several

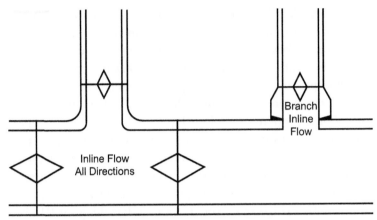

FIGURE 6.6 Schematic of inline and branch outlet differences

concerns and they are not fully resolved. Users of a proof-tested standard are cautioned and should examine the data and assure themselves of both the integrity of the submission and the validity of the test.

Generic Tests

An outline of a generic type of test is discussed here along with some comments on some of the known points of concern that have arisen. Any given proof test is to determine that the particular fitting being tested has at least as much pressure strength as the matching pipe for which it is intended. A test of an S40 fitting determines that the S40 fitting, in spite of any differences in thicknesses for the portions of the fitting to accommodate the increased stresses at the intersection, bend, or stress riser, is compensated for and does not result in a fitting that is less strong than its matching pipe. The major steps in developing a test are as follows:

1. The test assembly is important. The fitting being tested should be representative of the fittings produced by the manufacturer performing the test. The material used in manufacturing the fitting should include all the parameters, such as grade, lot, and any heat treatment performed on it. Since the test is usually for proof of compliance with the standard, dimensional verification is also required; all of the data should be compiled in the test data and be available for review by third parties as requested. Some standards explicitly require that the date of a particular test be kept at the manufacturer's facilities. Few require a specific report of the test, but many manufacturers do publish reports. There is little specificity as to what information the report must include. The latest

edition of B16.0 has added some reporting requirements. That is a step in the right direction. Repeatability of any test is an important element of that test's acceptability.

2. The assembly itself has to have a certain size. It is most important that the pipe extensions from the fitting be of a certain length. To make the assembly a pressure test chamber, some closure such as a pipe cap or ellipsoidal head must be attached at the end of the pipe. These heads can give a stiffening effect to the pipe, and the length must be enough that the attached pipe itself is not strengthened by this closure. Tests have shown that proximity to a pipe cap can actually nearly double the strength of the nearest section of that pipe. The standards give the required lengths.

3. The next step is to calculate the pressure—that is, the "target pressure" or minimum proof test pressure. This calculation may be the most controversial, or the part of the test that is discussed the least. It is also the part that differs according to what type of fitting is being tested. The Barlow formula is used:

$$P = \frac{2St}{D}$$

where

P is the specified minimum proof pressure. At one time the requirement was that this be exceeded by 5 percent to make the test acceptable. This is not always the case in every standard now. It is currently recognized by most that if the test assembly meets without bursting the theoretical pressure, it has passed the test. I recommend taking the test to burst if possible, as that will define an actual limit of the fitting. This gives some additional insight into the safety of the design.

D is the specified OD of the pipe.

S is the actual tensile stress of the test fitting (or in the case of branch fittings, the actual tensile stress of the run pipe being tested). The reasons will be discussed later. The tensile stress must test a sample of material that is representative of the fitting (e.g., the same lot) as closely as possible. This can be cut from the pipe for the branch fitting test. That material should be shown to meet the requirements of the specified material.

t is the nominal pipe wall that the fitting's marking identifies as the matching pipe for the fitting.

In the case of the stress, the actual stress one uses is a function of being sure that the test is not prejudiced in favor of the fitting. In the case of inline fittings, it would be possible to put a fitting made from a material that had a higher actual tensile strength, and if one then used a normal attaching pipe, the fitting would inherently be stronger regardless of the reinforcement. Recall that in performing pad-type calculations for that type of reinforcement, designers are not allowed to take into account any higher allowable stress in the calculations, but they are required to acknowledge the difference if the pad doesn't have an allowable stress which is as high as that of the pipe being reinforced. An adjustment must be made. If it is higher no credit can be taken for more strength.

The argument reverses itself in the case of branch outlet–type fittings. In this case the use of pipe of higher-strength material might bias the test in favor of the lower fitting. In either case a wise test developer will pick both pieces of pipe and fitting to be as close together as possible in terms of actual stress to eliminate any resulting disparities in results.

There is also some argument about whether to use the actual rather than the specified minimum material strength. Using the minimum would allow a biased test that happens to have material that was over the minimum strength. Using the actual rather than the minimum effectively eliminates the specified material from consideration.

Another argument is in regard to the use of the pipe nominal wall as opposed to the wall after the manufacturer's tolerance is eliminated. Some argue that it is not realistic that the code requires the reduction to that minimum wall, and that wall thickness should be the one used in the test. They say that one is never going to go that high. Those in favor have two reasons for this choice. First, in any specific piece of pipe you may or may not have that thin pipe, so the safe thing to do in a test environment is to assume that you will have that condition in the field and it should therefore be the field condition. Second is the possibility of field pressure variation and test variation and the fact that at present only one test is required. This weighs heavily in favor of some upward adjustment for safety.

This is one of the factors that will probably be addressed in the upcoming universal pressure test procedure. Most such procedures require something like a multiplication factor on the test results depending on the number of tests that are run. The new B16.9 test does have a set of multiplication factors when only one test is performed. It also allows and encourages the tests for given schedules to be run from different sizes within the range of sizes to help assure that the extrapolations allowed are covered. So, by comparison, reducing the denominator by a factor of 12.5 percent for the normal manufacturer's tolerance to account for statistical and test variation is not such a heavy price to pay.

Then there is the question of the applicability of the results of the tests. There are three common things that fittings standards apply: the size range, thickness or schedule range for a given size, and material grades that apply.

It is most common for the size range for a fitting to be from half to up to twice the size for similarly proportioned fittings. However, there are restrictions on reducing fittings. The standard in question remains silent on the issue of extrapolation as opposed to interpolation. Responsible manufacturers cover this by a test program that has overlap on the under- and oversize ratings. In developing this overlap program they can show interpolation through the standard sizes.

A close examination of the pressure calculations shows that the pressure rating of a particular size of pipe is dependent on its t/D ratio. So as the sizes change, the validity of the test is dependent to some degree on the t/D ratio. Most standards give a range of the ratio for validity. The most common range of that ratio is from 1.5 to 3 times. Some argue that as the ratio gets into larger sizes, the 3 times gets problematic. For instance, 3 times on a relatively large-diameter pipe can move the thickness into one of the thick wall regimes and might require additional consideration. In pipe sizes below 12 NPS this is not frequent.

The last range is the material grades the test is good for. The common argument is that the material is indifferent. Many tests are run on carbon steel. There is a significant difference in the malleability and/or percent expansion in a test of stainless versus carbon. Several argue that a test on carbon is not accurate for stainless. Certainly in this situation, increased distortion could be detrimental to the continued operation of the system. In setting allowable stresses for stainless, B31 codes often caution that one should use a different allowable stress where the distortion might cause a problem like a leak.

One of the more relevant concerns is one of horizontal data regarding tests. Currently, there is no explicit requirement for retesting the product over time. It may not be required where the fitting and its dimensions don't change. But there is also the question of the process that makes the product. That could make a significant difference. For instance, many of the proof-tested fittings standards do not specify dimensions that may be critical to the success of the fitting in resisting pressure, such as internal wall thickness. B16.9 recognizes this in tees and specifically states that if certain dimensions that are not controlled by the standard are important to users, they must specify this on their requirements. In effect, this may make that fitting nonstandard. When the universal test is published, the plan is to address the issue of horizontal data.

The final concern would be the record and/or report on the tests. There are some standards that require that the original test report be available for third-party examination at the manufacturer's facilities. This requirement is not in all standards or codes. In addition, there is no agreement on what specific form or what information should be made available in the report. At a minimum the reviewer should require some dimensional data on the assembly to ensure that the proper pipe lengths and other such dimensions were met.

The appropriate material data as specified, preferably for all components of the assembly, should be recorded. And a chart of the pressure that was produced as the test was conducted should be provided to ensure that the proper pressure was achieved. Some calibration of the pressure-measuring instruments would ensure that the pressures produced were accurate. These elements would be a minimum requirement, as well as a signature of the responsible party that conducted the test.

It is important to remember that a proof test is designed to prove something has, in this case, enough pressure retention capability to assure that it meets the margins of safety inherent in the codes. This means that it may subject a component to more pressure than the code would allow to help establish that margin. It is different than a hydrotest, which is a test to assure fabrication and design integrity.

This is a general description of a proof test with comments to help readers should they be required to approve a fitting that was justified as code compliant in this manner. The areas mentioned are subject to some degree of controversy in the description. The approving party has the ability to reject a test that violates any of the criteria that are not explicit in the test requirements in the standard.

Another way that some standards establish the pressure ratings of the fittings that are covered by them is to give the fittings a class and then relate that class to a specific pipe wall schedule. The most well-known standard is the ASME B216-11 standard, Forged Fittings Socket-Welding and Threaded. In this standard there are elbows, tees, crosses, and other shapes and fittings categorized in classes. The socket-weld classes are 3000, 6000, and 9000; and threaded fittings classes are 2000, 3000, and 6000. Each of the classes is matched with a schedule of pipe of the same size.

That schedule is different for the two different types of fittings. For example, a 3000 class socket-weld fitting is matched with S80 pipe. A threaded pipe matched to that same S80 pipe is rated as a 2000 class fitting. This can be confusing to people. They rightly assume that a 2000 class is not as high a pressure rating as a 3000 class. However, when one notes it is matched to the same schedule pipe, a question is at least suggested.

Before addressing that, we must discuss why a proof test is not directly applicable, as it is in a butt-weld fitting such as the B16.9 fittings. The answer is relatively simple. In pipe the OD is the standard dimension. As one who has worked with pipe knows, in sizes below DN 300 (NPS 12), the number does not describe the actual OD. Above that size the number and OD relate to actual inch dimensions and millimeter dimensions as if the millimeter was 25 mm/in. rather than 25.4. The point is that as schedules change the wall thickness changes. This change affects the welding process, whereas in the socket and threaded situation the pipe OD is constant for a given size and the wall thickness varies. However, it does not affect the fit of the pipe into the socket or the female thread receiver. This leads to the decision, for pressure purposes, to match a size of the B16.11 fitting to the schedule of the pipe.

Thus, the real question for this same S80 on two different classes depending on the type of fitting is, what is the pressure rating difference in a socket-welded pipe and a threaded pipe? The codes require one to reduce the allowable wall for pressure calculation by a mechanical allowance for things such as threads.

For example, a 2 NPS (50 DN) S80 has a minimum wall after a manufacturer's tolerance of 0.191 in. (4.85 mm). However, one must reduce that wall further by the thread depths. That reduction for pipe threads (Std B1.20.1) and the NPS 2 size is 0.07 in. (1.77 mm), making the final wall thickness for pressure 0.121 in. (3.08 mm). By manipulating the formulas one can calculate the pressure allowed for a threaded pipe of that size by assuming an allowable stress of 138 MPa (20,000 psi). One can also calculate the pressure with the Barlow formula as follows:

$$\frac{2St}{D} = \frac{2(20,000)0.121}{2.375} = 2038 \text{ psi} = \frac{2(138)3.08}{60.3} = 14.09 \text{ mpa}$$

If one uses other forms of the formula the specific answers will be different. Now if one uses the socket-weld wall thickness, he or she can find that the allowed pressure is approximately 1.6 higher, verifying that because of the threads the threaded pipe actually has a lower allowable pressure.

Does this mean that users must always stick to the pipe schedule that is listed for the particular class of fitting? In a word, no. The committee realized when making the schedule that they were not saying what is required to be used. Rather, they are saying that when the fittings conform to the dimensions given in this standard, that is the pressure that the fitting is proven to be good for. It should be noted that B16.11 is a standard that defined all the dimensions that are required to maintain a rating.

As to the correlation of fittings to the pipe table, the standard explains that the correlation does not intentionally restrict users to either thicker or thinner wall pipe. It further explains that the use for such thicker or thinner pipe may change which element of the assembly controls the final rating of the piping system at that juncture. It means you may use the fitting with such pipe, but the rating of the final assembly is a matter of a user's calculation. That calculation starts with the correlation of the table and the strength or rating of the attached pipe. One of the elements has the lowest rating. It may be the thinner pipe; it may be the fitting. It should be obvious that the thicker pipe might be the stronger one. However, that is assuming all of the materials have the same strength, but in some cases this may not be the case. This is particularly true for socket-welded fittings. Due to economic concerns it is not uncommon to use a lower-strength material for a small branch on a larger pipe. There can then be some calculation required in the nonstandard use of standard fittings.

The last way the standard rates the fitting is by establishing a pressure ratings table. The most common of these are standards for flanges. These standards include B16.5, B16.47, MSS-SP-44, and many others for different materials and classes and sizes. The B16.5 flange ratings are shared with B16.34 flanged valves, so there is a commonality of the valve standards and the flange standards. Those old enough might remember that B16.5 at one time included valves within that standard. It was determined there is enough difference between a flange and a flange valve that they needed separate attention. However, they quite often are bolted together, even now. There are several other flange standards in the United States and worldwide. The most common standards of fittings types, particularly in this discussion of flanges, are the DIN, which is a set of German standards; JIS, which is a set of Japanese standards; and ASME.

It is difficult to say which is the most used, but the United States is one of the larger economies in the world, and therefore probably wins the most used by default. The ISO organization has as one of its main goals to get a worldwide set of standards in use. They are recognizing that the DIN, JIS, and ASME standards all have a fairly large legacy (installed base), and that installed base has strong regional tendencies. So the default will probably be for many years to maintain the regional aspects. Since the B31 pipe codes are the basis of this book, we focus on the ASME standards. It is interesting to note that the ISO flanges standard 7005 in its newest edition includes B16.5 as an acceptable standard.

The B16.5 pressure temperature charts are very complete, but not entirely. There are no high-yield materials recognized in the B16.5 standards. This leaves room for the MSS-SP44 standard, which is designed for higher-yield and pipeline materials. The B16.5 size range is limited to 24 NPS (600 DN). This led to the development of B16.47 for sizes that are above 24 NPS (600 DN). That standard incorporates two older standard API 605 and MSS-SP-44 dimensions; it does not incorporate the high-yield materials.

The materials are important because the ASME charts tend to pressure rate the different class flanges over a range of temperatures. Different materials have different allowable stresses over those ranges. Some materials are feasible at higher temperatures, and some have lower stresses at the lower non-temperature-dependent stresses.

The technique used in establishing the allowable pressure temperatures by class is described in detail in the B16.5 standard. Basically, the first step is to group materials into similar allowable stresses over the temperature range for that material. Then there is a calculation that includes the material class as part of the calculation. Finally, there is a rating ceiling that includes consideration of the distortion that the pressure would make to a flange of that class. The calculations are based on the lowest-strength material within a particular group.

Having completed the calculation, the temperature-pressure chart is established and interpolation between temperatures, but not between classes, is allowed. Currently, there are charts for Celsius and pressure in bars, as well as charts for the USC units of pound per square inch and Fahrenheit. As an example, assume the material is A350 LF2, the temperature is 225°C, and there is a pressure of 42.9 bars (see Table 6.1). The question is whether class 300 flanges can be used.

Example of the Interpolation

Interpolating between 200°C and 250°C, the difference is 50°C; at 225°C the difference is 25°C. Therefore, the temperature difference ratio is 25/50 or 0.5 between the two rating temperatures. The pressure difference is (43.8 − 41.9) or 1.9 bars. Using the temperature ratio the pressure at 225°C would be

0.5 × 1.9, or 0.95 bars. The rating pressure would be 43.8 − 0.95 = 42.85. Your pressure is 42.9 bars; therefore it is not acceptable to use class 300. However, class 400 has a rating of 55.9 bars at 250°C, so it is acceptable.

Naturally, this example was chosen to show the possibilities one could pick with a different material group, such as group 1.2, which would accept the slightly higher pressure. One could pick the class 400 flange, which is not a common class and might be hard to find in stock. Or one could make a special flange that might have higher cost due to being a one-off construction as opposed to the mass-produced standard flange.

One technical problem with choosing the class 400 flange may not be a real problem; in fact, it may be a solution to another technical problem. The flanges in B16.5 and other standards are designed for pressure only. Flanges may have moments or forces imposed on them from the thermal expansion of the pipe and other external loads. A question then arises: How does one handle this problem?

There are several ways to handle the problem. Probably the most conservative way is with equivalent pressure. This method simply calculates an equivalent pressure from the moment and/or axial force and adds it to the system pressure. It then checks to see which flange class can handle that equivalent pressure. The method utilizes the following formula:

$$P_e = \frac{16M}{\pi G^3}$$

where

P_e is the equivalent pressure that is added to the actual system pressure
M is the moment in the appropriate units
π is the constant
G is the diameter of the gasket load reaction and a function of the flange facing and type of gasket (see Figure 6.7).

In the preceding example, say you have a moment M of 100,000 lbs, and G is 12 in. Then,

$$p_e = \frac{16(100,000)}{3.14(12^3)} = 294.7 \text{ psi}$$

Table 6.1 Excerpt from Table 2-1.1 of the B16.5 Standard

Temperature, °C	Class 300 Pressure, Bars	Class 400 Pressure, Bars
200	43.8	58.4
250	41.9	55.9

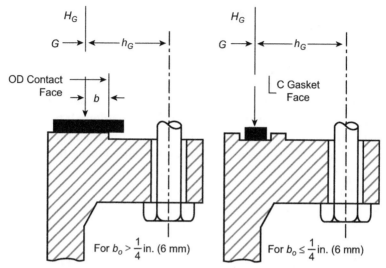

FIGURE 6.7 Gasket reaction diameter *Note: b_o* is defined as *N/2*, where *N* is specified in Section VIII, Division 1, Appendix 2 of the ASME boiler code. It can be loosely interpreted for defining purposes at this stage of the calculations as the gasket width that is within the raised face.

or 20.3 bars. Adding this to the 42.9 bars of the system, one gets a pressure of 63.2 bars, which is clearly over what a class 400 flange would take.

Once again, one could go up to a class 600 flange where there is ample pressure retention capability even when the equivalent pressure is added. However, recall that this method of equivalent pressure is considered conservative. This implies that there is another method available to check to see if it is possible to use the class 400 flange. Remember that the class 400 flange may not be economically available, but that is not a technical decision.

Unfortunately, the less conservative methods require essentially as much computational effort as the design of a flange using the design method given in ASME Section VIII, Division 1, Appendix 2, and so one might as well just design a flange for the system in question if for some reason the change to a class 600 is not an acceptable alternative. This type of flange design is best performed with a computer program and there are several available on the market. Or one can be written with minimal effort using a spreadsheet. If one is only doing an occasional flange design, using a spreadsheet might even be acceptable.

The previous example used 12 for G, which is not exactly correct, because it is actually a computed factor and rarely comes out with even numbers. However, the three parts of Figure 6.8 shows the results of the closest commercial class 400 flange with an approximate G of 12 and the 100,000 in.-lbs external moment.

Note the change from the SI version during the discussion of moments. This is because the codes are still in B16.5 USC dimensions. It turns out that as previously noted the B31 codes have not yet converted their allowable stresses to SI units. The program used is in USC units. Finally, for calculation purposes and demonstration of what the calculations may do, the units are not significant. The commercial programs often have automatic conversion buttons in their systems where the mathematical conversions are as simple as clicking a button or setting that preference. This is not true for the program I developed.

When reviewing Figures 6.8A, B, and C, note several things. First, the dimensions of the flanges do not change drastically within a certain size

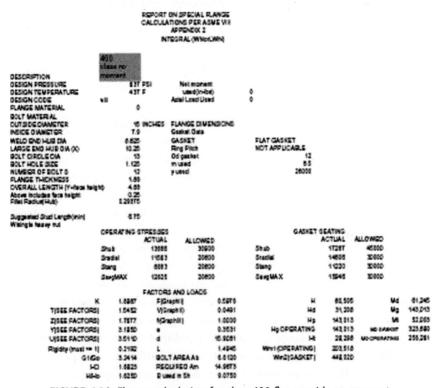

FIGURE 6.8A Flange calculation for class 400 flange with no moment

REPORT ON SPECIAL FLANGE
CALCULATIONS PER ASME VIII
APPENDIX 2
INTEGRAL (WNorLWN)

DESCRIPTION		Sample with no moment			
DESIGN PRESSURE		697 PSI		Net moment	
DESIGN TEMPERATURE	Atmos	F		used (in-lbs)	100,000
DESIGN CODE	VIII			Axial Load Used	0
FLANGE MATERIAL		0			
BOLT MATERIAL					
OUTSIDE DIAMETER		15 INCHES	FLANGE DIMENSIONS		
INSIDE DIAMETER		7.9	Gasket Data		
WELD END HUB DIA		8.625	GASKET	FLAT GASKET	
LARGE END HUB DIA (X)		10.25	Ring Pitch	NOT APPLICABLE	
BOLT CIRCLE DIA		13	OD gasket		12
BOLT HOLE SIZE		1.125	n used		6.5
NUMBER OF BOLTS		12	y used		26,000
FLANGE THICKNESS		1.93			
OVERALL LENGTH (Y-face height)		4.87			
Above includes face height		0.25			
Fillet Radius (HUB)		0.29375			
Suggested Stud Length (min)		8.75			
With single heavy nut					

OPERATING STRESSES	ACTUAL	ALLOWED	GASKET SEATING	ACTUAL	ALLOWED
Shub	14,941	30,900	Shub	11,187	30,900
Sradial	12,545	20,600	Sradial	9455	20,600
Stang	8652	20,600	Stang	7276	20,600
SavgMAX	12,693	20,600	SavgMAX	10,321	20,600

FACTORS AND LOADS						
K	1.8987	F (Graph)	0.5981	H	69,770	Md 61,245
T (SEE FACTORS)	1.5452	V (Graph)	0.0492	Hd	21,209	Mg 89,614
Z (SEE FACTORS)	1.7677	f (Graph)	1.0000	Hp	109,919	Mt 55,414
Y (SEE FACTORS)	2.1250	*	0.2934	Hg OPERATING	109,919	Mo GASKET 209,652
U (SEE FACTORS)	2.5410	d	15.9713	Ht	22,562	Mo OPERATING 279,147
Rigidity (must <= 1)	0.2393	L	1.4954	Wm1 (OPERATING)	157,599	
Gt/Gs	2.2414	BOLT AREA Ab	6.6420	Wm2 (GASKET)	225,959	
HO	1.6929	REQUIRED Am	12.0384			
H/Ho	1.6191	B used in Sh	9.0750			

FIGURE 6.8B Flange calculation sample with no moment

between classes, with the exception of the thickness of the flange and the size of the bolts. These changes are the result of careful manufacturing planning several years ago to minimize tooling, as well as other practical reasons.

The bolts, in combination with the gasket, are the mechanism to seal the flange tight. The tightness and evenness are important. For that reason ASME developed standard PCC-1 to describe the appropriate way to perform this assembly. In the design of the flanges the bolt pattern is kept even in groups of four to obtain, as nearly as possible, a uniform chordal distance between the bolts. The size of the bolt is changed to allow the proper tightening torque and therefore the pressure holding the flanges together and squeezing the gasket without overstressing the bolts. That is one of the reasons the conversion to SI units kept the USC bolt sizes. They are not exact SI equivalents. For reference, there is a chart in the Appendix that compares the SI bolts to the USC bolts. This allows knowledgeable users to select alternate sizes on their own.

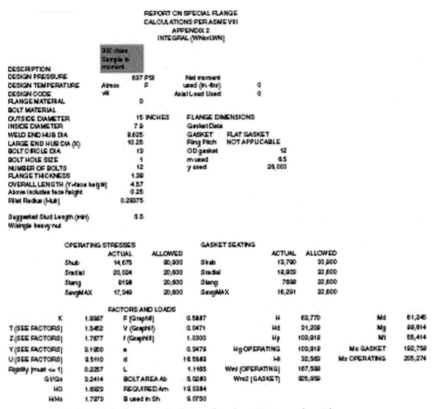

FIGURE 6.8C Flange calculation for class 300 sample with moment

The calculation also provides a sample showing the 100,000 in.-lbs moment added and the more accurate, less conservative calculation (Figure 6.8A). Figure 6.8B illustrates a flange calculation with no moment and Figure 6.8C represents the class 300 flange. It was determined by interpolation that the example was very slightly overstressed at that temperature. Several things should be noted in observing these calculations:

1. The stresses for the gasket seating case or condition are the same for both. These stresses are calculated at the appropriate condition when the flanges are bolted together and the proper amount of torque (bolt load) has been applied to seal the joint and gasket. Naturally, there would be no moment at that point, unless of course the assembly wasn't properly aligned.

2. The operating stresses have changed from the condition of no moment to the condition of the applied moment. The stresses in the no–moment case are quite low. This is because the conservative check said that it the class

would need to change from 300 to 400, which resulted in the allowed maximum pressure being much higher than the system pressure. The moment application (Figure 6.8B) shows that the stress is much higher in the radial case. In fact, it is 61 percent of the allowed stress, but the conservative equivalent pressure method said it would not meet the requirement. This is an indication of the amount of conservatism in that method. A moment that passes using that method should be quite safe to use.

3. Note that four different stresses are checked in each case. They are tangential, radial, and hub, plus the highest average of the hub and the radial and the hub and the tangent. Note also that the allowable hub is higher than the other allowable stresses by a factor of 0.5. The averaging ameliorates this higher allowable stress.

4. Note that the major change in the stresses is caused by a change in the moment. A moment is calculated from the pressure alone, and when a moment is translated from the pipe load it is converted to the moment that is felt at the gasket reaction diameter G and added. This of course doesn't happen in the gasket load case.

5. Note that in this report there are several factors reported that are internal to the calculation. Some of these are basically integration constants. Some are ratios and some are calculations that are used to calculate G. There is also the total bolt area and other factors. They are in the report to allow a knowledgeable check of how the stresses were calculated.

6. There is also a rigidity index, which is discussed in detail later.

The calculation in Figure 6.8A shows that the flange could in fact take the 100,000 in.-lbs moment. It is reasonably distant. There should not be any trouble using that moment, provided the calculations are documented.

The calculation shows that the high stress is very slightly under the allowable stress. At 97 percent it might be an engineering judgment call to use it. Again, it would have to be documented if it were. One might ask why it passed. The most logical explanation is that the A35 LF2 is in a group and that group most likely has a spread of more than 3 percent in the allowable stresses. One should be cautioned that B16.5 flanges are not individually designed. A calculation might show the opposite. They have stood the test of "grandfather time" and are acceptable when used per the standard without calculation.

The rigidity index is another nonmandatory check. This comes from the fact that the flange method of calculation may not produce a flange sufficiently rigid with respect to flange ring rotation. Excess rotation may affect the gasket seating and efficiency. Therefore, ASME developed a method, in a nonmandatory Appendix S, to check that the rigidity is within acceptable limits. That check is applied in the report. Its formula is

$$\text{Index} = \frac{52.14 M_o V}{LEg_o^2 h_o K_I}$$

All of the factors except E, which is the modulus at the design temperature, and K_I are in the report. When K_I is set to 0.3, the maximum index must be equal to or less than 1. These K's are different for different flange types and loose-type flanges have a smaller K. All K's are based on experience, and thus the formula is basically empirical.

There is then the problem of how thick a blind flange must be for the pressure. This is a different problem as it is a flat plate subjected to uniform pressure across its area and is bolted down as a corner factor. The pipe codes reference ASME Section VIII, Division 1, UG-34. These are closures and have different edge-holding methods. The most applicable one to blind flanges are items j or k of Figure UG-24 of that paragraph, and they set a constant C of 0.3 for the calculation and show a diameter d that basically goes through the center of the gasket or the pitch diameter if a ring-type gasket is used. Given this, the equation for the thickness is

$$t = \sqrt{\frac{CP}{SE} - \frac{1.9 \, Wh_G}{SEd^3}}$$

In this case E is the quality factor. For a forged or seamless blind flange that would be 1. The other factors are shown in the report for the flange in question. It should be checked for both the gasket seating and operating case since the bolt loads and operating stresses may change in opposite directions and give differing answers. Naturally, the maximum thickness is the one to use. Some standards such as MSS-SP-44 give thicknesses for blind flanges; others don't.

The design of flanges is codified and there are several sources that can design flanges. They are a unique type of fitting. The flange one buys is not the complete joint—there are bolts that are required. A gasket is required also. The B16.5 standard has recommendations for the use of gaskets and bolts, but they are not requirements since the flange manufacturer would have no control over which bolts or gaskets are actually used in making the final assembly. In addition, the amount that the bolts are tightened influences the actual tightness of the final flange. ASME has recently published a standard entitled "Guidelines for Pressure Boundary Bolted Flange Joint Assembly" that addresses the actual putting together of the flanges. It is given mention in B16.5 and is gaining mention in other standards and codes as the publication cycles allow.

This bolting problem also makes B16.5 somewhat of a hybrid standard in its metrication. There are no exact equivalents in SI or metric bolting to USC bolting, which affects bolt-hole sizes as well as circles, both of which affect the gasket dimensions for that area. Consequently, the committee decided to keep the USC bolt sizing for the metric or SI version of the standard.

A short digression into the factors of flange design might help to explain the concerns. It is important to remember that flanges are an analogy to Newton's law that states that every action has an equal and opposite reaction. Flanges inevitably come in pairs. As such, to tighten them the holes must somehow line up and the hole in each flange must accept the specified bolt. To ensure as tight a seal as possible some gasket must fit between the two flanges to ensure that the best possible seal is achieved. This basically puts the flanges in two states. In one, when not in service the bolts are tightened to an amount that makes the gasket seal leak-tight. Then when the service reaches temperature and pressure, that pressure tends to create a change in load, which tends to loosen the gasket seal pressure. The bolt must be strong enough to withstand all of the forces that those two conditions create and still stay leak-tight. These states are addressed in the design calculation methodology. ASME has a methodology that is universally accepted. There are others, for instance DIN EN 1591 and EN 13445-3, but these methods are mathematically different and the goal is to solve the problems mentioned here.

It must be pointed out that this discussion relates to the most common flange in piping, which is the raised-face type of flange. This type is technically one where the gaskets are entirely within the bolt circle and there is no contact outside that circle. There are flanges called flat-face flanges that are permitted by B16.5. It is assumed by that permission that the gaskets remain inside the bolt circle and touching does not occur. ASME BPV Section VIII, Division 1 does have a nonmandatory appendix that addresses the calculation methods to use for flat-face flanges that have metal-to-metal contact outside the bolt circle. Other special-purpose flanges, such as anchor flanges, swivel flanges, flat flanges, and reverse flanges, are discussed in Chapter 9 on specialty components.

When studying the flange dimension charts for B16.5 flanges, note that the bore of welding neck flanges above a certain size is specified by the user. The standard points out that it is based on the hub having a wall thickness equal to that of pipe of a specific strength. Therefore, they express a formula to determine the largest bore, which is not to be exceeded if the ratings of the standard are used. If that bore is exceeded, the ratings do not apply or the flange is not a standard flange. That formula is

$$B_{max} = A\left(1 - \frac{C_o p_c}{50,000}\right)$$

where

A is the tabulated hub diameter at the beginning as listed in the charts

C_o is a constant 14.5 when p_c is in bar units and 1 when p_c is in psi

p_c is the ceiling pressure as listed in Annex B, Table B1 or B2, in B16.5. This is useful for a manufacturer that suspects users are specifying too large a bore, or for users to see how large a bore can be specified for that size flange.

Not all pressure-temperature rating charts in standards are quite as complex as the ones in B16.5. They all serve the purpose of letting users choose a fitting from the standard that will work in their service. Note that if interpolation is allowed one can tell if a rating is applicable to the intermediate service listed. If interpolation is not allowed the next higher service must be chosen.

As mentioned before, one might be tempted to use metric bolts with these B16.5 flanges. While it is true that there are no exact equivalents, there are ones that are close. Some charts on bolting in the Appendix give comparisons for users. The previous discussion about the differences in bolting makes it a user responsibility if such bolts are used. There are subtle changes in the calculated results that may cause problems in the service for which this substitution is made.

CHAPTER 7

Piping Flexibility, Reactions, and Sustained Thermal Calculations

Contents

OVERVIEW

In Chapters 5 and 6 we discussed the ways to calculate what the pressure design requirements are in a piping system. However, that discussion covered only part of the requirements. Many pipe systems operate over a range of temperatures or at least at a temperature other than the installation temperature. In addition, all piping systems have longitudinal stresses that occur. These come from the pressures acting along the pipe's longitudinal axis, as well as loads from the fluid the pipe carries, the weight of the pipe itself, and any insulating or other coating material. These are also exacerbated by loads from external events such as wind, earthquakes, and other natural elements.

All of these stresses interact together in some manner. The wind, earthquake, and similar loads are generally classified as occasional. The weight and pressure loads tend to act continuously. The ranges of temperatures, while ideally both constant and continuous, are not always perfect. For reasons of simplicity, we discuss various stresses separately here. As usual, the discussion will focus primarily on the B31 codes with an occasional diversion to other subjects where there might be other methods that are equally acceptable alternatives.

Piping and Pipeline Calculations Manual
ISBN 978-0-12-416747-6
http://dx.doi.org/10.1016/B978-0-12-416747-6.00007-3

EXPANSION AND STRESS RANGE

We start with expansion stresses, which come about with temperature change. Normally, the temperature change would be from the installation or nonoperating temperature to a higher operating temperature. However, the cryogenic industries would consider the nonoperating temperature hot and the operating temperature cold. There might also be operations cases where there is more than one operating case or temperature. For this variety of temperatures, the difference involved is really one of direction of the expansion only.

Readers should assume that the piping is metallic and that the metal expands with an increase in temperature. Interestingly, if the pipe is not constrained in any way, this expansion causes no mechanical stress. However, no pipe system is totally unconstrained. As the temperature increases, stress is put on the pipe by the constraint. If the pipe has a change in direction, stress is also established by the connection involved in that change in direction (see Figure 7.1).

This figure is, of course, a very simple example, but it illustrates that the expansion in different directions leads to stresses. One can imagine the complexity that occurs with more than one change in direction and/or three-dimensional changes in direction.

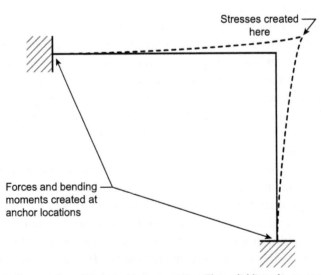

FIGURE 7.1 Changes in a simple *L* pipe connection. The solid line shows ambient pipe and the dashed line shows expanded pipe.

One of the difficult things to grasp in this is that the change is first of all a range of stress states rather than a single one. Also, it is a secondary stress, not a primary stress. A secondary stress is simply one that is created by a constraint, which as mentioned is exactly what an expansion stress is—without a constraint there is no stress. This means that the secondary stress is self-limiting. This is most likely because the temperature has been reached. It is true that there are more precise and complex definitions, but this one serves to explain what generally happens.

This brings us to the general discussion of what the allowable stress range will be. The fundamental equation for the B31 codes is

$$S_A = f(1.25S_c + 0.25S_h) \qquad (7.1)$$

with an alternate form under a certain condition (when the hot stress is greater than the longitudinal stress, one can utilize that difference). This alternate form is represented by the following formula:

$$S_A = f[1.25(S_c + S_h) - S_L] \qquad (7.2)$$

where

S_A is the allowable stress range

f is a stress range reduction factor (discussed later in more detail)

S_c is the cold stress

S_h is the hot stress

S_L is the longitudinal stress

The first question that might come to mind is: Where did the equations come from? Since we are discussing a range of temperatures, the c(old) and h(ot) subscripts refer to the extremes of the temperature range. Remember that when one is working in a cryogenic-type application the cold and hot subscripts would have different meanings. However, since we are talking about an additive expression of positive numbers, it is only necessary to remember that the last expression could be named the operating condition rather than the hot condition; the cold condition is the starting or ambient condition.

B31.3 has an appendix that addresses this issue as an alternative way to evaluate the stress range. It uses the term *operating conditions*. It points out the changes that might occur as the system moves from one operating case to another. It also addresses other issues and was designed to incorporate calculations that are made easier by the use of some of the current computer stress-analysis systems available commercially.

The choice of constants regarding the two end stresses is one place where the range concept is often confusing. First, it is important to remember that one of the ultimate philosophies of establishing stresses is not to exceed the yield stress of the material at any temperature. This is of course not including the time-dependent stresses like creep. When one is working at those temperatures other concerns must be addressed. Second, the allowable stresses are basically established at ⅔ the yield when in normal conditions. Finally, for steels a complete cycle from yield to yield constitutes a total of twice the yield. It then becomes simple to convert the thought that the allowable stress is equal to ⅔ the yield, so that yield then would be ³⁄₂ or 1.5 times the allowable stress. So the expression is $1.5(S_c + S_h) = 2$ (combined yield). Giving one a margin of something like 20 percent for all the things that one doesn't know, the 1.5 factor becomes 1.25 (i.e., $1.5 \times (1 - 0.2) = 1.25$).

Why then is Eq. (7.1) of the form it is where the 1.25 only applies to the S_c and 0.25 is applied to the hot? The simple fact is that the longitudinal analysis is based on this design or operating condition and in the worst case has used up 100 percent of the S_h. Only 0.25 of that stress is allowed in the range calculation, which leads to Eq. (7.2).

The simple and conservative way to analyze the entire system is to use the result of Eq. (7.1) as allowable for every place in the system. However, if the system is complex, the longitudinal stresses might vary across sections of the system. It is possible these sections do not use all of the available longitudinal stress. If the calculated longitudinal stress is less than the allowable stress, then one can use that excess available stress in determining the magnitude of the stress range for that section of the system where this is true. Therefore, Eq. (7.2) lets one use all of the S_h provided that the calculated longitudinal stress is subtracted.

Since Eq. (7.1) is considered the simple and conservative method, Eq. (7.2) is often called the liberal method. It does take a little more caution because the longitudinal stress will quite often vary from section to section of the piping system, so one will have different allowable stresses in those sections.

The next factor to consider is the f stress range reduction factor, which is established at an assumed 19- to 22-year life. The common lore says that this is roughly equivalent to one full cycle a day for a 20-year life. The formula given to compute the factor is $f = 6(N)^{-0.2}$, which gets a value of 1 when the number of cycles is 7776. The B31.3 committee had many requests from users to consider fewer cycles in the design life of a project and thus expand the f to more than 1. An allowed limited expansion was developed based on some experiments. They limited the value to 1.2, and there were limits on S_c and S_h, the service temperature, and the ultimate strength of the material. These limits were placed on increasing the f factor because the experiments did not show that higher tensile stress was the only factor contributing to fatigue strength.

The reduction was also allowed to go to the lower limit of 0.15 for f. The intention here was to allow users to go to an indefinitely high number of cycles. Using the 1.2–0.15 range, the number of cycles ranges from 3125 to 102 e6. This range is considerably higher than the original range, which was basically from 7000 to 250,000. The increase in range is not adopted for all of the codes. The research on the extension continues as ASME looks for ways to give pipe designers a method to handle high-cycle, low-amplitude situations. The original cycles were based on thermal growth, which can be characterized as low-cycle, high-amplitude cycles. Since those original experiments, the usage has expanded to things like floating platforms subject to wave patterns that have immeasurably high cycles and, absent a sea storm, relatively low amplitude.

There are other techniques in this area that call the method into question regarding the mathematics of determining the slope of the reduction factor, as well as the ambiguity about tying this to allowable stress only. The methodology of piping code continues to evolve.

The equations cited here refer to full or complete cycles. But it is easy to foresee that temperature cycles do not always go from ambient to working and back to ambient. Many processes have cases where they operate at intermediate levels. This also would apply in counting the deviations that might occur from wave height. All beachgoers can note that all waves do not achieve the same height, however ceaseless they may be.

For these partial cycles ASME gives a method of calculating the equivalent full cycles in systems with varying cycles. The first step is to determine which of the varying cycles creates the maximum stress range and record that number as S_e as well as the number of cycles N_e for that maximum stress range. The next step is to compute the various stress ranges as S_1, S_2, and so on, until you have all the varieties. Then compute the ratio (r) of each lesser stress to the maximum stress range (S_1/S_e) and the number of cycles for those. The equivalent full-stress range is then computed by the formula $N + N_e + \Sigma(r_i {}^5 N_i)$ for $i = 1, 2, 3, \ldots, n$. Lets do an example.

It is important to remember that the different stresses to determine equivalent cycles are based on the calculated S_E for the different load (or cycle) cases that are involved. This means that one has to do some analysis before one can calculate the equivalent cycle range. As the discussion above states you do all of the cases, choose the maximum S_E, and then calculate the ratios of the other cycles.

For this example a simple piping system was postulated between two anchors. This system had two long-radius elbows. One was horizontally arranged and one was vertically arranged, which made the system a 3D system. One could just as easily set up an even simpler system, such as is shown in Figure 7.1. The 3D system was used in conjunction with a pipe stress program. Figure 7.1 would be easier to calculate by hand. Since we haven't discussed the details of any type of calculation, this example will not go into the details, but will express the results.

Different cycles at different temperatures were established on this system. The material was carbon steel pipe and elbows. The temperatures and number of cycles are shown in the following table along with the S_E values obtained for the various calculations. It should be noted that the length and size of pipe between the elbows and the anchors as well as the pressures would affect the results; however, they are not germane to what we are demonstrating. As college professors are apt to say, "the calculations of a sample are left to the student."

Temp, °F	Cycles	S_E (Max), psi	Ratio to Max S_E	Ratio R_i^5	S_L
750	5000	21,630	1	1	2805
650	5000	17,935	0.83	0.39	2805
500	3000	12,735	0.59	0.07	2805

Note: For this high S_E, the S_c is 18,300 psi and the S_h at 780°F is 13,000

Note that the total number of cycles is 13,000, a relatively high number of raw cycles, but using the equation above that relates the equivalent cycles to the cycles for the max S_E and the ratios to the fifth power of the other cycles it reduces to

$$N = 5000 + 0.39(5000) + 0.07(3000) = 7160$$

This is only a little over half of the raw cycles and is very close to the standard default number of cycles of 7000. So the user could decide not to take advantage of B31.3, allowing some small increase in the S_A through the formula that says

$$f = 6(N)^{-0.2} \leq 1.2$$

In this case that would calculate to 1.02. The reader can check that both formulas for S_A would be sufficiently high at 1 to make the system code compliant. It then becomes one of those judgment calls the designer has to make when things get close. We know that it is all a judgment. It should be pointed out that this procedure is a variation of the cumulative damage rule known as Miner's rule. This rule could be subject to some controversy.

FLEXIBILITY ANALYSIS

What then does one do with this computed allowable stress range? This brings us to what the ASME codes call *flexibility analysis*. We all know that pipe, especially metallic pipe, isn't really all that flexible. However, if you manage to pick up a long stick of a reasonably sized piece of pipe, say a 2 NPS (DN 50) S40, and hold it in the middle, the ends will sag some from the weight of the pipe. That is part of the flexibility.

Another way to start the description of the pipe flexibility requirements is to imagine the requirement to connect a pump to a tank. For simplicity, assume that the pipe is always at the same level—that is, the pump is elevated or the tank is lowered and the straight-line distance between the two connections is 37.2 ft (11.3 m). Further, assume that the two pieces are arranged in such a way that the flanged connection is such that the pipe is perpendicular to the flange. We then have a straight section of pipe that is subjected to a temperature increase from the pumped fluid of 600°F (315°C). The pipe is constrained so the expansion causes stress. For a pipe such as A106 B, that expansion would be 1.7 in. (43 mm). The formula that explains the thermal (axial, in this case) force is

$$\text{Thermal force} = E(\text{strain due to expansion})(\text{metal area})$$

If the pipe we discussed was 6 NPS (DN 150) S40, it would have a metal area of 5.58 in.^{2}, which is 0.0035999 m^2 (say 0.0036). The E in USC at 600 is 26.7 lbs/in.^{2} and in SI that would be 18,631.3 kgf/mm^2 (to make the units consistent). Note that due to the difference in scales there can be some difference in the final answers due to things like accuracy and significant figures. Let us run the examples.

$$F = 26.7e6\left(\frac{1.7}{37.2 \times 12}\right)(5.58) = 567,375 \text{ lbs in USC}$$

$$F = 18,631.3\left(\frac{4.3}{11.3 \times 100}\right)(3599.99) = 257,102 \text{ kgf in SI units}$$

Note the units cancel out, which is important, so be sure to do a unit check when working in an unfamiliar system no matter which direction you are going. In any system one can see that the stress is high. In the USC system it is 567,375/5.58 or 101,680 psi. It is well above any code–allowable stress for that pipe.

Note that, if the analyst was interested in the stress, the elimination of multiplying by the metal area would give the stress directly. In many of those cases the interest is to know the pounds of thrust that the expansion is producing on the end or anchor point. So from that calculation one can get both answers by the manipulation or use of the metal area.

The thrust or force would additionally cause a bow in the pipe that would also cause bending stress. In short, the pipe would fail in some more or less disastrous manner. This is, of course, assuming the pump and its support were strong enough to hold that flange rigidly. A word here is appropriate about how one knows that this 37.2-ft pipe expanded to 43 mm (1.7 in.). It is a function of the linear thermal expansion coefficient. This little creature is not impossible, but fairly hard to find. The way we were taught in physics is to use the thermal expansion coefficient, usually called α, often expressed in length/length per degree. It turns out the coefficient, if used with the appropriate and consistent units, is measuring-system indifferent.

The formula for this is fairly simple: $\Delta = \alpha L_o \Delta t$. The L_o can be either in inches or meters, the Δt must be in compatible degrees Kelvin for metric or Fahrenheit for inches. It happens that $1°$ Kelvin is the same amount as $1°C$, so that isn't a problem. However, the Δt might be. You must know the base from which the α is derived. Often tables are in different bases. Generally it in inches per inch in U.S. customary but some give inches per hundred feet or the equivalent in metric. The linear expansion tables in the Appendix list some commonly used α's, and show that there are several different temperature ranges. This is because the actual thermal growth rate is different for different temperature ranges, and the coefficient given is essential for the mean or average rate of growth for that range. For our straight piece of carbon pipe, α ranges from 70 to 600 and is 7.23 e-6 for inch/inch.

Now here is something to remember if one is converting to metric. A metric degree is 1.8 times larger than a Fahrenheit degree, so the same metric α would be 13.014 e-6. Thus, the calculation is as follows: $\Delta = 0.00000723(37.2 \times 12)(600 - 70) = 1.71$ in. Now the metric is a little harder. We have already converted α, so the calculation is $\Delta = 0.000013014(37.2 \times 0.3048)(530/1.8) = 0.04344$ m, which is of course 43.44 mm. Using metric tables would eliminate the need for the conversion. As one can see there is a fair amount of converting to accomplish a relatively simple calculation.

LINEAR EXPANSION DUE TO HEAT

ASME thought so also, so they did some converting to something called total linear expansion between 70°F and the indicated temperature/100 ft. The set of linear expansion tables in the Appendix have been converted to that total per foot in both inches and millimeters. Because the base table is in °F, the °C column is mathematically converted. The added advantage is that ASME indicated the coefficients for several commonly used pipes and alloys of pipe, making the calculation somewhat more accurate for different pipe material.

The use is simple: Choose the page that has your pipe material, go down the appropriate temperature column until you find the temperature that is closest to the temperature you are working with, move to the right for the coefficient in inches per foot, and multiply that by feet. If you are working in SI, move to the left. If one needs super accuracy for some reason, linear interpolation is acceptable. Certain table values were computed by the same method. The table also moves from cryogenic temperatures to very high temperatures and has no values or zero where that material is not acceptable at those temperatures.

70°F (21°C) was chosen because it is commonly considered the temperature at which the installation is made and from which the temperature change starts. When one is working between two other temperatures, there is more math involved, but one can just read the difference between the coefficients of the two temperatures and use that difference as the coefficient for that temperature range.

We return now to the discussion of what one does when a straight line fails. In fact, these are the types of actions to take for any configuration that fails by whatever means. To remedy this two things are done. Often the pipe is made longer by adding bends or elbows to help create flexibility. It is well known theoretically that a curved piece of pipe ovalizes to some extent while being bent. This ovalization causes the elbow to bend a different amount than a steel bar beam would be predicted with curved beam theory. This gives rise to what is known as a flexibility factor to account for the difference in bending and therefore resulting stresses. This applies in some manner to all types of fittings.

So far we have illustrated that the flexibility analysis required by all codes in some manner is nothing more than a specialized form of structural analysis of a given pipe configuration. There are very few portions of a piping system that are straight pieces between fixed points other than between elbows. For one thing, pipe as a fluid transport medium often connects things like pumps or compressors and storage tanks, reactor tanks, or turbines or other rotating equipment. The piping has to take a circuitous route around equipment and the openings are often at different levels. Pumps often sit on the ground, while it is best to pump into a tank from the top to reduce the need to overcome the fluid head in the tank.

In addition, equipment can be damaged by excessive end reactions from the pipe termination point. And although not quite so pronounced, there could be a problem with the equipment reacting back on the pipe. Newton did say there were equal and opposite reactions.

In addition to the reactions, there is the possibility of the stress range from a thermal excursion being in excess of the allowable range that was calculated earlier. Another concern is that undoubtedly there will be joints somewhere in the piping layout that might start to leak from the moments and other forces that the thermal excursion creates. It is easy to understand why this analysis can be very important.

REQUIRED FLEXIBILITY ANALYSIS

But the question remains, how do we know when that analysis is required? Not all piping systems need a formal or complete analysis. For instance, if a new system is a duplicate of another where no significant changes have been made, and that previous system has a successful service record, there is no need for another analysis.

The B31 codes in general offer a simple analysis for a two-anchor system of uniform size that can suggest to you whether a more formal analysis is required. The codes take great pains to point out that there is no general proof that the system will yield accurate or conservative results. It is looked upon by many current analysts with disdain; however, experienced analysts can use it for systems that do not experience severe cyclic service. It also is a great way for a novice to get a feel for what might be needed in a piping layout.

The simple program the codes give that has no general proof may indeed not be accurate. However, that program generally has proven to give a conservative result for several of the more common pipe configurations. One of the convenient results of the piping experience is that there are a few simple configurations that are repeatedly found. This has led to the chart-type solutions where one can choose a configuration and look up predesigned and precalculated parameters, multiply them by the specifics of the condition, and get an answer. The codes' equations tell whether one has to go that far.

But first we should look at those common configurations. There are eight that cover the most common configurations and are the basis for the empirical development of the formula. They are shown in Figure 7.2.

The formula given in Eq. (7.3) has been shown to be conservative for these configurations; shape 8, the unequal U bend, is the least conservative. Note that the corners are square in the diagrams in the figure. These shapes are considered the most conservative approaches and they also are the easiest to calculate. The figures imply only 2D piping runs; there are of course 3D configurations. Each additional dimension and branch or constraint adds complexity to the computational process. Methods used in modern computer programs are calculation intensive to adapt to bends and elbows. We discuss one of the methods later when we talk about stress intensification factors.

At any rate the code formula can be utilized as a first cut. It should be noted that the more experienced the analyst, the less he or she might agree with the usefulness of the formula. That formula is

$$K_1 \frac{S_A}{E_a} \geq \frac{Dy}{(L-U)^2} \tag{7.3}$$

where

K_1 is a constant; depending on the measurement system, it is 208,000 for SI and 30 for USC

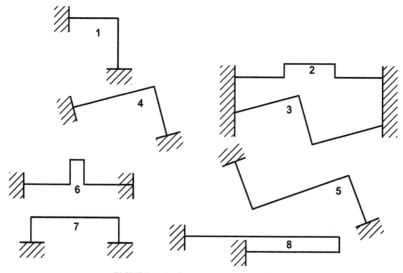

FIGURE 7.2 Common piping shapes

E_a is the reference modulus of elasticity in the appropriate units, MPa or ksi

S_A is the allowable displacement stress range calculated in Eq. (7.1), MPa or ksi

D is the OD of the pipe (which must be the same throughout the run), in. or mm

y is the resultant total displacement strain to be absorbed by the system, in. or mm (This is usually simply ΔTemp(length), but attention must be paid to whether there are any reverses that might take back some increase from the straight run.)

L is the developed length of the piping between the anchors, ft or m

U is the anchor distance, which is the straight line between the anchors (This would have been the distance in the previous straight pipe example.)

Note that in effect this places a weight on the difference between the straight pipe previously described and the developed length of the piping shape that is used between the anchors. As pointed out, we were axially and in other ways overstressing the straight pipe.

For example, let us take a simple 90° shape (diagram 1 in Figure 7.2). The data might be as follows:

- Horizontal leg, 22 ft; vertical leg, 30 ft.
- Straight-line distance between the anchor points, 37.2 ft (try Pythagoras).
- The developed length of the pipe, 52 ft.
- Reference modulus, 70° 29.5 e6.
- Thermal expansion, $\alpha = 4.6$ in./100 ft at 600°F; from the straight line we calculated the total growth as 4.6(52/100) or 2.39 in.
- The pipe is 4 NPS (4.5-in. OD).

Using an f of 1, the S_A where the cold S equals 20,000 and the hot (600) S equals 17,000 is 29,250 psi. The left side is equal to

$$30\frac{29.25}{29.5E3} = 0.0297$$

The right side then equals

$$\frac{4.5(2.39)}{(52 - 37.2)^2} = 0.0491$$

The right side is clearly larger than the left, so this would not pass the first test. Some formal analysis is required. We mentioned that the empirical relationship is based on square corners and the type of configuration we just checked is the simplest, so it is most probably very conservative. In fact, one of the studies done in developing this relationship puts this square–corner configuration as the lower margin of the formula.

One who experiments with this example will find several interesting things about it. First, one might notice that another configuration could get a longer developed length, say a loop type (diagram 2 or 6 in Figure 7.2). For example, make the loop configuration so that the developed length is 57.2 ft and the left side magically (mathematically actually) becomes 0.0268. Thus, it is smaller and passes this test.

If one has access to a computer program that does the calculations, he or she can quickly input the simple 90° configuration just discussed. In doing that it is quite likely that one will find that the configuration does not overstress the piping. At least it didn't when I checked it. This is a reflection of the conservatism in the simple check. As a point of reference, in talking with many experienced analysts on the subject, a majority would say that the test is not necessary. However, as discussed in the following, it is a good way to begin to get an understanding of the flexibility problem. For that reason, it is discussed here.

One should also notice that Eq. (7.3) is not sensitive to the wall thickness of the pipe. Clearly, the left side only varies with the S_A, which would vary with temperature. Raising or lowering the temperature would affect the result. Another point with the calculations is that the ΔT considerations were ignored in this example. This is an approximation or test; if the result were very close to the pass/fail criteria one could reduce the expansion by 70°. However, many engineers and experienced analysts would say we should look at another solution or decide that a more complete analysis is required.

Naturally, as the pipe size changes, the change in OD will result in a change in the left-side results, and at some point the configuration will change from a pass status to a nonpass status or vice versa. This of course depends on the direction of the change.

If one changes the leg lengths the results will change considerably. The ratio of direct anchor distance to developed length is very sensitive. It is a proof of the flexibility changing with the layout by fairly drastic amounts. Lastly, if one makes an adjustment in the developed length for a long-radius elbow, there is a slight reduction in the value of the right side because of the shorter arc length versus the straight corners. These are not all the differences one would find in an exhaustive study. However, it begins to show that elbows are more flexible than square corners.

Experimenting with this simple formula and changing the mentioned criteria, such as pipe size, temperature, leg lengths, configuration shape, and even materials, will give an insight into what happens in pipe flexibility. Knowing full well the results are based on ratios and do not contain all the variables that can and do affect the flexibility, one begins to get a grasp of what happens. For further information, see the chart of the various results and their changes in the Appendix.

Various Methods of Flexibility Analysis

Many prefer to do quick checks of the system with the older methods of analyzing pipe systems. These methods vary extensively and, excluding the high computational time, some are quite accurate. It should be noted that the results of such systems can be quite varied. In the 1950s when the systems were all that was available, *Heating, Piping, and Air Conditioning* magazine ran a series of articles where many of the then-competing systems were pitted against a common piping system to compare the results. A brief discussion of the results follows, and for those more interested, there is a more detailed discussion of these hand-type calculations in the Appendix.

There were 12 different methods employed. The average result was a 3062-lb force with a standard deviation of 391 lbs and a range from the lowest result to the high of 1040. The percent deviation represents a ± of 20 percent from the average. That is not all that impressive but it is about what the state of the art was at that time. Given the many variables that any model might require in a complex system, it might be quite good enough.

In an interesting book, *Introduction to Pipe Stress Analysis*, author Sam Knappan uses a graph (see Figure 7.3) that shows the variety of results from the various pipe flexibility systems available in the 1980s. While the principles remain the same, the actual code values, formulas, and rules have had several opportunities to change in the last three decades. It is advisable to consult the current codes for specifics.

It should be pointed out that there were no PC-type flexibility programs available when Knappan's book was written and the computer analyses cited may no longer be extant. In fact, some of the "hand" methods cited may not be available either. The current crop of computer programs gives comparable results.

One of the difficult elements in any engineering system is to determine the degree of accuracy that precision gives as opposed to the reality that precision accomplishes by precise calculations. Assumptions and variables often are the more critical elements.

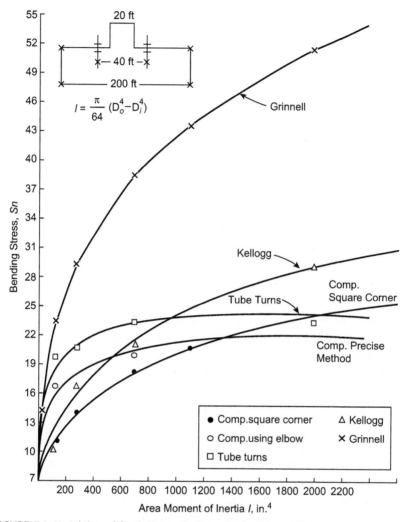

FIGURE 7.3 Variability of flexibility analysis methods *(Source: From Kannappan,* Introduction to Pipe Stress Analysis; used with permission.*)*

As was pointed out, this is really a rather sophisticated ratio analysis. Therefore, it does not give supposed absolute values, which would be required in any formal code check. We will use the same configuration with a more accurate method known as the Spielvogel method. This method was used in the early study, and it was well within a standard deviation of 0.6 of the average of the methods used in the articles discussed earlier. I might add that the average is a little suspect as many of the methods have been abandoned. This is not totally true of Spielvogel. It is known that some testers use it to check new releases of current laptop programs. They also use it for simple configurations.

The method is relatively simple to explain. Step by step the process is as follows:

1. Assume the two-anchor situation (use previous example). The remaining steps are mathematical thought experiments.
2. Remove one of the anchors to allow free thermal expansion. One of the advantages of computer programs is that they do not remove anchor systems.
3. Find the centroid of that configuration. This is another thing the computer programs do "behind the scenes."
4. Attach a perfectly stiff connection between the centroid and the free end. This is an imaginary thought experiment.
5. Calculate the expansion distances and therefore the location of the end point at the end of the expansion.
6. What forces in the x and y direction are required to move the centroid back to its original position can be shown and calculated. Once the forces are calculated, the resulting moment based on the offset dimensions can be calculated.
7. That having been said, the actual mathematics, even for a simple $90°$ bend, can be daunting.

The mathematics involve the moment of inertia of the lines and the product of inertia of the lines.

When one computes for the 4 NPS pipe in the previous example, the moment at the bend corner creates a 7000-psi stress in that corner. And the stress from the reacting moment is a little over 10,000 psi. Both are well within the allowable range of 29,250 psi. This is another indication of how conservative the simple check is. One should be cautioned that the large disparity will not occur in all pipe shape configurations.

Another benefit of this system is that by computing the centroid and the x and y amount of force on the configuration, one can determine the thrust line. This is the line at which there is a zero moment. This is known variously as the neutral axis and sometimes as the wrench axis. By knowing the distance of any point on the system from that thrust line, one can compute the moment of that system. This is also the line where it is prudent for analysts to place any heavy components, such as valves, as close as possible because of that relationship. The work is tedious, even more so when one is working with a 3D system. That is where the pipes change not only in the x and y directions, but also in the z direction; one has to do all of the work in three different planes and then combine the results.

Another relatively easy check is the guided cantilever method. It can be described most simply by referring to the configuration 1 in Figure 7.2. This figure has two legs. For purposes of discussion call leg 1 h (for horizontal) and leg 2 v (for vertical). A more detailed picture of what happens in an expansion is shown in Figure 7.1. Here is what happens at the point where Figure 7.1 says stresses are created.

Leg h is bent upward by the amount that leg v grows; call that Δv. Leg v is bent by the amount that leg h grows; call that Δh. Now admittedly this calculation will be conservative because it ignores the rotation and other niceties that occur and a probable elbow. However, using cantilever straight beam theory one can get an idea of what length the legs need to be to determine if they are flexible enough.

The formulas for a guided cantilever assuming a thin-walled pipe, which is most schedule pipe, can be reduced to

$$ L = \sqrt{\frac{E}{48S}} \sqrt{\frac{D\Delta}{}} $$

where
  L is the acceptable length in ft for the specified stress S
  E is the modulus of elasticity of the chosen pipe material
  D is the outside diameter of the pipe
  Δ is the appropriate expansion calculation
If you use MPa for E and S and millimeters for D and Δ one must use a huge constant instead of 48 (actually 330,000) to get an answer in meters. As my math teacher used to say "the proof of the metric constant is left to the reader." It depends on what units you use and what you want.

The approach would be to make the layout the natural way one would from the plan or project requirements and check to see if that leg is of a length that will be within the stress you have used in the calculation as presented.

It should be obvious, but like all things that should be sometimes they need a nudge. The length L that is calculated is the length that gets the stress that is used. If the length is less, the actual stress would have to be higher, and if the length is longer the actual stress would be less. One remembers this is a conservative approach and a detailed analysis would give a lower stress at the point in question.

If one wants to estimate the stress from a given length, with algebraic manipulation one can derive a version of the equation that calculates the stress generated in that leg from the Δ generated by the opposite leg.

If one wants to use a consistent material (in USC) one can substitute something like 29e6 for E and 20000 psi for stress and the first square root reduces to 5.5. Those numbers are pretty good for carbon steel at room temperature.

Modern Computer Flexibility Analysis

Computers began to emerge as a method of analyzing pipe structures. At first, it was mainframe and card-type input, which was time consuming but much faster, giving answers that were more accurate. Now, there are many PC- and laptop-based programs. That accuracy basically comes from the ability to include more variables because computers can rapidly repeat thousands of calculations. The spread in the test mentioned was mainly because some of the systems took calculation shortcuts to eliminate the long and tedious calculations required. On the other hand, some of the methods that we discuss later in a little more detail are still useful when one doesn't need a full-blown calculation of the entire system.

Suffice it to say that currently most pipe analysis systems can be run on laptop computers. Even in those cases one needs to be very knowledgeable about the computer program settings. If the settings are changed, those changes will cause significantly different results. In one system I am familiar with, the setup options have as many as 66 detailed settings, most of which have to do with tolerances of the calculation results and may affect the outcome. This book will not discuss in detail how to operate these programs. Software manufacturers have courses and manuals that do that sort of thing far more expertly. What one program does is not exactly the same as another, so detailed discussion is next to impossible if it is not on a program-specific basis.

Most, if not all, programs use the theorems of Castigliano or the energy considerations as the basis for their calculations. At least one newer program from the Paulin Research Group uses a finite-element analysis approach. Many use a spreadsheet approach for their inputs to set up the lengths of pipe and other project-specific data. The most useful ones allow you to switch between the idiosyncrasies of the various code rules. They also usually have an extensive database of the more common materials, fittings, pipe sizes, and other such useful data. In the calculation procedures they work along the pipe section in a finite-element way to get the final answer.

As such, two different programs would give a similar answer for a similar problem.

These programs are much closer than the ± 20 percent of the previously cited study for the other methods. However, this is only if the operators understand the program and use compatible settings. In one case, in a code committee project where we were trying to get some sample baseline programs to show people how to set up a piping problem, two skilled analysts who were using the same program got significantly different answers. Upon investigation it was found that they had, for very supportable reasons, selected different settings for the program, which led to the different results. Neither was wrong, just different. When they found this and made appropriate adjustments the answers were the same. An entirely different program with compatible settings got results well within 5 percent of the other program's answer. This is to say a modern-day analysis by a competent analyst can be expected to give the right answer in a significantly shorter time than in the days of the 1950s study.

Why then even worry about the other methods of analysis? Well, the answer is that one problem with a computer is, how does one know the answer is correct? You put the data in, and the computer blinks and gives you an answer. Did you put all the data in properly? Were all the decimal places in the correct location? Do all of the settings, including choice of code, make sense for the problem at hand? In short, you need to have some sense of the answer to know if the answer is sensible. Thus, there is the need for some sort of hand checks. There have been tales of many horrendous results from computer analysis. These stem not from the computer, but from the inputs or the settings. The Spielvogel method is detailed, along with those pesky line moments and other details, in the Appendix.

It should be noted here that in doing a complete piping system analysis one should endeavor to use one of the modern computer systems. Computer programs may have a steep learning curve; however, once learned they eliminate a considerable amount of hand work. For reference, the programs mentioned most often by those who write the ASME codes are, in no particular order, Caesar II (Coade), AutoPipe (Bentley), CAE Pipe (SST), and Triflex (Piping Solutions). As with all programs, they have differences due to their authors' approach.

In addition, many of the popular FEA programs can be used for analysis of components and some have the capability to work a piping system. The program Fepipe has the additional convenience of automatically converting the FEA-calculated stresses into ASME stresses. This can be helpful because most FEA programs use distortion energy theory while ASME codes are written around maximum shear stress. The two methods can give different numerical answers. The techniques and math procedures here are more suited to small problems that might be found in the field or in a particular problem, where setting up a full-blown analysis may not be feasible. Having said that, the use of laptop programs continues to increase and a knowledgeable analyst can perform many simple analyses rather easily.

Stress Intensification Factors

The next major piece of the flexibility puzzle is the stress intensification factor (SIF). This may be the most important part of flexibility computation. At the same time, it may be the least understood. There are two reasons for the misunderstanding. The first is that SIFs are a mixture of empirical testing results blended with some theory. The second is that the basis of SIFs is the fact that empirical data relate to a piece of butt-welded pipe rather than the "polished specimen" data from the more conventional methods of developing a stress–cycle curve, which is known as an S–N curve.

There is a third anomaly that is more or less easily explained. SIFs are based on a fully reversed cycle. That is to say, bend the metal forward and then completely back. This is opposed to the more conventional S–N curves, which are based on amplitude or one-directional offsets.

It is also true that all of the data are based on low-cycle, high-amplitude systems, which are characteristic of the heat up and cool down cycles that most pipe is exposed to. There are increasing efforts to utilize these data for high-cycle, low-amplitude phenomenon, such as vibration. That effort creates some problems and is under study by the ASME piping codes. It is discussed more thoroughly in Chapter 10.

The rationale for the other differences of this second "misunderstanding" of SIFs is partly historical and very much practical. A. J. Markl did most of the empirical work in the 1950s with his team of engineers at Tube Turns. Their first set of experiments was directed at finding how many cycles a joint of butt-welded pipe could be bent before it would fail. From that they developed the expression for the failure stress as $iS = 245{,}000N^{-0.2}$ where the i, based on his tests, for straight pipe can be set to 1.

In Markl's original paper it was pointed out that the actual exponent for N varied from 0.1 to 0.3 but most values were within 20 percent of the 0.2. There might also be some argument regarding the constant 245,000; subsequent tests indicate it might be related to the modulus of elasticity. The new B31-J code establishes a constant C to cover that eventuality. We discuss this code later in this chapter. It is important to note that the S or stress that is utilized in these formulas is the nominal stress that is created by the moment that is invoked on the pipe or fitting. That is, of course, the stress as computed by dividing the moment invoked by the section modulus, or

$$S = \frac{M}{Z}$$

It then becomes a simple matter to run a test on a particular fitting, and given that one will control the moment by the displacement and count the cycles, one can then rearrange the formula and create an SIF for that geometry. The rearranged formula is

$$i = \frac{245000N - 0.2}{S}$$

From those tests done by Markl and blending the data with some theory involving a flexibility factor directly related to a factor h, known as the flexibility characteristic, the formula can be written as

$$h = \frac{tR}{r^2}$$

where

t is the wall thickness (for code purposes this is a nominal wall)
R is the radius of curvature of the centerline of the pipe
r is the mean radius of the pipe

This characteristic remains untouched for bends or elbows, which are what the theory is based on. For other fitting configurations where the i factor is determined directly, the h can be back-figured to match the experimentally determined SIFs and fit to the form of the equation for stress intensification factors that the code developed, which is

$$\frac{c}{h^x}$$

where

C is some constant
x is some power of h

Alternatively, some other experimental form may be used from the empirical data.

It is from that arduous methodology and many tests that the codes developed their SIF appendices. In general, they are all based on the same data. Some, like B31.1, establish that one uses the highest factor developed whether doing analysis in an in–plane or out–plane direction. Others, like B31.3, give both the in–plane and out–plane figure so one can apply it to the proper moment when doing the analysis. Appendix D from B31.3 can be found in this book's Appendix.

An example of working up an SIF is given in the discussion of the B31-J book later in this chapter. First, it is prudent to discuss the terms *in-plane* and *out-plane*. In-plane bending is the bending that is in the same direction as the axis of the header pipe. Out-plane bending is at a 90° angle to the axis of the pipe. Figure 7.4 shows the two types of bending.

In the figure the opposite directions on opposite ends are the plus and minus directions depending on the sign convention used. Also note that there is the expression for torsional moments, which represent a somewhat less direct computational requirement.

It was noted before that it is a simple matter to run a test on a fitting to determine an SIF for that fitting geometry. This would require some kind of test rig and test procedure. A picture of one test rig that has been used for several years to develop or confirm the SIFs for proprietary fittings is shown in Figure 7.5.

The rig in Figure 7.5 was first used to test WFI fittings and is now used by the PRG research organization to test and verify analytical configurations

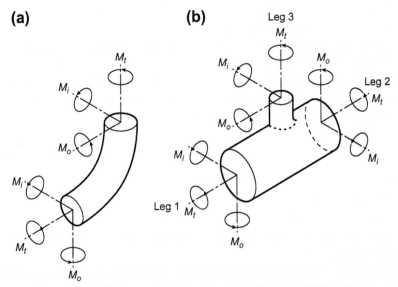

FIGURE 7.4 In-plane (a) and out-plane (b) moments for fittings *(note: i for in-plane, o for out-plane, and t for torsional)*

FIGURE 7.5 SIF test rig

in ongoing research on the SIF questions for ASME and other organizations. It is one of the configurations used to test such fittings.

SIF DEVELOPMENT METHODOLOGY AND B31-J

The test procedures are quite involved and that is the reason ASME has published B31-J as an approved test methodology. The methodology can be summarized as follows:

1. The fixture is mounted in the appropriate place (e.g., see Figure 7.5). The test sample should have water with a head of at least 12 in. (300 mm) so that when a through wall crack occurs it can be determined by the water coming out.

2. The first step is to apply calibrated force (within 1 percent) and measure the deflection. This is done in both the positive and negative directions from the zero point. It is noted that the definition of SIF includes fully reversible displacements.

3. This manual testing of force versus displacement is continued until it is completely obvious that the line is no longer linear. Once that point is reached the load is unloaded in the same steps as it was loaded and the process is repeated for the opposite direction. The minimum number of steps in this should be five to the maximum displacement and five back in both the positive and the negative directions.

4. The loading/unloading data are plotted and the best-fit straight line is used in subsequent determinations of the SIF.

5. A suitable displacement of the fully reversible cycling from the displacement-controlled device (usually hydraulic) is chosen. This displacement should be set so as not to cycle faster than 120 cycles per

minute and the displacement should be picked in such a way that minimum fully reversed cycles should be 500 or more.

6. The action should be started by counting the cycles and watching carefully at the area of anticipated leakage. When leakage is established, the cycling should be stopped and the distance from the point of application of the load displacement to the leak measured.

The follow-up procedures are then completed.

The moment at the leak point is determined by multiplying the force times that measured distance. The size of the force is determined from the previous load displacement chart that was made in the manual load/unload cycle step. Should there be a change in the displacements, the equivalent cycles can be calculated using the following formula for the number of cycles:

$$N = N_j + \sum (r_i)^{\frac{1}{b}} \times N_i \quad \text{for} \quad i = 1, 2, \ldots$$

where r_i is the ratio of the displacements, taking care that the $r_i < 1$ and b is the material exponent 0.2 for metals. Naturally, the N_j is the cycle for the largest displacement, and essentially that is the force used in the moment calculation. That moment is calculated into a stress by the previously mentioned formula

$$S = \frac{M}{Z}$$

where the Z is of the pipe intended to be used with the component.

The SIF can then be calculated using a variation of the formulas previously discussed,

$$i = \frac{C}{SN^b}$$

where b is the material constant 0.2 for metals, C is a constant that is 245,000 (1690 MPA) for carbon steels and adjusted as discussed later for other materials, S is the nominal stress at the leak point just calculated, and N is the cycles.

To be valid there needs to be at least four independent tests. If there are four tests the average i can be used. If there are fewer than four, that average needs to be multiplied for the factor R_i as given in Table 7.1.

For variations in materials the following formulas are used for adjusting C in the SIF formula:

$$C = \frac{245,000 \times E}{27,800,000} \quad \text{for USC units}$$

Table 7.1 Multiplication Factor for SIF

Number of Tests	Test Factor, R_i
1	1.2
2	1.1
3	1.05
≥ 4	1

$$C = \frac{1690 \times E}{192,000} \text{ for SI units}$$

In addition, a comprehensive report is required that is certified by a registered professional engineer or person of equivalent experience. It should be pointed out that this summary is not complete in the details. Before one actually performs a test of this nature it is strongly suggested that he or she get a copy of the current B31-J and use it.

For complete understanding, here is a short example. A tee-type fitting is to be tested; the size is 4 NPS standard weight, carbon steel. We will simulate the actual load/unload test. We set up the test rig with a 24-in. extension for the driving mechanism for the cycling. We have produced a sample load/unload chart that is a manual load and measures the displacements (Figure 7.6). The example test data in the chart is purely arbitrary and is not from any actual tests. Most existing tests are proprietary information.

After examining the chart, one determines that the straight-line portion begins to enter the yield domain of the material and test specimen. Then

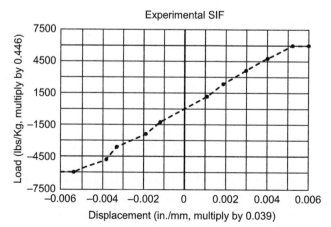

FIGURE 7.6 Example load/unload SIF chart

one picks a displacement of 0.005 in., which is slightly below this yield, and one that we can expect to last for more than 500 fully reversed cycles. The adjustments are made to the limit stops on the machine and the automatic cycling is started.

Note that there was a check to see whether there was enough water in the test specimen and the process of watching for the leak, assuming some automatic device is not extant on the test rig. The allowed cycles are as high as 120 cycles per minute. However, experience with the machines has shown that the number of cycles per minute is often more on the order of 20 cycles per minute. So this is a period that can be described as slow.

For purposes of our thought experiment we will assume that the leak appears at 845 cycles. Again in practice sometimes the original choice of displacement is too low and a change is made to increase the displacement, or it could be too much and the failure would occur before 500 cycles. This would cause extra experimental work. So often one starts out at a lower displacement, and as the number of cycles increases toward the minimum, one can reset the cycles to a higher displacement. This creates the need for more calculations to arrive at the equivalent cycles.

In the thought experiment it worked without the adjustment of cycles. The leak occurs and a measurement from the load application to the point of the leak is taken. That measurement shows the leak was a little higher than the actual surface intersection and was actually 23.75 in.

Going back to Figure 7.6, we find that the 0.005 displacement equates to a load of 5500 lbs (in SI units, that is 0.125 mm and 2495 kg). We can calculate the moment as $M = FL = 5500 \times 23.75 = 130,625$ in./lbs (2.33 e6 kg/m). The Z of NPS 4 (DN 100) S40 is 3.21 in.^{3} USC (34,237 cm^3 in SI). Therefore, the stress is equal to

$$S = \frac{M}{Z} = \frac{130,625}{3.21} = 40,693 \text{ psi in USC (2881 Mpa in SI)}$$

This sets up the calculation for the experimentally determined SIF. The equation is

$$i = \frac{C}{SN^b} = \frac{245,000}{40,693 \times 845^{0.2}} = 1.56$$

What can be determined from this? Since this is a thought experiment and can't be repeated, the multiplier 1.2 would have to be applied to the 1.56 to make a usable SIF of 1.87. If in actually doing this, the average result of four tests was 1.56, that would be the applicable SIF for that fitting.

Recall that this is a test of a tee-type fitting. The SIF appendices for B16.9 tees in B31.3 and B31.1 indicate that the basic formula for a tee this size and schedule would be 1.82, for the one test example. This is a good time to point out that it is always good to read the applicable notes in these code figures and tables. For B16.9 tees, note 8 points out that if the crotch thickness as well as the tees' crotch radius meets certain criteria, the flexibility characteristic, and thus the SIF, could change. If the crotch thickness met those criteria, the SIF would reduce to 1.45.

This could give an indication of whether to perform further tests. If the crotch thickness of the tested product was approaching or better than the thickness of the criteria in note 8, it could be beneficial to perform more tests to establish a lower SIF for this product. If the thickness were intermediate or less, it would become a judgment call.

The lowest allowable SIF is 1. This is because that is the basic stress multiplier for the welded piece of pipe and is therefore the baseline stress multiplier. As a historical note, in the original code equations the lower SIF was the default calculation. That was because at the time of the original tests, the crotch thicknesses of production tees were on the order of the thickness specified. Ongoing testing in the 1980s showed two things: (1) many production tees no longer had that thicker crotch, and (2) this had a significant effect on the tested SIF. The committee further investigated and found in fact that the preponderance of the tees available had the thinner crotch, and they then adjusted the default calculations to reflect the actuality. They maintained the lower SIF option through note 8, which states that if you have this desirable condition you may use the lower calculation.

The current default SIF calculations of the B31 codes are reproduced in the Appendix. These are usually found in an appendix to a particular code book. Within the body of the book these are invoked as the ones to be used in the absence of other objective evidence. By publishing B31-J, ASME has offered a path to develop the other objective evidence. Using an analogy of the geometrical configurations, developing an SIF for a geometry that is not offered in the code default calculations is also allowed. This, of course, requires either some testing or theory with a great deal of experience in that field.

There has been an extensive research project within ASME to develop new formulae for SIF. These have been done with extensive review of literature and computer analysis of various configurations. A new set of formulas has been developed. The intention is to place them in an appendix of B31 J and again allow the various code books to adopt the new methodology as they see fit for their particular applications. They reflect refinements in both the slope of the base curves and other data as the research has pointed to the need for change.

The elements are now in place to conduct a flexibility analysis of a piping run or system. But first we must talk about an inherent part of the piping system in Chapter 8.

CHAPTER 8

Pipe Support Elements, Methods, and Calculations

Contents

OVERVIEW

After designing and laying out a piping system and analytically taking it to the temperature that the intended system will have, we could have a mess. A piping system is essentially an irregular space frame. It is often quite slender and, when at a high temperature, could be compared to a tangle of spaghetti unable to support itself in its intended shape.

This leads to a pipe stress analyst's dilemma—where and how to support the system. This includes issues related to how and where to restrain the pipe. We also want to know if restraint of movement is needed anywhere. If it is, there is a question of which direction or rotation would be the most beneficial. The question of bracing most often comes into play when thinking of nonconstant loads.

While it may not be obvious, similar problems occur in pipelines. Certainly there could be the problem of earthquakes. Also, no matter how buried the pipeline may be it has to come up to surface occasionally to go through a pumping station or a terminal or a pigging point. There are the problems related to when the pipeline has to turn (some sort of blocking might be needed). There would be river crossings, street crossings, and even occasionally crossing the path of another pipeline. These might not require pipe hangers like industrial piping but they do require thought and supporting mechanisms of some sort.

The ASME codes refer to nonconstant loads as occasional loads. These loads include such things as earthquakes, winds above some level, and snow. There may be other such occasional loads that can occur, but it is not

Piping and Pipeline Calculations Manual
ISBN 978-0-12-416747-6
http://dx.doi.org/10.1016/B978-0-12-416747-6.00008-5

specifically known as to when they would occur. This would include dynamic loads from a pressure relief upset or other such loads that are not specifically prepared for but are known to potentially happen over the intended life of a system.

An analyst's problem is that the supports, restraints, and braces interact with the reactions of the pipe. We learned that in simple analysis there could be two anchor points. In most systems those anchors would be something like a flow generator (pump) and maybe a storage or pressure vessel. There could be all sorts of other anchor types, but the essence is that usually the anchor points are equipment. That equipment has limitations on what kinds of forces and moments it can take from the piping thermal movements, as well as the other types of loads that may be developed during operation. The various supports, restraints, and braces will affect the size of the forces that are developed.

This results in what can be described as the piping designer's conundrum. Changing the type and location of the supporting and restraining devices will change the resulting forces. There are limits to the forces because of the equipment. In most cases there are also limits to the ways the piping can be routed.

The location of the equipment is quite often limited by the process or site needs, or other restrictions imposed for external reasons to the piping flexibility. We discussed in Chapter 7 that one of the ways to increase or decrease flexibility to change the stresses the thermal growth will generate in the system comes from adding elbows, loops, or lengths and/or moving the piping system closer to the thrust line. We know that the supporting system has to have some external structure to support the supports.

SUPPORT DESIGN

The question of how to bring a system into flexibility compliance at first glance seems to be insurmountable, but this is not necessarily so. For one thing, experience will quite often tell one where the supports, and so on, will be needed. There are some simple starting rules and ways to resolve these problems. That is, a designer uses experience and simple checklists to select the initial locations and type. Then the required analysis of the results can be made. For an experienced or fortunate designer the results put the reactions within the allowable parameters as well as the resulting stresses. Otherwise, some changes have to be made like adjusting types, changing locations, and changing components until an acceptable combination is found.

Fortunately, the Manufacturers Standardization Society (MSS) has developed a set of standard practices that helps with the decision process. Those standards are SP-58, SP-69, and SP-127. They are helpful and do give some allowable stresses and maximums for things like spacing, which have been accepted by ASME and others to eliminate further calculations.

It is important to point out that this is an area where the use of flexibility software will be of great service. Each program has a built-in methodology to do the type of formulas introduced later in this chapter by eliminating tedious calculations. Again, for an individual case the manual calculations can be utilized. Understanding what goes into them is also important to a designer.

For purposes of this chapter a set of general rules is set out to help in the determination of the location of the original supports based on some rudimentary rules. Then calculations that go into determining the details of those types of installations are presented. As stated in Chapter 7, the calculations need to be tested by the flexibility analysis to establish that there are no overstressed points and that the final reactions are within the limits that the anchoring equipment can accept.

Referring back to flexibility analysis, it is also good to run through some of the manual calculations both in thermal movement and adjusting shapes, etc., to gain some theoretical understanding of what is really happening in this spaghetti-like frame. This might be a good time to interject that a complicated system can be broken into simple systems and then analyzed for the support issues with an equilibrium-type solution. One of the complexities that is hard to discuss but easy to visualize is that at the anchor point, not only does the pipe move thermally, but most likely, unless the anchor point is something like the Rock of Gibraltar, the anchor point will also move, however slightly. In some situations it will move in concert with the pipe, and in others its movement will oppose the movement of the pipe. This is another situation where experience, actual or theoretical, is most helpful.

For all practical purposes the codes do not cover pipe support systems design. Some, like B31.1, do give guidance regarding support design. Since the MSS series has been upgraded and combined, MSS SP-58 is the *de facto* international standard. The codes will also allow or point to the MSS standard practices mentioned previously.

B31.3 also lists some objectives to meet in designing such support systems. They can apply to any piping system. In summary they are the following:

1. Prevent stresses in excess of permitted stresses.
2. Prevent joint leakage.

3. Prevent excess forces on equipment to which pipe is connected.

4. Prevent excess stresses in the elements of the support system.

5. Avoid resonance from vibrations.

6. Avoid interference from the thermal movement with other pipes or structures.

7. Prevent excessive sag or distortion, including pipe in a creep condition.

8. Shield from excess heat, which could overstress the supporting components.

These seem obvious at first glance. However, there have been times when a designer did not allow for things such as subsequent maintenance of nearby equipment due to locating support elements in such a manner as to obstruct access. Other such unintentional errors from not checking for these not-so-obvious occurrences can be mitigated by following such a checklist.

There are a few other guidelines that can be helpful, including the distance between supports. This is a function of sag from such things as weight, locating the supports near heavy point loads such as valves or risers, locating the lines near structures or where a structure can be provided, and avoiding locating items like those mentioned in changes of direction like an elbow. For economic reasons one should group piping together, such as on pipe racks that one sees all over a plant.

It is time to work on some of the calculations that will be encountered in the design. The first is the spacing of supports. This can be a source of controversy. The spacing is of course dependent on the weight of the piping being supported. MSS SP-58 provides an acceptable table based on a pipe size and material filled with either water or vapor. It is essentially the same spacing as the B31.1 table, but has more materials and variations. Since it is specific to certain wall weights and other criteria, this book will lead you through the underlying calculations.

A straight portion of pipe between two supports is a simple beam. Therefore, the stress calculation becomes one of determining the end connections and calculating the stress. There are two possible end configurations: pinned or fixed. Those stress formulas are, respectively,

$$S = \frac{WL^2}{8Z} \text{ for simply supported}$$

$$S = \frac{WL^2}{12Z} \text{ for a fixed end beam}$$

where

S is the bending stress of the pipe

W is the weight per unit of the pipe

Z is the section modulus (in.^{3}, mm^3) of the pipe

Experience has shown that neither configuration is entirely correct. So there is a convention in piping to use the formula

$$S = \frac{WL^2}{10Z}$$

which is a compromise between the two end conditions. This can be considered the standard calculation and the simply supported end for a conservative calculation.

To get an understanding of what this means it is useful to compare the calculations with the MSS chart for a 6 NPS (DN 150) standard-weight water-filled steel pipe. That factor's accepted length is 17 ft (5.2 m). By rearranging the equation using the compromise factor of 10 and setting the allowable bending stress at 15,000 psi (103 mPa), we can get the following results for the calculation:

$$L = \sqrt{\frac{10 \times z \times S}{W}} = \sqrt{\frac{10 \times 8.5 \times 15,000}{31.48}} = 16.77 \text{ ft } (5.213 \text{ m})$$

This is a reasonable comparison to the MSS chart for that size water-filled pipe. For reference, a pipe that size weighs 18.97 lbs/ft and the water in it would weigh 12.97 lbs/ft, giving us the 31.48-lb weight used in the calculation.

A careful reading of the allowable stress charts will show that there are not a lot of pipes that have exactly 15,000-psi allowable stress in the ASME codes. This is especially true of B31.1. If one uses A53 grade B and calculates those two separately and averages, one gets a figure of 16.96 ft (5.27 m), which shows that the MSS chart is reasonable for using the 15,000-psi figure for water-filled pipe.

However, that doesn't mean one is home free. An adjustment to the weight could be needed if the fluid used is changed or the schedule is different from standard. For example, the MSS chart says that the same pipe filled with a vapor would have a maximum distance between supports of 21 ft (6.4 m). This implies that the vapor has a weight of 1.1 lbs/ft (0.498 kg/ft). It is also obvious that using piping made of other material or operating at a higher temperature reduces the allowable stress and the calculated maximum L becomes less.

Once that maximum *L* has been established, it becomes a simple matter to place the fixed supports along the line at that or some smaller distance, taking into account the other restrictions. If one has a concentrated load, say a valve somewhere in the line, that changes the calculation also. There is one other consideration in this placement. Most pipe, as we have learned, is not in a single straight line. There are often changes of direction. It is advisable, when there is a change of direction between two supports, that the maximum span dimension calculated be reduced by a factor of 0.75. Given this information, the following is a simple example showing how the process works.

First, we posit a small horizontal piping layout in Figure 8.1. The supports are labeled *A*, *B*, *C*, and *D*. They have been located following the guidelines previously suggested. The line is a horizontal line of size 6 NPS, 150 DN, standard weight, and it is assumed to be filled with water and have no insulation. At the valve end is the first anchor point (or nozzle), and at the 10 (all given dimensions are in feet) end is the other nozzle. Recall that the maximum span is 17 ft (5.2 m) for straight pipe, and through a change in direction that span is 12 ft (3.6 m). The layout is dictated by the site requirements. No decision has been made as to the type of support. The supports are considered to be rigid and could be hung from above or supported from below. The line is assumed to be at some elevation above

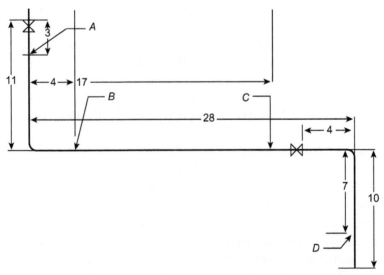

FIGURE 8.1 Pipe support diagram. *Note:* all dimensions are in feet

grade, which does not enter into the support loads. All line loads are gravity or down loads, and the support loads are therefore acting up. The salient factors are listed in Table 8.1.

$$CGD = \frac{R(1 - \cos\theta)}{\theta}$$

where θ is in radians and R is the centerline radius.

These factors will be used for the development of the loads on the supports. This will be done through the use of equilibrium calculations using conventional static analysis (i.e., there are no dynamic loads or other loading considered).

The procedure is to break the piping into sections that are statically determinate, make free bodies of those sections, and solve for the forces resulting from the estimated weights in Table 8.1 and the dimensions from Figure 8.1. Then resolve the entire system by superposition as required.

On examination there is an unencumbered straight section between supports B and C that is 17 ft (5.2 m), and so the force on each of the supports will be

$$\text{Force} = \frac{17(31.48)}{2} = \frac{535.2}{2} \quad \text{or} \quad \text{Force} = \frac{5.2(47.1)}{2} = \frac{244.9}{2}$$

$$= 267.6 \text{ lbs} \qquad\qquad = 122.5 \text{ kg}$$

For the next section consider using the section from the valve to support B and create the free body from the valve through support B. Examination shows two unknowns; however, if one sums the moments around the pipe through the valve and support A, support A is eliminated and the only

Table 8.1 Salient Factors for Figure 8.1

Element	Size (Description)	Weight, lbs (kg)	Other Information
Valve near A	Flanged gate valve	190 (86.1)	150 class
Valve near C	Flanged check valve	150 (68)	150 class
All elbows	Long radius	28.8 (13)	CG 0.5 ft★ (0.15 m)
Pipe	6 standard	18.97/ft (28.3/m)	No insulation
Water	N/A	12.51/ft (18.8/m)	N/A
Water in elbow	N/A	7.36 (3.3)	Based on centerline length

Note: All 6 NPS standard weight.
★Center of gravity distance (CGD) from the end of the elbow in the direction of the attached pipe centerline based on the formula

unknown is support B. That calculation then becomes, using the axis y, $\Sigma M_y = 0$, which is

$$0 = -150 \times 0.5 - 31.48 \times 3.25 \times \left(0.5 + \frac{3.25}{2}\right) + 4B$$

Support B is then 73.1 lbs (33.2 kg).

Having determined the load on support B the next step will be to sum around the nozzle at the end of the valve to determine the value of B. The formula is

$$\Sigma M_x = 0$$

$$0 = -190 \times 0.5 - 9.5 \times 31.48 \times 5.75 - 10.5 \times (28.8 + 7.36)$$

$$-11 \times 3.25 \times 31.48 + 73.1 + 3A$$

Support A is then 9738 lbs (4417 kg) from this free body.

The next free body comes from the end nearest D and goes back to C. Once again we do the calculations in two steps, first summing around the y axis, thus eliminating all D and that portion of the pipe. The calculation in SI units is

$$0 = -(13 + 3.3) \times 0.15 - 47.1 \times 1.1 \times \frac{1.1}{2} - 1.5 \times 47.1 \times 6 + 8C$$

Support C from this free body is 125.4 lbs (56.9 kg).

Now we can sum moments about the support C check valve axis, eliminating those contributions:

$$0 = -(13.3 + 3.3) \times 0.15 - 47.1 \times 2.9 \times \frac{2.9 + 0.15}{2} + 7D$$

So support D has a load of 66.22 lbs (31.1 kg).

The remaining step is to combine the loads that are taken from both the first span and the two free-body sections. The first load calculation on both supports B and C was 267.76 lbs (122.4 kg). To support B we have to add the free-body load of 73.1 lbs (33.2 kg), making the total load on support B 340.7 lbs (155.6 kg). The comparable total load on support C would be 393 lbs (179.3 kg).

It is noted that the load on support A seems quite high at 9738 lbs (4417 kg). At this point some changes may need to be made. An intermediate support before the first elbow, thus dividing the load on support A between two supports, is possible. Or a special support might be utilized on the valve, eliminating it from the piping support. Again, these are areas where experience and skill would come into play.

A flexibility analysis has not been run. This may add some concerns that are not apparent. Since no insulation was posited the temperature was not extreme, but if the temperature had been higher, insulation would probably be considered. As the allowed stress went down with the temperature, the maximum span would have been reduced, calling for more support, which might have reduced the load now apparent on support *B*.

Last, there were no vertical riser pieces of pipe in this layout. There are no spacing or span rules on risers, as essentially they do not have the sag problems of a horizontal run. There, expansion can cause movement problems on the attached horizontal pipe. They do have weight and must be supported in some manner, so they require attention in piping and support design.

It was mentioned that the type of support was not decided in the exercise. Each manufacturer of piping supports and hangers/supports includes in its catalog the several types it manufactures.

As mentioned, there are three main categories: support, restraint, and brace. Within each category there are some subsets. Some generic types of these are pictured in the MSS documents and are shown in the Appendix of this book. It may be helpful to discuss some of the subsets in the three generic categories at this time.

The following are some of the subsets under *restraints*:
- *Stop.* A device that permits rotation but prevents movement in one direction along one axis.
- *Double-acting stop.* A device that prevents movement in both directions along one axis.
- *Limit stop.* A device that permits limited movement.
- *Anchor.* This device is essentially a rigid restraint; however, it is also often considered as a piece of equipment that would accept without harm only a limited moment or force.

The following are some of the subsets under *supports*:
- *Hanger.* A support that does so by suspending the pipe from a structure.
- *Guide.* A device that prevents rotation about an axis.
- *Resting or sliding support.* A device that provides the support from beneath the pipe and offers no resistance except friction.
- *Rigid support.* A support that provides stiffness in at least one direction.

- *Constant effort.* The most common type is a spring support that is intended to supply a constant supporting force through a range of movement.
- *Damping device.* Commonly called a snubber, which acts as a shock absorber in its efforts.

Braces do not have the same number of common subsets. They are employed to act as restraints for forces that generally do not come from sources such as thermal expansion or gravity. MSS SP-127 offers guidelines on bracing.

The next issue for a designer is to select the appropriate type of support for whatever is being designed. In the previous example, the entire line was specified as horizontal. So there would be no normal up or down thermal movements. Rigid hangers would be appropriate. Also, there was no insulation, indicating no high temperatures. However, for the sake of discussion, a high temperature can be posited to demonstrate how one might determine how much the rigid hammer could be expected to sway or swing as the pipe is moved from side to side from thermal expansion.

As an example, think of a 90° right-angle turn where the legs are 50 ft (15.2 m) in length. The line is a 10 NPS (DN 250) carbon steel pipe operating at 650°F (343°C). The expansion factor in inches per foot for that line would be 0.512 (4.27 mm/m). Therefore, the total expansion of any one leg would be as follows:

$$\text{Expansion} = 50 \times 0.512 = 2.56 \text{ in. for UCS units}$$

$$\text{Expansion} = 4.27 \times 15.5 = 64.9 \text{ mm/m for SI units}$$

This expansion would be at the point of direction change and would cause the right-angled line to move to the side that much. Imagine that line to have a perfect pivot on the opposite end from the corner. This would cause that line to form a straight line between the pivot and the corner. One can readily see that this would cause a proportional sideways displacement anywhere along that line (see Figure 8.2).

Note that this is a simplification that ensures that anyone using this method will have a conservative estimate of the displacement at any point. In the real world both lines would expand and interact, causing some displacement on each. This would in effect cause the displacement on either not to be purely proportional. This is one of the many differences between a rigorous, computer-type analysis and the manual, field-type calculations expounded in this book.

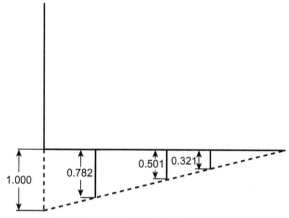

FIGURE 8.2 Proportional movement

Given the 2.56 in. (64.9 mm) deflection it is simple to note the amount of sway that a hanger would have hanging anywhere along the line. Then, depending on the length of the hanger, one can calculate the amount of arc lift that would be imposed on the pipe. As a rule a maximum angle of 4° is the point at which another design might be required from a conventional rod–type hanger.

Interestingly, as well as logically, this same proportional sort of analysis can be used to determine the distance a rigid support can be placed away from a change in direction. It is simple to rotate the corner from Figure 8.2 in such a way that the thermal growth is vertical. The arguments for proportional displacement don't change just the directions.

In the case of vertical growth, there is a formulaic way to determine the minimum length along this line that a rigid support can be placed that will not cause excess stress beyond an arbitrary stress that is established in calculating that minimum length. The formulas are

$$L = \sqrt{\frac{\Delta D(10^6)}{1.6S}} \text{ in USC}$$

$$L = \sqrt{\frac{\Delta D \times 0.62}{S}} \text{ in SI}$$

where

  L is the distance to first rigid restraint, ft or m
  Δ is the displacement to be absorbed, mm

D is the pipe's outside diameter, in. or mm

S is the allowable stress of pipe, psi or MPa

Using the 50-ft (15.2-m) example, we find

$$L = \sqrt{\frac{2.56 \times 10.75(10^6)}{1.6(10,000)}} = 41 \text{ ft}$$

For the SI system we find:

$$L = \sqrt{\frac{64.9 \times 273 \times 0.62}{68.9}} = 12.62 \text{ m}$$

The use of 10,000 psi is standard in many of the charts made before modern calculators were available. Then, conversion factors were provided to convert the chart factor for 10,000 psi for other stress values. This was to avoid the tortuous hand calculations to extract the square roots. Modern calculators make it simpler to just use the square root or 0.5 power function and calculate for the specific stress required.

The distance calculated establishes the point where a rigid holder or restraint would create the established stress from thermal movement.

With experience and skill, analysts can use this concept to do some preliminary flexibility analysis. The concept is to assume that the system is a two-anchor system and that the anchors are immovable (not realistic, but small anchor movements might be acceptable). Then calculate the movement inward along the axis of each leg. Establishing one leg's movement makes the assumption that the other legs will absorb the movement. They have a total length of some amount. Then calculate for a chosen stress the length that would be needed to absorb that stress. If the total absorbing leg length is longer than that needed from the calculation, then one can assume that it is an acceptable layout/length combination. After having done this for all legs and all axes of the layout, if the result is always more absorbing pipe length than required, one can make the tentative assumption that the layout will pass.

It should be pointed out that this method has many underlying assumptions of the system and is not for amateurs; as the saying goes, "Do not try this at home." For that reason it is not delineated here. The reason it is noted as a possibility is that for knowledgeable analysts, it is a "quick and dirty" analysis that might be useful in determining a potential source of a problem in the field, or a potential solution might even be suggested by going through the calculations. For someone just learning the business, it is an exercise that might give insight into what happens with complex stuff.

NONRIGID HANGERS

To move to the next item on the agenda one will surely find that there is somewhere in the system where the movement is such that a rigid hanger is not advisable and some other sort is required. As was noted in the discussion of subsets, there are two "nonrigid" types, the constant force and the spring hanger. The spring hanger is the more common.

As was indicated, the determination is that the location of the hanger is one where, by calculation, one has determined that a rigid hanger is not adequate. So the first step will be to determine the range of movement for that location. It is conservative to use the proportional method previously described for the calculation if a more rigorous method is not available. The MSS hanger documents break this into four ranges, which vary from ¼ in. to 3 in. Then they recommend both a load variability and a hanger type for that specific range.

The standard types of charts and the spring hanger/constant force hanger are replicated in the Appendix. Each spring manufacturer has a similar table for their particular models, which is more useful because the applicability of a particular design is based on the actual spring rate and load capability for the situation, and the loads imposed on the movement that the spring hanger is designed to control.

The following is a generic discussion of what the spring hanger will do. It follows that as the spring is loaded or unloaded it will impose a variable load on the rod of the hanger, and that load is eventually imparted to the piping system. This is based on the fact that the piping weight doesn't change as it moves up or down. The convention in calculations is that the operating or hot condition is considered the base load, which then becomes the load at the neutral position of the spring. This is considered the hot load. This implies that as the system is cooled down the spring is collapsed and, based on the spring's K or spring rate, adds load. This makes the cold load the higher load. That, plus the weight load calculated for that position, is the total load on the system. And that load, since it is higher, changes the system stresses. However, the mathematics is the same with either convention, and if one is working with a system where the movement is down, it may be more convenient to ignore the conventional considerations (see Figure 8.3).

Some sample calculations to illustrate are in order. Let us posit a system where the load is calculated as 2500 lbs (1134 kg) at the position where we need a hanger, and it has a movement of 1.2 in. (30.5 mm) at that location. We want the difference between the loads to be less than 600 lbs (272 kg).

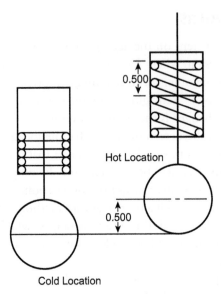

FIGURE 8.3 Spring hanger hot and cold positions

We have a spring hanger that has an acceptable range within our movement range. It has a spring rate of 450 lbs/in. (10 kg/mm). Will the hanger meet the load difference requirement?

$$1.2 \times 450 = 540 \text{ lbs—meets the 600-lb criteria}$$

$$30.5 \times 10 = 350 \text{ kg—does not meet the 600-lb criteria}$$

This is one of the vagaries of standardization: If there is not a spring hanger that has an acceptable range and a spring rate of 8 or less, some other solution is required.

Normally since the hot load has a lower allowable stress than the cold load this will not cause a problem. However, the standard is to keep that variability between loads within a certain percentage. The MSS documents set that percentage at 25 percent. It is not uncommon to reduce that to something like 10 percent when the system is deemed critical. The calculation of that variability is

$$\text{Variability} = \frac{\text{cold load} - \text{hot load}}{\text{hot load}} \leq 25\%$$

In the case of the USC measurement, the cold load would be $2500 + 540$ or 3040 lbs, so the actual variability would be

$$\frac{3050 - 2500}{2500} = \frac{540}{2500} = 22\%$$

which is under the 25% standard limit.

One can use that variability to calculate a desired spring rate. When using a chart or standard-type catalog choice the calculation of the spring rate is not necessary, but it can be useful to back-check any chart-type selection method. That calculation is

$$\text{Spring rate} = \frac{\text{variability} \times \text{hot load}}{\Delta}$$

where Δ is the movement. In this case it did not meet the requirement, so for the spring rate that was required the desired variability is

$$\frac{272}{1134} = 24\%$$

Thus the needed spring rate is

$$\frac{24 \times 1134}{30.5} = 8.9 \text{ kg/mm}$$

This leads to the 8 or below previously noted.

There can be a situation where one cannot find a standard spring hammer, or for one reason or another, variability of the load has to be tighter than the 25 percent or even the 10 percent criterion mentioned for critical systems. In those cases a constant force–type support is required. While there are some standards, according to MSS they are generally in the 6 percent variability range. Anything more critical would most likely be special.

Counterweight-type arrangements are probably the most consistent in terms of least variability. As noted, the weight of the piping does not change if one gets an opposing or counterweight mechanism. So barring some reduction in the weight of the counterweight mechanism due to wear, corrosion, or lack of maintenance, when properly installed counterweight arrangements would be as close to zero variability as possible. They are also the most special and require the most space and the most continual attention to be an effective solution. For these reasons they are avoided when possible and not discussed further here.

The ordinary method of constructing a constant force hanger is to utilize the helical spring and interpose a variable crank between that spring and the rod transmitting the force. The variability of the lever is on the bell

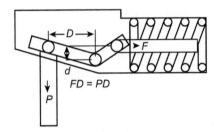

FIGURE 8.4 Constant force hanger arrangement

crank, which offsets the change in load due to the spring constant rendering the load imposed to the piping as nearly constant as possible (see Figure 8.4).

Since the force is designed to be as constant as possible the only calculations needed have already been described. They are the range of movement at the location and the calculated load, which should be constant. There are no further calculations remaining.

RISER SUPPORT

There has been little discussion of supports for risers. This is because they have no "span sag" requirements. However, this is not to say that there are no requirements. The first and most obvious requirement is that something is needed to hold risers erect. Quite often this can be a rigid holder somewhere in the riser itself. The holder has to hold not only the weight of the riser, but any weights that are inflicted on the riser, because the system is not supported in a balanced manner.

In the discussion of spring hangers, the notion of variability was introduced. One result of the fact that there is variability is that some "imbalance" is being introduced to the system. Those imbalanced loads create stress somewhere and the intervening anchors or restraints have to be able to withstand them.

Another interesting aspect of the support system for a riser is that it can change the direction that the thermal growth moves. Take a simple example where a rigid anchoring device is placed at the middle of the riser. This has the effect of causing the growth above this rigid point to be in the upward direction. Conversely, the growth in the portion below the rigid point will be in the downward direction. This has no physical effect other than changing the signs. Analysts must attach proper signs to the growth while performing the balancing static analysis.

It should be pointed out that this result occurs when any rigid anchor is placed in a line, and if it is rigid enough, the direction of the movement might be changed. In fact, occasions can occur where only a certain amount of movement/rotation can be allowed or no movement/rotation in one direction is allowed. These are occasions where stops, double-acting stops, or limit stops, as well as guides, can be used. When used, they will affect the magnitude of the stresses or loads on any given piece of equipment.

As an example, if a system approaching an equipment-type anchor has an excess load in one axis of the end stop, that load can be reduced by putting a stop on that line that will limit the movement and thus reduce the load on that axis. It will, however, change the loads and stresses in other axes. To determine if this is a problem, we need to consider more rigorous analysis.

As was stated earlier and emphasized as we go along it is important that the close design work be done by one of the computer modeling systems. They simply eliminate much tedious work and as a result they also reduce the potential for calculation errors.

It is also important to note that during the installation process there needs to be a rigorous walk-down of the final installation. Unfortunately the location on the drawings is not always followed closely for many unspecified reasons. As one gets familiar with the total analysis, one will find that in some cases minor changes in hanger location can result in significant changes in the stresses created during operation. It is always best to avoid such potential problems.

This discussion of hand methods is meant to familiarize the reader with the process the computer systems use and also to present a means to do a "field check" when the need arises.

Specialty Components

Contents

EXPANSION JOINTS

One component that is also used in compensating for thermal expansion is an expansion joint. There are basically two types of expansion joints: a slip joint and a bellows. The argument in favor of expansion joints is that they take up less space than a pipe loop, which is one of the ways to add flexibility to a run of pipe. This also can save material. The other limitation of both the pipe loop and the slip joint is that they only offer compensation along one axis of the pipe in which they are used.

The slip joint is a theoretically simple device; however, it is, practically, very difficult to work with. A slip joint is essentially a sleeve over two disconnected pieces of pipe that allows the two pieces to move toward or away from each other as they expand or contract. There of course must be some sort of sealant between the OD of the pipe and the ID of the slip joint to keep the fluid from leaking out. This also would require some amount, if not an excessive amount, of maintenance. For those reasons the slip expansion joint has lost its popularity and is used less in more recent piping.

The bellows expansion joint is the one that is most often used currently. There is a certain disdain among experienced pipers who take the position that the use of an expansion joint to reduce reactions at a particular equipment anchorage or for other similar reasons is an admission of lack of skill or planning. This may not be exactly true. It is entirely possible that such an expansion joint would be the most economical solution. Expansion joints may be used for any of several reasons besides space saving. They reduce expansion stresses, they reduce the pressure drop when used in place of elaborate looping and other flexibility–increasing layouts, and they also reduce mechanical vibration.

Piping and Pipeline Calculations Manual
ISBN 978-0-12-416747-6
http://dx.doi.org/10.1016/B978-0-12-416747-6.00009-7

In the following, we discuss the bellows expansion joint only. The term *bellows* is probably the most generic term, as there are several fundamental types of bellows that are employed. The use of a specific sort of bellow is a function of the bellows manufacturer. Each manufacturer has its own tooling that is used for the common types of materials, and as the materials change, the manufacturers may change the type of bellows to suit their design expertise. These fall into two basic categories of bellows: those that are formed and those that are fabricated. Within each category there are four shapes, shown in Figure 9.1, that are recognized by the Expansion Joint Manufacturers Association (EJMA).

In addition, there are several fundamental designs of complete bellows. These different shapes afford different degrees of freedom and are applicable in different situations requiring the flexibility afforded the bellows. An expansion joint can work with three fundamental movements:

1. *Axial movement.* This is the movement that occurs along a straight piece of pipe that has no interactive pipe in other directions.
2. *Lateral movement.* This can be in any direction that is perpendicular to the axis. If there is movement at other than 90° from the pipe's axis it can be resolved into its resultant movement by

$$\sqrt{x^2 + y^2} = \text{resultant}$$

or vice versa.

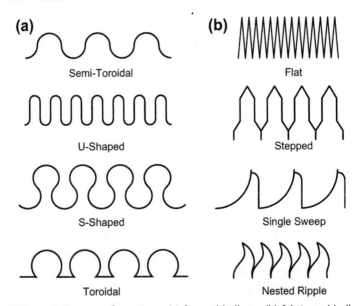

FIGURE 9.1 Bellows configurations: (a) formed bellows; (b) fabricated bellows

3. *Angular rotation.* This is the bending of the pipe's centerline.
It should be noted that expansion joints have very little or no torsional resistance, and should it be present in the system, special considerations would be required.

The joints can be reduced to eight fundamental types of expansion joints. They are described in Table 9.1.

This is a good point to refer to Appendix X in B31.3, "Metallic Bellows Expansion Joints," which outlines the requirements used for this chapter that are compatible with EJMA standards. The general chapter, 300, states that it does not specify the design details, but rather assigns the design details of all elements to the manufacturer of the joint. Expansion joint design requires significant knowledge and testing and this is deemed an appropriate assignment. It also assigns the designs of the main anchors and the intermediate anchors to the pipe designer.

In addition, it imposes factors of safety, places limits on the design stresses, and has rather detailed requirements for fatigue analysis and testing. It should be noted that this type of combined responsibility on the final design of the expansion joint and its final assembly does not lessen the responsibility of a pipe designer. One could make the argument that it adds to that responsibility in that a designer has to find ways

Table 9.1 Expansion Joint Types

Name	Description of Usage and Rationale
Single	Absorbs all movements of pipe section it is installed in
Double	Two single joints with a common connector rigidly anchored
Universal	Two sets of bellows acting as one joint to accommodate lateral movement larger than a single unit (where a double acts as two different singles)
Tied universal	Absorbs pressure thrust; will absorb no lateral movement external to the tied length
Swing expansion	Will absorb lateral and/or angular rotation in one plane only
Hinged	A single bellows to permit angular rotation in one plane only; note these should be used in double or triple combinations
Gimbal	Designed to permit angular rotation in any plane by use of a gimbal
Pressure balanced	Designed to take angular rotation and lateral movement while restraining the pressure thrust force.

Note: It is always good to determine the proper type to use in conjunction with the manufacturer.

to ensure that the detailed design is done properly and all EJMA requirements are met.

One thing to remember is that giving the manufacture all the information required to properly design the expansion joint means somewhat more detailed information than the simple design pressure and temperature. As mentioned earlier, quite often the material is set before the piping design is started and a designer's job is one of making the proper stress evaluations rather than focusing on the details. If a designer is working with one of the software systems that has built-in catalogs for expansion joints, those questions might be an integral part of the inputs.

However, the bellows themselves are made from thin material. Often this material is less than 0.125 in. (3 mm) thick. Many times the material is stainless steel, but sometimes a different material is required to handle the increased probability from the somewhat different flow patterns inside the joint. This may be true even if liners are used to reduce that probability. Suffice it to say, the inputs required by the joint manufacturer will be more specific than one would have to furnish if one were providing, say, a standard B16.9 tee. The process requires much more collaboration and effort.

As was noted, one of the ways that expansion joints absorb multiple thermal movements is through their thin convolutions. Because they are thinner and not circular, the stress equations are much different than the hoop stress used in many piping equations. It stands to reason that certain shapes and methods of bellows manufacture would have different fatigue results for the same fundamental set of thermal cycles. It should come as no surprise, then, that fatigue is an important consideration. B31.3 says that a fatigue analysis should be done and reported for all cyclic conditions. It further requires that the analysis should be in accordance with EJMA standards.

Since this book is about pipe and piping calculations, a detailed methodology is not discussed here. However, to give readers a better understanding of the complexity, we discuss the fatigue test requirements as described in Appendix X of B31.3, which follows EJMA standards. Fatigue tests are required for both new and different materials and for new manufacturing methods. These are separate test requirements because the ultimate goal is to develop a factor that relates the specific test to the manufacture of the bellows. The tested expansion joint must be a minimum of 3.5 in. (89 mm) in diameter, and it must have at least three convolutions.

A manufacturer must qualify the manufacturing process used with a minimum of five tests for unreinforced and a minimum of five tests for reinforced bellows in the as-formed condition and manufactured by the organization making the tests. These tests are to be for austenitic stainless steel. Then if the manufacturer wants to use a material other than the as-formed austenitic stainless steel, they must perform a minimum of two bellows fatigue tests with a difference of at least a factor of 2 in the stress range. Heat treatment after forming is considered a different material. This test must use the appropriate manufacturing factor.

This can be confusing and it may clarify things to run through the nomenclature and the formulas that are used in the testing. The key is the minimum number of tests that have to be available for any material and manufacturing process.

The X factors developed are ratios to a lower-bound set of tests that are the reference tests used to develop the EJMA design fatigue curves.

$$X_f = R^f_{min}$$

$$X_m = K_s R^m_{min}$$

The sub- and superscripts refer to fabrication tests and/or to material tests. The K_s is a statistical factor based on the number of tests (N_t) and is calculated as follows:

$$k_S = \frac{1.25}{(1.470 - 0.044N_t)}$$

The R_f and R_m are the minimum ratios of the test stress ranges calculated by the EJMA formulas and divided by the reference stress ranges as listed in the following for each test. Those reference ranges for unreinforced bellows are:

$$\frac{58 \times 10^3}{\sqrt{N_{ct}}} + 264 \text{ MPa for SI}$$

$$\frac{8.4 \times 10^6}{\sqrt{N_{ct}}} + 38,300 \text{ psi for USC}$$

The following equations are for reinforced bellows:

$$\frac{73 \times 10^3}{\sqrt{N_{ct}}} + 334 \text{ MPa for SI}$$

$$\frac{10.6 \times 10^6}{\sqrt{N_{ct}}} + 42,500 \text{ psi for USC}$$

It should be pointed out that N_{ct} is the number of test cycles to failure, which is a through-thickness crack. Reinforcement of a bellows is generally considered to be a hollow tube or a solid rod placed in the bottom of the groove formed by the convolute.

It should further be noted that X_f cannot be greater than 1 and that X_m is not allowed to be greater than 1 unless five or more such tests are conducted on the same material.

A great number of tests are required of a bellows manufacturer, and as that manufacturer adds materials and manufacturing methods to their product line, the number of tests increases.

This is a testament to the seriousness that the expansion joints manufacturers and the B31.3 code place on the establishment of reasonable certainty that installed joints will have an adequate service life for which they are intended. It might also serve as a gentle reminder to users of such specialty components that it is unwise to deal with someone who cannot make the same kind of assurance that the joints' service life will be adequate.

This sort of understanding can help a pipe designer who is using these assurances for a proposed expansion joint. For instance, assume that you have a project that has the standard 7000 cycles during the service life. Further, suppose that you are intending to use an austenitic stainless steel expansion joint that is unreinforced. From the reference equation for such a device we can calculate a stress range of 138,706 psi (957 MPa). This seems like an extremely high number, but it goes back to the EJMA stresses and is a lower-bound stress. As you read the manufacturer's report, it is a base from which to start.

A piping designer does have responsibility for the layout, anchors, guides, and supports. We address here the differences among these items and supports for other types of equipment, at least for calculation considerations.

In preparing the piping arrangement so that an expansion joint will operate properly, there are three basic concerns that should be considered:
1. The friction force that the sliding pipe creates
2. The spring force that the bellows that act as a spring generates
3. The pressure thrust force that the expansion joint generates

The symbols used in the following discussion are:

E is the modulus of elasticity of the pipe material (taken at 70°F (21°C)), psi or MPa

I is the moment of inertia of pipe, in.^{4} or m^4

P is the design pressure, psig or MPa

f is the bellows initial spring rate per convolution, lb/in. or kg/mm

e_x is the axial stroke of the bellows per convolution, in. or mm

A_e is the effective pressure thrust area

C is a constant (0.131 for USC units and 15.95 for SI units)

The first force is the friction factor between the pipe and the supports or guides that are recommended to be placed between the main anchors. It should be noted that the joint and connecting pipe make the equivalent of a column, and a relatively weak one at that. As such, it is subject to buckling. For this reason the EJMA standards recommend certain guides along the pipe to keep this buckling from happening. The number of guides on either side of the joint is a function of the system and the particular joint configuration, so there is no fixed guide number required.

There is, however, a set of spacing rules where D is the pipe OD as follows:

- The first guide is placed 4D from the joint.
- The second guide is placed 14D from the first guide.
- Any subsequent guides are placed at no greater than

$$L_{max} = C\sqrt{\frac{EI}{(PA_e \pm fe_x)}}$$

The friction force, then, is equal to the weight of the length of pipe times the friction factor between the guides (supports) and that weight.

The spring force is equal to the spring rate of the joint times the displacement of the pipe over the total length of pipe. It should be noted that this is the spring rate of the bellows as opposed to the spring rate of the convolutions. The expansion joint manufacturer would supply information as to what each of those spring rates would be. But in any event it is the expansion rate for the temperature and that spring rate.

The pressure thrust force of an expansion bellows is different and sometimes difficult to completely understand. A piping designer knows that there is a horizontal force in a section of pipe coming from pressure. Assume a section of pipe capped at each end. This is known as longitudinal stress and is defined as

$$S_L = \frac{\pi P D_{internal}}{4}$$

This, of course, is from the design pressure P only. In normal pipe this is handled by the pipe in the pipe wall, which absent excessive pressure is stable. Note that effectively it is half the hoop stress and therefore the pipe should fail in the burst mode first. However, split the pipe into two pieces and insert something in between those two pieces that has only a modicum of axial and/or lateral and angular resistance, and you do not have a stable system. The pressure would begin to move the two sections apart. In effect, there is nothing to keep the two sections from being forced apart by the pressure until something tears apart. It is these three forces—friction force, spring force, plus the pressure thrust force—that the main anchors at either end of the section need to resist.

The magnitude of the pressure thrust force can be surprising. It is more than the force when one calculates the longitudinal force using the pipe ID. Recall in Figure 9.1 that the small diameter of the bellows is equal to the pipe ID. The bellows extends some amount beyond that ID. That extension has a height h. By geometry the ID + h would constitute a mean diameter of the convolute. It is this mean diameter that is the effective area A_e. That area times the design pressure then becomes the pressure thrust. It is best to run through an example.

Set up a 14 NPS 350 DN standard wall pipe that has an unrestrained expansion joint midway between two main anchors that are 210 ft (64 m) apart. The system is properly supported and guided (see Figure 9.2). The design conditions are:

- Temperature is 450°F (232°C)
- Bellows spring rate is 10,000 lbs/in. (178.72 kg/mm)
- P (pressure) is 215 psig (1.5 MPa) gas
- Weight of pipe + insulation = 65.3 lbs/ft (97 kg/m)
- Mean diameter (from manufacturer) is 18 in. (457 mm)
- Friction coefficient is 0.3
- Expansion rate for carbon is 0.0316 in./ft (2.63 mm/m)

We start with the friction force. Calculate the weight of the pipe:

$$\text{Weight} = 210 \times 65.3 = 13,713 \text{ lbs in USC}$$

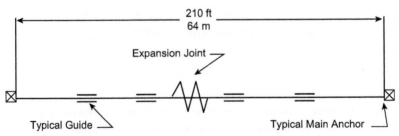

FIGURE 9.2 Expansion joint support calculations example

$$\text{Weight} = 64 \times 97 = 6208 \text{ kg in SI}$$

Multiply by 0.3 to establish the friction force at

$$\text{Force} = 0.3 \times 13,713 = 4114 \text{ lbs in USC}$$

$$\text{Force} = 0.3 \times 6208 = 1862 \text{ kg in SI}$$

The spring force is a little more difficult. First we must calculate the expansion:

$$210 \times 0.0316 = 6.64 \text{ in. in USC}$$

$$64 \times 2,63 = 168.3 \text{ mm in SI}$$

The spring rate force F_s is then

$$F_s = 6.64 \times 10,000 = 66,400 \text{ lbs in USC}$$

$$F_s = 168.3 \times 178.72 = 30,078 \text{ kg in SI}$$

The final calculation is the pressure thrust force, PA_e and A_e, which is defined as the area of the mean diameter or

$$\frac{\pi D_m^2}{4}$$

So we get

$$PA_f = 215 \times \frac{\pi 18^2}{4} = 54,710 \text{ lbs in USC}$$

In this case MPa must be multiplied by 10^{-1} for the computation to be on the same order of magnitude:

$$PA_f = 1.5 \times 10^{-1} \times \frac{\pi 457^2}{4} 24,604.5 \text{ kg in SI}$$

Note that as in all conversions the final answer is not the exact equivalent. It is always better and less frustrating to work in one system or the other.

So now the forces are summed to get the total force A_f. Remember that the friction force will be split between each of the two main anchors. This is because the expansion joint is midway in the pipe. However, they will each receive all of the spring rate force and pressure thrust force. That anchor force is

$$A_f = 0.5(4114) + 66,400 + 54,710 = 123,497 \text{ lbs in USC}$$

$$A_f = 0.5(1862) + 30,078 + 24,604 = 55,613 \text{ kg in SI}$$

The last calculation of this exercise is to locate the guides. In this case we will assume that there were ten convolutions and therefore the per-convolution expansion would be 0.66 in. in USC and 16.8 mm in SI, and that the spring rate per convolution would be equally divided by 10. The calculations are shown in USC units. The L (in ft) would be

$$0.131\sqrt{\frac{29 \times 10^6 \times 372.8}{(54,710 + 1000 \times 0.66)}} = 58.5 \text{ ft}$$

The equivalent length metric can be calculated by substituting the metric values calculated; for the E and I of the pipe posited it should come to 17.8 m. The exercise is to familiarize those who are used to working with USC units with the intricacies of converting to metric in the calculations. It is difficult to know precisely what units one will be given in the problems or reports for which one checks. It is a true learning experience to gather the skill to convert even within the metric system to make the units compatible with the inch–pound system. For those fluent in metric it is almost as difficult. The problem is the myriad of varieties as one crosses disciplines.

About the only concern left for expansion joints is that Appendix X in B31.3 has a leak test requirement that requires a 10-minute duration. It allows the adjustment of the test pressure to the ratio of the modulus of elasticity of the test temperature to that of the design temperature. It also allows a combination of hydrostatic and/or hydrostatic–pneumatic tests that must be in accord with the test in the main book.

I hope that there isn't disappointment that there was little actual design calculation guidance for establishing the stress levels in the various convolutions. The intent of this chapter is not to turn readers into accomplished expansion joint designers, but to give them the skill and understanding required to work with various expansion joint manufacturers, and to install an understanding of what their role is in the process of working through a project that includes a need for an expansion joint.

ANCHOR FLANGES

In the following, we discuss another specialty component that has a relationship to expansion joints—some of them require flanges at their ends to attach to a piping system. These flanges may require special considerations. In that sense they are not the standard flanges from sources such as B16.5. There is another type of flange that is rarely if ever mentioned in the codes: the anchor flange. Its use is quite common and it is certainly not a standard flange.

ASME B31.1 Appendix VII discusses buried piping. This subject is quite extensive and that appendix refers to several sources for further information. When one is talking about pipelines, the simple fact is that they in general cover miles of territory and that converts to several different types of soils. The B31.1 appendix suggests that readers should consult the project geotechnical engineer for assistance in resolving uncertainties about certain critical soil parameters.

This is not to say that there is nothing to say about the anchors, especially those at a building or equipment structure. The B31.1 appendix also defines the location of what is called a virtual anchor. It is defined as the point or region where there is no relative motion at the soil/pipe interface. It is not a leap of imagination to assume that this virtual anchor would rarely come at the point of penetration for the building. Unfortunately, the B31.1 appendix usually shows that anchor location as the typical straight line with diagonals indicating the fixed location.

One of the more common ways to accomplish anchorage is to use an anchor flange. Alfa Engineering, an anchor flange manufacturer, accurately describes anchor flanges on its website as being designed to restrain pipeline movements and spread the pipeline axial forces throughout the foundation in which the flange is anchored. Restraining the pipeline movement ought to be taken into account, particularly at points of directional change, interconnection spots, river crossings, and so forth. Most commonly, they are embedded in a reinforced concrete block.

An interesting thing about Alfa Engineering's statement is that most commonly anchor flanges are embedded in a concrete block. In my personal experience with anchor flanges, 99 percent of the time there are three issues. First, a block buried in soil is of course subject to frictional restraint. A cursory search of literature shows that the friction factor varies from 0.3 to 0.7 depending on the soil. Normally the friction acts on the opposing surface depending on the direction of the force. In the case of gravity, that is relatively easy to compute. It would be the weight of the block times the friction factor. In the case of horizontal or semi-horizontal forces, the resisting force would presumably be the force of the pipeline acting on the side surfaces. And in reality it would be both. Second, the anchor flange literature only defines the force that comes from the piping reaction on the flange embedded in the block. The known specifications just ask for the force and do not ask for the sizing of the block.

The third issue is the bearing stress on the concrete. ACI-318 is the American Concrete Institute's code on concrete design. It defines the bearing strength as

$$P_{bearing} = 0.85 f_c A_{bearingarea}$$

where f_c is what is normally called the concrete strength. This is often set as 3000 psi (21 MPa) in the United States and is usually specified in that manner. This same code recognizes that the footing area may be larger than the bearing area and allows a multiplier factor of

$$\sqrt{\frac{A_2}{A_1}} \leq 2$$

where A_2 is the larger area. This is due to what is called the stress cone—as one applies a force to the bearing area, the imposed stresses progress through the concrete in a conical shape.

If one assumes that the block is buried and as such acts as a horizontal footing, this multiplier can be used. The dilemma comes when one does not know how large the block area may be. However, one can rationalize that the 0.85 multiplier can be nullified by this multiplier as shown in the previous equation.

Assume an area of the bearing stress is 100 in.^{2} (64,500 mm^2). Note that the component of concern is a flange and a circle. To nullify the 0.85 multiplier to 1, the square root would have to result in 1.18, which means the area would have to be 139 in.^{2} (89,655 mm^2) larger. Assume the area is a complete circle. The diameter would be 11.28 in. (286 mm). To achieve the larger area, the diameter would increase to 13.3 in. (338 mm). This is an increase of slightly over 1 in. (25 mm), which is easily rationalized as a safe assumption. So a designer can derate the concrete if the desire is to be conservative. However, not derating can also be justified.

This brings us to the discussion of calculating the dimension of the actual anchor flange. One might first ask, what does it look like? Readers probably have a mental picture of an ordinary flange, which is probably accurate. However, this flange requires no bolts and actually is welded to a pipe on either side. So one description would be a pair of flanges somehow melded together without bolts. Refer to Figure 9.3, which also gives the dimensional notations for the ensuing calculation.

The computation follows the procedure outlined in ASME Section VIII, Division 1 for flange design without the bolt portion of the calculations. Perhaps the best way to demonstrate the calculations is to set up a problem and go through the calculations step by step. The problem is as follows:

- The pipeline is 16 NPS (400 DN) XS wall, 0.5 in. (12.7 mm)
- The operating pressure P is 1200 psi (8.27 MPa)
- The installation temperature is 60°F (15.6°C); the high temperature is 100°F (37.8°C)
- The minimum temperature is 32°F (0°C)
- The concrete strength is 3000 psi (20.7 MPa)
- Both the pipe and flange material have an allowable stress of 20,000 psi (137.9 MPa)

The symbols used are:

A_m is the area of metal in the pipe

A_{ID} is the area in the ID of the pipe

α is the linear coefficient of expansion in μin./in.°F from Table C-3 of B31.3, 6.13 e6

α_m is the linear coefficient of expansion m/m/°C using the conversion of 1.8 times the B31.3 factor, 11.03 e6

N is the centroid of the annulus formed between the OD of the pipe and the OD of the flange. In Figure 9.3 it is the annulus formed by the diametric dimensions A and C. For those who don't have access to these

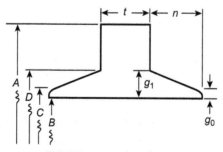

FIGURE 9.3 Anchor flange

arcane formulas they are repeated here for convenience. The N diameter is equal to the following formula:

$$N = \left(\frac{2}{3}\right)\left[\frac{A^3 - C^3}{A^2 - C^2}\right]$$

The load will be calculated first. When the temperature is increasing, both the pressure load and thermal load are included. When the temperature is decreasing, only the thermal load needs to be calculated. The hydrostatic pressure load only acts in tension in that case. Note that it would then be the rarest of cases where the load from the thermal decrease is the higher of the two loads. The higher load is the one that would govern.

Before we can calculate the loads, we need calculate the metal area in the pipes:

$$A_m = \frac{\pi 16^2}{4} - \frac{\pi 15^2}{4} = 24.35 \text{ in.}^2 \text{ in USC}$$

$$A_m = \frac{\pi 406.4^2}{4} - \frac{\pi 381^2}{4} = 15,708 \text{ mm}^2 \text{ in SI}$$

Note that in both formulas the first factor is dimension C in Figure 9.3 and the second is dimension B. Each of these dimensions will be used as we proceed through the calculation.

The load for the temperature increase is therefore

$$L = \alpha(\Delta T)EA_m + \frac{\pi B^2}{4}(P)$$

$$= 6.13 \times 10^{-6}(100 - 60)(29e6)24.35 + \frac{\pi 15^2}{4}1200$$

$$= 385,206 \text{ lbs in USC}$$

$$L = 0.177\alpha(\Delta T)EA_m + \frac{\pi B^2}{4}(P)$$

$$= 11.03 \times 10^{-6}(37.8 - 15.6)(2.e9).0157 + \frac{\pi 381^2}{4}8.27$$

$$= 174,727 \text{ kg in SI}$$

Once again, note the factor, in this case 0.177, in the SI equation, which is often necessary to convert a USC formula into a compatible answer when working with metric units.

Since the force when computed for the temperature when it decreases to the freezing point is obviously less for the reasons mentioned before, we will not do that in this exercise. This is not to say that the decrease in temperature will not always need calculation. That would depend on several factors in the design regime. Often the calculation can be eliminated by examination of that design regime, as happens here.

Having assured ourselves that we have the load that the flange needs to resist, we have to select an OD. One option is to use the standard diameters of flange forgings that are in B16.5 or some other flange standard. Here, we will use B16.5, which has metric sizes, and note that the pressures are in bars rather than the MPa we specified at the outcome. Fortunately, the SI system allows us to merely slip the decimal point to the right and the 8.27 MPa becomes, for all practical purposes, 82 bars. We find this in the tables and that pressure requires a class 600 flange. For the NPS 16 size specified, the OD is 685 mm. With the pipe ID of 381 mm the total area of the supposed annulus is

$$\text{Area of annulus} = \frac{\pi\left(685^2 - 381^2\right)}{4} = 254,520 \text{ mm}^2$$

Is this going to be enough area for the concrete bearing stress? We choose to believe the block will be large enough that we can use all of the stress, so the bearing stress could be the full 20.7 MPa. The computed load is 724,727 kg/254,520 mm^2, which translates to a far lesser load than the 20.7 MPa allowed, so this anchor flange could be much smaller from the standpoint of bearing on the concrete. In this case, even using the conservative 0.85 multiplier would still make the concrete acceptable.

A smaller OD is better because it would shorten the moment arm and thus reduce the stresses in the ring and allow a thinner ring thickness. The advantage of using a B16.5 forging is that it may be cheaper; however, it may not have a thick-enough ring and a new forging would be required in any case. For purposes of this exercise we will set the OD at 581 mm (22.875 in.).

Having decided on the OD for dimension A in Figure 9.3 we can proceed with the remainder of the calculations. First, we calculate N:

$$N = \left(\frac{2}{3}\right)\left[\frac{581^3 - 381^3}{581^2 - 381^2}\right] = 487.9 \text{ mm in SI}$$

The comparable USC calculation is 19.6. This establishes the load circle diameter so the moment arm can be calculated. It is half the distance from C to N and the symbol is h_g:

$$h_g = \frac{N - C}{2} = \frac{487.9 - 406.4}{2} = 40.75 \text{ mm in SI}$$

$$H_g = \frac{19.6 - 16}{2} = 1.8 \text{ in. in USC}$$

Given the moment arm, it is a simple matter to convert the loads to moments to calculate the stresses:

$$M = \frac{Lh_g}{B}$$

$$= \frac{174,727 \times 40.75}{381} \ 18,688 \text{ kg/m per mm of pipe diameter in SI}$$

$$M = \frac{385,520 \times 1.8}{15}$$

$$= 46,262.4 \text{ in.-lbs per inch of pipe diameter in USC}$$

For the next steps we have to compute the various shape factors, which all start with the K factor, defined as

$$K = \frac{A}{B} = 1.525$$

Since this is a ratio it is the same in SI and USC units. This is one of the few times it is not difficult to convert to metric from USC and vice versa.

The factors in Table 9.2 all come from the ASME charts in ASME BPVCode, Section VIII, Appendix 2. Before we can compute all of them one has to determine how long to make the hub and at what taper. For this exercise we will use an approximate taper of 14°. Further, we will set length h at 3 in. (76.2 mm); this makes the dimensions in Figure 9.3 labeled g_1 and g_0 18 in. (30 mm) and 0.5 in. (12.7 mm), respectively, by setting the pipe wall at those dimensions.

Given those dimensions we can calculate the shape factors. We will not repeat the graphs. There are a set of formulas to calculate those and they are in B16.5. For this exercise we have precalculated the basic formulas. There are a few that include some more calculation, which are provided in Table 9.2.

Table 9.2 Anchor Flange Exercise Factors from ASME Graphs or Equations

Factor Symbol	Value
F	0.701754
V	0.109415
f	1
T	1.699841
U	5.246574
Y	4.774393
Z	2.508722

There are two that we need to calculate from data that we have here; they use the factors as well as input data. The first is the ratio g_1/g_0:

$$\frac{g_1}{g_0} = \frac{30}{12.7} = 2.36$$

This is another ratio and therefore the same in all units.

The second is the symbol h_0, which is

$$h_0 = \sqrt{Bg_0} = \sqrt{15 \times 0.5} = 2.73 \text{ in.}$$

The metric calculation needs to be multiplied by 0.0393 to use the ASME method:

$$h_0 = \sqrt{Bg_0} \times 0.0393 = \sqrt{381 \times 12.7} \times 0.0393 = 2.73$$

Then there is factor e, another ratio, which is

$$e = \frac{F}{h_0} = \frac{0.702}{2.73} = 0.26$$

Finally, factor d:

$$d = \frac{2U}{V} h_0 g_0^2 = \frac{2 \times 5.25}{0.109} \times 2.76 \times 0.5^2 = 65.74$$

Again, the g_0 is multiplied by 0.0393 for the SI version to work with the ASME method.

After this rather rigorous and time-consuming method of calculating the various shape factors, one can calculate some other factors and finally calculate the stresses that will arise. But first, one has to assume a thickness of the ring. There is no real way to determine the ring thickness without further calculations. In general, it requires an iteration of at least one step.

Even if the first estimate passes, it is a good idea to estimate one slightly smaller thickness to determine if there is a possibility of making a thinner one and therefore a shorter forging. In fact, as in all designs one should try to establish a working range to determine if the design is optimized. The shape factors are used to calculate some stress constants and then one can directly calculate the stresses.

Those stresses are the hub stress, radial stress, and tangential stress. The hub stress is allowed to exceed the allowable stress because it is basically a bending stress, and the calculated stress in bending is a maximum at the extreme fiber and in fact reduces to zero in the center. This of course shifts some of the stress to the ring and therefore the practice is to limit the average of the calculated hub stress and each of the radial and tangential stresses to less than the allowable stress. This procedure, which is used in ASME design, is to assure a designer that the shift is not excessive.

In the case of the anchor flanges there is a need also for the shear stress calculation to assure that the force does not actually push the hub through the ring. And of course one makes a final check to ensure that the bearing stress of the concrete is not exceeded.

It should be pointed out that this discussion has been centered on the use of a concrete restraining block. This is because, as pointed out at the beginning of the discussion, it is the most common means. It is not the only means. Sometimes the resisting structure is some kind of metallic frame. In those cases there are two major differences. First, it is unlikely that the restraint will bear on the complete annulus from the pipe ID to the flange OD. Most likely it will be only on the flat surface of the ring. This requires some adjustment regarding the amount of bearing area and certainly the bearing capability of the structure or frame that is used to resist the flange. It also requires that one take into consideration whether or not there is any movement of that frame either from deflection or in some cases a planned movement. This affects the total amount of expansion to resist. It is specific to the situation; readers are cautioned that there may be other considerations in such a situation.

The stress constants and the actual stress calculations are provided in Tables 9.3 and 9.4. This is possible because at this point in time the unique items have been calculated and the process is merely one of running a calculator or a spreadsheet. Tables 9.3 and 9.4 show the name or symbol, the formulas, and the results for a given assumed thickness and a second thickness where one can get a feel for the differences. Also, the acceptance stress criteria are repeated in the stress table.

Table 9.3 Stress Constants for Anchor Flange Exercise

Constant Name	Formula	Calculated Value
For Thickness 1.75 in. (t)		
Alpha (α)	$te + 1$ (factor e)	1.458
Gamma (γ)	αT (factor T)	0.852
Sigma (σ)	t^3/d (factor d)	0.082
Lambda (λ)	$\gamma + \sigma$	0.0934
For Thickness 2.125 in. (t)		
Alpha (α)	$te + 1$ (factor e)	1.544
Gamma (γ)	αT (factor T)	0.908
Sigma (σ)	t^3/d (factor d)	0.146
Lambda (λ)	$\gamma + \sigma$	1.054

TABLE 9.4 Stress Calculations for Anchor Flange Exercise

Name	Formula	Calculated (psi)	Allowable (psi)
For Thickness 1.75 in. (t)*			
Hub stress, S_h	$S_h = \frac{fM}{2\lambda g t^2}$	17,905	30,000
Radial stress, S_r	$S_r = \frac{\alpha M}{\lambda t^2}$	23,808	20,000
Tangential stress, S_t	$S_t = \frac{MY}{t^2} - ZS_t$	13,546	20,000
Average of S_h and S_t	$\frac{S_h + S_t}{2}$	15,725	20,000
Average of S_h and S_r	$\frac{S_h + S_r}{2}$	20,856	20,000
Shear stress	$\frac{load}{\pi(B + g_t)t}$	4056	12,000
Bearing stress	$\frac{load}{0.7854(A^2 - C^2)}$	1845	3,000 (per spec)
For Thickness 2.125 in. (t)			
Hub stress, S_h	$S_h = \frac{fM}{2\lambda g t^2}$	15,851	30,000
Radial stress, S_r	$S_r = \frac{\alpha M}{\lambda t^2}$	15,242	20,000
Tangential stress, S_t	$S_t = \frac{MY}{t^2} - ZS_t$	11,456	20,000
Average of S_h and S_t	$\frac{S_h + S_t}{2}$	13,653	20,000
Average of S_h and S_r	$\frac{S_h + S_r}{2}$	15,546	20,000
Shear stress	$\frac{load}{\pi(B + g_t)t}$	3340	12,000
Bearing stress	$\frac{load}{0.7854(A^2 - C^2)}$	1845	3,000 (per spec)

* *Note*: S_r and the average of S_r and hub fail M are the moment.

Note that the tables are for the thicknesses in inches. This is another example of the need to have in this case the thickness in millimeters, converted by multiplying by the 0.0393 factor, and then comparing the converted stress calculated in psi to the allowable stress in whatever units the particular specification requires.

One can find very little of such data and is forced to use complex calculations where the literature quite frequently uses USC units. Such is the engineering Tower of Babel that has been created by not converting completely.

There are many other specialty types of components, but they are less amenable to discussion than expansion joints and anchor flanges because these components follow much more conventional dimensional stability and therefore lend themselves to discussion.

BLOCKS FOR ANCHOR FLANGES

Now that we have designed an anchor flange, the natural next question is what do we anchor it to, or as my English teacher would say, to what do we anchor it. We do have a clue that it has something to do with cement ,since one of the data checks was to be sure the flange didn't exceed the bearing stress of the concrete. In fact it is quite common in pipelines to use concrete blocks to restrain pipe.

The purpose of the anchor flange is to restrain the pipe as it goes into a pumping station so that the reaction from the transverse stress loads and any growth do not cause damage to the equipment by having too high a reaction load. In fact, the pipeline might need the same type of blocking to keep the movement from a horizontal bend or forces from a reduction in size and/or going up and down a hill from causing damage. Even though the pipes are often buried, they move around quite a bit under the ground. Not as much as an earthworm, but one might be surprised.

One of the fortunate things that occur when one designs an anchor flange is that a great deal of the required information to design the anchor block has already been calculated or assembled to design the flange. It should be noted here that the method described in this book is not the only method of designing such a flange. It is possible to design the flange while designing the block. However, as one might suspect that would require much more complete information about the pipeline system than is usually available to the flange manufacturer. So the technique described could be considered a bit on the conservative side.

The issue mostly revolves around the soil conditions in which the anchor block will be placed. It requires knowledge of the soil friction angles and the soil density, both of which vary with the type of soil at the exact site. While it is true that one can generalize those types of figures, they come from Mother Nature, not the steel mill from which the flange and pipe material arrive. As such they are far more variable than the rest of the data. If one were to check with a skilled geologist on what to use he would get an answer somewhat like this: "That requires an expert and should not be tried at home."

One of the general themes of this book is that we are guiding you and giving you enough information so that you will be able to tell if the "expert" and you are in the same ballpark so you will be able to ask the questions that are needed to increase the probability of success. If you want to become an expert, we understand there are several good universities teaching the subject.

We will discuss the issues and give some general parameters because they will be necessary to demonstrate the process. The careful reader is cautioned that the numbers used in this context are somewhat like reading a historical fiction. They may be real, but they may not be what is appropriate for your specific design project.

The first issue is the parameter of soil friction angle. This is the angle that one uses to calculate friction factors between the buried soil and the pipe and the concrete. Another parameter is the density of the soil. Both are a function of the soil type and we all know that can vary from rock to sand. Table 9.5 is a small table of soil types to show you that variability.

For purposes of this discussion, we will use 120 and 1920, respectively. We will also use 25° as the soil friction angle. Each code has some regulation about how much soil cover must be on the pipe. It varies by code and location; we will use a cover depth of 48 inches because that is the one the li1uid code B31.4 requires for normal excavations.

Table 9.5 Soil Types and Representative Values

Soil Type	USC, lbs/ft³	Metric, Kgf/m³
Sandy soil	112	1800
Gravelly soil	125	2000
Silty soil	131	2100
Clay soil	119	1900

Note: The soil friction angle is more variable because most soils are a mixture of the four types. The range of the soil friction angles in these soils, depending on the percentages, varies from a low of 20 degrees to a high of 35 degrees.

Table 9.6 Summary of Design Dimensions for Anchor Flange Example and Listing of Block Elements that Need to be Determined

Element Name	USC Dimension	Metric Dimension	Source
Flange OD	22.875 in	581 mm	Previous design
Flack thickness	2.125 in	54 mm	Previous design
Pipe metal area	24.374 in^2	15,708 mm^2	Previous design
Axial load	385206 lbs	174727 kg	Previous design
Soil weight	120 lbs/ft^3	1930 Kgf/m^3	Specified
Soil angle Φ	25°	25°	Specified
Coefficient of friction, pipe to soil	0.22	0.22	Tan(1/2 sin Φ)
Weight of pipe	84.35 lbs/ft	122.98	Calculated (table)

We mentioned that the calculation of the flange itself was somewhat conservative, especially in calculating the flange OD. This comes about because the conservatism revolves itself around the fact that in the development of the flange design technique the Poisson effect was not built into the calculations. For those who are deep into the mathematics of stress and such calculations, the Poisson effect is a ratio which describes the change in a transverse direction in relation to the change in an axial direction.

In the case of steel this ratio is often taken as 0.3 In the case of the anchor flange, whose purpose is to eliminate the effect of the expansion of the material of the pipeline due to temperature change, it is possible to limit the change in the expansion along the axis of the pipe by using the shrinkage that would occur due to the expansion of the diameter due to hoop stress. The mathematical techniques also can be affected by the soil conditions, which affect the block size and other elements in the system design. So as was mentioned before, when all the factors are not known it is prudent to take a more conservative approach. It is possible to say that the approach is too conservative and some will. But that is just a version of the following question: How safe is this product? We all know that things like conservatism and safety are not precise, so we have invented the concept of design margin or in a politically less correct world, safety factor.

Having discussed the unknowns it is now time to summarize what we do know and what we need to find out for the block design. It may be best for convenience to summarize the factors and list what we need to do with them as well as the unknowns and how we decide them. Below is a summary table of the design we developed previously. Only the necessary elements will be repeated in this summary. We have little need to use the flange design factors again.

From this point we do a few calculations that are more complicated so we will show the calculations. The first thing is to calculate the friction force on the pipe from the soil. In doing so we will use a nominal two feet of pipe as our basic length to receive force. The calculation is

$$pipefriction = .22 \left[\frac{\pi OD(soilwt)(cover)}{144} \right] + 2(pipewt)$$

$$= .22 \left[\frac{3.14(22.875)(120)(48)}{144} \right] + 2(84.35)$$

This calculation equals 669 lb/ft.

Next we calculate what is called the length of pipe moving at the free end ; it is based on the previous calculation:

$$L = \left[\frac{A_m E \Delta}{6f} \right]^{.5} = \left[\frac{24.347(29e6)(1)}{6(669)} \right]^{05} = 420 \, ft$$

Then we calculate the total friction force by multiplying the two calculations together:

$$F_f = pipefriction(L) = 669(420) = 280,980 lbs$$

Voilà, we have a figure that the block must overcome and it is the anchor force less the friction force just calculated. To put is succinctly,

$$Blockforcerequired = 385,206 - 280,980 = 104,226$$

This doesn't seem like much and in reality it isn't a big force for the block to overcome. A big force often takes iterations to get a satisfactory solution. There are still several things left to consider.

In calculating the above forces we used a cover depth of 48 in. In calculating the actual height of the block we have to decide whether the block will be completely buried. For simplicity we will assume the top of the block is all 1.67 ft below grade. We should know that the cover depth is to the top of the pipeline so that comes into the actual block size. It is best to set the pipeline and anchor flange in the middle of the block so that gives us a starting point. We know that the flange size does not exceed the bearing stress of the concrete, but we are not sure yet that it won't shear or punch itself out. We should also check to see that the moments created have sufficient stability to not cause the block to tip. Finally the block must have enough soil resistance and/or soil friction to accomplish the required additional forces.

The three block dimensions we must establish are the height h from top to bottom, the width W, which is along the axis of the pipe, and the length L, which is the distance transverse to the axis of the pipe. As mentioned we start from the center of the block. It is important to mention that by convention there is a 3-in. free band, or at least half a foot added to each of the dimensions around the outside of the block. This is to allow for the corners wearing off and to give space for any reinforcing bars that may be required in the concrete.

Attack the height h first. We recall that the OD of the flange is 22.875 inches. We will use 2 ft or 24 in. as a guide. It seems a little silly to hold cement to within thousands of an inch when one thinks about it; the surface is not that smooth. For starters we will suggest that the height should be at least three flange ODs high. This is flange on the top and the bottom of the embedded flange. When we recall the 3-in. band, we might make the band a full foot instead of the minimal half a foot. We positioned the block to have a six foot h or height. We must check to see if the cover of the pipe is actually the 48 inches we specified. The pipe and flange are situated in the center of the six foot block. So the dimension from the centerline of the pipe to the grade line will be the three feet from the center of the block plus the 1.67 feet from top of block to grade or 4.67 feet (56 inches). However, the cover is defined as from the top of the pipe so we must subtract the 8 inch radius of the pipe and the result is the specified 48 in cover.

Another factor is the total depth to the bottom. It is called H and is the cover depth which we set at 48 in. + ½ pipe diameter + ½ h. This dimension in feet is $4 + 0.67 + 3 = 7.67$.

Now since we already know we don't need a tremendous amount of additional resistance, for the first iteration of the length L we can go for symmetry in that direction and establish that dimension as 6 ft also.

The width W, which is along the axis of the pipe, is somewhat more tricky. Recalling that we need to check the viability of the shear cone, we need to do a little more calculation. Shear cones by their nature start from the OD of the flange and progress at larger than a 45° angle to the outside of the block. For many reasons, most of them intuitively obvious, it is best to make the shear cones large and within the 3-in. band of the block. So let's do a little math here. The Block is posited at 5 ft square (minus the band) which means that the large end of the cone should be no more than

4.5 feet or 54 inches. Now we use the actual OD of the flange, 22.875 inches. Take $54 - 22.875 = 31.125$ in., which is how much the diameter could grow. Round that down to 30 in. That means it can grow 15 in. radially. One of the nice things about the 45° angle is that the cone will grow that amount in the same 15 in. Recalling that the flange is 2.125-in. thick we can say that the W dimension needs to be $2.125/2 + 15$ in. from the exact center of the flange ring or 32.125 in. overall. Once again we could round it down to 32 in. or 2.66 ft; for checking resistance we will use 2.5 ft. We know from our process that the large end of the cone will be inside the band.

In checking the resistance, we do need a little more geological data. Fortunately our friend the solid angle helps here. K_p, the coefficient of passive solid pressure, is defined as the following:

$$K_p = \frac{1 + \sin \Phi}{1 - \sin \Phi} = \frac{1 + \sin 25}{1 - \sin 25} = 2.464 \text{ use } 2.5$$

There are three resistances to check. First is the soil resistance from the anchor trying to push the block forward. That formula is

$$F_R = \frac{pf(H)^2(L+h)K_p}{2} = \frac{120(7.67)^2(6+6)2.5}{2} = 105,892 \text{ lbs}$$

The reader will notice that this is more than was needed. It appears that one does not need to go any further to achieve enough resistance. It is usually the case that you have to fiddle with the block dimensions to get an appropriate amount of resistance. This factor is usually the largest, so one generally starts with it and adjusts dimensions like L and W. The h dimension requires some adjustment to get the amount of cover and the amount you want the block buried. In some cases the block is left with some of the top exposed for location purposes; all these factors make h the trickiest to set. The W dimension is affected mostly by the shear cone and often can only be increased a small amount before the cone goes outside the block and causes other, more critical, design problems. The L dimension does affect how much earthwork is required in the right of way, which may limit it, so it is a like a high–wire balancing act.

At any rate there are two other resistance factors. We will just state the formulae here and the additional work is left to the student as an exercise in problem solving.

The first of the calculations is the friction between the bottom of the block and the soil due to the weight of the block and the friction of the soil on concrete. For reference is it quite common to use 150 pounds per cubic foot for concrete. The friction factor of the concrete and a solid is commonly set as tan $2/3\Phi$ and called k. The formula is

$$F_b = k(L)(W)(h)150$$

The last resistance factor is that between the sides of the block and the fill material. Here we use K_A, which is the active soil pressure because we are moving past the soil, not pushing it as it resists. K_A is defined as $1/K_P$ or, in this case, $1/2.46$. The formula is

$$F_s = K_A(pcfsoil)(H/2)(W)(h)2k$$

The total resistance is the sum of all three: $F_R + F_b + F_s$.

This little exercise has not made anyone an expert. However, one can rest assured you can talk with an expert and he will not notice a massive amount of egg on your face. We haven't detailed anything regarding the tipping or the stability ratio. This is a function of the nearness of the pipe's force to the center of gravity of the triangle of the effective retaining wall of the block. It once again delves into the soil conditions, which requires more specialized information. We have not discussed the reinforcing bars needed. These last two elements are not within the scope of this section.

It is important to understand the difficulties that the block designer faces and to have a feel for all the work that goes into it. It is also good to have this type of information should one be thrown into the arena to face such problems.

CHAPTER 10

High-Frequency versus Low-Frequency Vibration Calculations

Contents

OVERVIEW

Many folks think that the parts of the B31 codes that discuss thermal expansion and stress ranges also deal with vibration analysis. It is true that there are similarities, and in recent years there has been an extension of the number of cycles in the displacement range calculations to 10^9 cycles, which makes it appear so. One should note that there was also a reduction of the number of cycles from the traditional 7000 down to a maximum factor of 1.2, which translates into 3125 cycles during the expected life.

This came about because there was ample evidence that in many industries they do not plan to have the nominal one full cycle a day. The limit of 3125 cycles effectively says that for the same life it is one cycle approximately every two days. Obviously for an industry that plans more cycles, like an industry that plans batch runs rather than continuous runs, there would be a higher number of cycles for the same life.

One of the reasons behind the extension of cycles came from the increased use of floating platform ships to process or store offshore oil. These have a very high number of cycles. The current way this is handled is by using the DNV (Det Norske Veritas) system. That method uses an S-N slope of $-\frac{1}{3}$, and is in agreement with the general approach used in

Piping and Pipeline Calculations Manual
ISBN 978-0-12-416747-6
http://dx.doi.org/10.1016/B978-0-12-416747-6.00010-3

Europe. It contrasts with the slope of $-\frac{1}{5}$, which the ASME piping codes adopted from the work of A. R. Markl. It has proven successful with the low-cycle work for which it was intended. The proper approach for higher cycles is examined within the piping codes to accommodate the growing need for high-cycle analysis.

Platform ships have extensive piping and the wave action is significantly more frequent than one wave cycle a day. So there is a need to increase the number of cycles and in effect reduce the stress range as the stress reduction factor goes down drastically. Fortunately, for that industry there is rarely high temperatures and thus lower stress is involved. However, if one does a little checking, the highest number of cycles, 10^9, still translates in a 20-year life to something like 1.5 cycles per second, which is lower than expected vibration from an electric motor.

This is to say that it is not vibration as we have come to know it. The most common vibration that might be encountered is from an electric motor that is slightly imbalanced and has somewhere around 3600 RPM (revolutions per minute). This translates into something like 1.9 e9 cycles per year. That is a far cry from even a 20-year life. In fact, it is relatively safe to say that without proper protection, most things are subject to vibration failure much sooner.

The B31 codes approach vibration in an indirect manner. They recognize what is called severe cyclic conditions. It is defined in B31.3, which is the code that seems most involved with cyclic loading, as any cycle that produces a stress range in excess of 0.85 times the allowable stress range, S_A. This can be taken as a working definition of a vibration load. For example, let's calculate a stress range reduction factor for a vibration from a 3600-RPM motor.

Example Calculations

For 3600 RPM for a life cycle of 6 months (this assumes that there would be periodic inspections that would allow corrective action at that time), the number of cycles according to the design from the vibration would be 3600 × 60 min/hr × 24 hrs/day × 182 days/6 months = 9.4 e8 cycles. Plug that into the equation $f = 6(N)^{-0.2}$ and you find a factor of 0.096.

For a calculated S_A of, say, 12,000 psi (82,737 kPa) multiplied by the factor, you get a range of 1152 psi (7948 kPa). Then 0.85 of that would be 979 psi (6749 kPa).

The question then becomes: What kind of moment from vibration creates that stress in the component? We will discuss that in more detail later.

SEVERE CYCLIC SERVICE

What do the codes say regarding accepting severe cyclic service as an analogy for vibration and establishing what that might be? First of all, we will be mostly discussing what is said in B31.3, because it addresses this aspect of the vibration problem most completely of all the ASME piping codes.

There are numerous references to severe cyclic service regarding what not to use for a component or feature of a component in a particular service. There are some specific items that say what type of material, such as piping, one may use in that service. There is also a table in the code that addresses the acceptance criteria for types of welds in severe cyclic service. While these admonitions may not be explicit in the other books, they can be used as good guidelines in any application.

All books in their sections on pressure design assert in some manner that the rules are for loads from pressure only. Any external forces from things, such as thermal expansion and contraction, live loads, and other special considerations, shall be given so the designer can make that connection withstand such loads.

Again, B31.3 specifically addresses vibration in this manner. The piping should be arranged and supported to eliminate excessive harmful effects. The code points out that vibration may come from impact, pressure pulsation, and turbulence in the flow; resonance with other external sources such as pumps and compressors; and wind. An earthquake is a shake or rattle event, but it is a subject unto itself.

So in essence the codes give guidance and admonish one to consider higher-frequency vibration. So a design analyst is left with the question: What do I do?

The business of vibration requires a great deal of expertise to be well versed in it. Here, the intent is to give readers enough of a feel for the requirements and rudimentary elements of the subject that one can do elementary things in it, and have a filter for when dealing with experts. The idea is that piping analysts should know enough to know what they know and what they don't know. When one is aware of what one doesn't know, then, as Lao Tzu said long ago, "When you know what you don't know you have genuine knowledge." That is, if you are fully aware of the limit of your knowledge you will ask questions.

TYPES OF VIBRATION

For the purposes of piping and pipelines there are two major categories of vibration: mechanical and flow-induced.

Mechanical Vibration

The first category of vibration is the one we have been talking about, which is generically called mechanical vibration. Within that there are two major divisions:

- *High amplitude–low frequency.* This is the one usually handled in the codes with thermal expansion.
- *Low amplitude–high frequency.* This is typically the type of vibration we were discussing in the example of an electric motor at 3600 RPM. There are of course other sources of such vibration.

There is no definite line between the two. For instance, take the wave action that was one of the reasons for extending the frequency chart for the stress reduction factor. In a nominally calm sea, the amplitude of the wave action is certainly low, and it has a relatively low frequency. However, in a hurricane or other storm the wave amplitude can be extremely high, and we certainly hope for a low or at least short-acting frequency. The major differentiator must be the judgment of the amplitude. Typically, a low-amplitude vibration would be so small it would be hard to see and would have to be measured by some instrument or felt by touching. In some cases those types of vibrations can also be heard as a buzzing sound. Higher-amplitude vibration can be seen.

One time in a test to determine the stress implication of an attachment welded to a large pipe, we were using amplitudes of a significant portion on an inch. It was amazing to see the pipe wall ripple like a wave in the ocean. It was a shame that a video wasn't taken of that test. It certainly destroys the common knowledge that steel is completely rigid. With the proper power, steel will ripple like flapping a quilt to shake the dust out.

Flow-Induced Vibration

The other major category of vibration for piping purposes would be flow-induced vibration. Again there are two major types:

- *External flow.* This is something like wind. Probably the most famous example of this is the "galloping Gertie Bridge" in the Tacoma Narrows where the wind caused the collapse of the bridge fairly shortly after it opened. If you haven't seen that video, the Internet has excellent examples and film clips of it. If you aren't familiar with that bridge you should type "galloping Gertie" into Google and watch the YouTube video. It is a revelation.

- *Internal flow.* All pipes have internal flow at some time. In some circumstances, that flow goes past a branch opening that is closed at the other end and can cause a vibration. That can be likened to blowing across a soda bottle and hearing a sound. The sound is a vibration that is at a frequency in the audible range. This often happens in cross-flow heat exchangers and safety relief valve installations.

Degrees of Freedom

A *degree of freedom* can be described as how much information is needed to describe a system. For instance, in a system with one degree of freedom, such as a single spring in a constrained environment so that it can only move vertically, the position can be described by one dimension—the distance from a fixed point to a point on the spring.

It is incorrect to determine from this example that systems with one degree of freedom are necessarily simple. Consider an automobile engine as a unit—that is, separate from the car. When it is running it has many moving components and can certainly be called complex. However, by asserting that the moving components like the crankshaft, pistons, valves, head, and others are rigid, the position of each component is described by the position of the crankshaft. That would make it a system with one degree of freedom.

Mount that engine in a car with motor mounts and the degrees of freedom would increase. For instance, they could go to seven. In our 3D world there are six potential degrees of freedom for anybody. They are the three xyz dimensions and the three rotational dimensions, which, if we were talking forces, would be moments. In our example, the seventh would be the crankshaft rotation. Put the car in gear and let it move and more degrees and complexity are added.

The degrees can become infinite, which this book will not explore. Generally speaking, we limit the discussion to one-degree-of-freedom systems. As degrees of freedom are added, additional considerations are required, basically additional mathematical crunching. There are complete books that do nothing but show methods to handle these computations. And of course there are computer programs available that can deal with more advanced calculations. These programs can handle the complexity of many degrees of freedom much easier than the simple calculations. The basic concepts are discussed here in more detail than they are in the piping codes. Always keep in mind that there are more comprehensive methods available that should be employed when one gets to the edge of this basic knowledge.

WORKING WITH VIBRATION

Regardless of the type of vibration or the degrees of freedom there are certain things that are common to all types. We limit the discussion to linear and simple harmonic vibration, which are the most common vibrations encountered. For discussions on nonlinear and nonsimple harmonic vibrations, there are more complete sources and references available.

The basic language of this discussion is straightforward. The first concept is that the motion repeats itself in a specific time period, called T. In simple harmonic motion, that means a sinusoidal pattern. The amplitude of that motion is x, and the home position is x_o; the formula then becomes $x = x_o \sin t$. The period is usually measured in seconds, so the frequency is measured in cycles per second, which is

$$f = \frac{1}{T}$$

In vibration analysis it is common to use the symbol w, which is known as circular frequency and is normally in units of radians/second.

A sinusoidal function actually repeats itself in 2π radians. This means that the expression $wt = 2\pi$ would substitute for just plain t. The basic vibration equations then become the following:

$$T = \frac{2\pi}{w} \text{seconds} \tag{10.1}$$

$$f = \frac{w}{2\pi} \text{cycles/sec} \tag{10.2}$$

$$f = \frac{30w}{\pi} \text{vibrations/min} \tag{10.3}$$

There are two major concepts that are most important to understand regarding vibration. The first is natural frequency. Natural frequency is often called the resonant or resonance frequency. This leads us to the second major, but related, concept—the concept of resonance. It is important because it is the place where it is safe to say that one does not want to operate.

Both concepts can be described by an analogy before we get into a calculation discussion. Many have struck a bell and heard the resultant gong or peal. You may have noticed that different bells of different materials, sizes, or different shapes have a different tone. You may have seen a commercial on TV or a demonstration where a glass is struck with an instrument and there is a resulting tone (natural frequency). Then someone reproduces that tone, say an opera singer, and the glass shatters (resonance).

As noted, as the size, shape, or material of the device changes, the tone changes. Each object has its own natural frequency. Within limits, it can be calculated. Limits are easily calculated for simple shapes or configurations. However, as shapes or configurations become more complex, the calculations become more complex. It is not inaccurate to say geometrically more complex. One reason is that as one gets a more complex system, the degrees of freedom of the system increase. This calculation is one that easily lends itself to computer analysis, providing you have the proper software.

Before we can actually calculate a natural frequency, there is a "spring constant" that must be calculated. The symbol for this is k and it can be defined as the load per inch of deflection. Following is a formula for a cantilever that is in terms of deflection. Note that it is a rearrangement of a cantilever beam formula:

$$k_c = \frac{3EI}{L^3} \qquad (10.4)$$

where

E is Young's modulus

I is the moment of inertia

L is the length in consistent units with the E and I

Given the spring constant, we can now compute a natural frequency for a mass at the end of a cantilevered shaft. This particular configuration was chosen because in piping we quite often can model something as a mass at the end of a cantilever—for example, a drain valve on the end of a branch. Eq. (10.5) is based on the mass of the valve, M, and the mass of the beam, m:

$$w_n = \sqrt{\frac{k}{M + 0.23m}} \qquad (10.5)$$

where w_n is the natural frequency. The mass of the beam is a small portion of the equation. This can be attributed to the fact that the beam or pipe is not vibrating as much at the "anchored end" as it is at the free end.

Given these equations and establishing a piece of equipment vibrating at 3600 RPM, we can calculate how close the actual vibration is to the natural frequency of this system.

Example Calculations

We start with a valve of 50 lbs at the end of a 7-in. 2 NPS S80 pipe. The source of the vibration is the 3600-RPM equipment. What is the constant k of the pipe valve system? What is the natural frequency? What is the ratio of the frequencies?

- The moment of inertia is 0.868 in.^{4}
- The pipe is 2 NPS (50 DN)
- The modulus of the steel pipe is 29 e6 psi
- The mass of the valve is

$$\frac{50}{32.2} = 1.55$$

- The mass of the 7-in. pipe is

$$\frac{\frac{7.0}{12}(5.022)}{32.2} = 0.090$$

- The constant k is

$$\frac{3(29e6)(0.868)}{7.0^3} = 220,163$$

- The natural frequency is

$$w_n = \sqrt{\frac{220,163}{1.55 + 0.23(0.090)}} = 374.5 \text{ radians/sec}$$

- The frequency of the 3600 RPM is

$$\frac{3600}{60}(2\pi) = 377 \text{ radians/sec}$$

- The ratio is

$$\frac{w}{w_n} = \frac{377}{374.5} = 1.007$$

What is the ratio all about? The answer is relatively simple: the things that happen when that ratio is 1—that is, when the forced frequency and the natural frequency are the same—are not good. There is a lot of math associated with the development of this force multiplier, but the simple result reduces down to a simple equation for the multiplier:

$$\text{Multiplier} = \frac{1}{1 - \left(\frac{w}{w_n}\right)^2} \tag{10.6}$$

Examination shows that when the ratio is 1, the divisor is 0, and the answer is indefinite, or infinite. Further examination shows that when the ratio is less than 1 the answer is positive, and when it is more than 1 it is negative. That negative number represents a change in the phase angle of the resultant wave. For all practical purposes that has no real effect on the resulting increase or multiplication. The multiplier can be treated as if it were the absolute value. The base graph showing that relation is in Figure 10.1. So, for our example, the force multiplier is

$$\frac{1}{1 - (1.007)^2} = 71.18$$

The moment of a 50-lb valve on a 7-in. pipe force going sideways would be 350 in.-lbs. Multiply that by 71.18 and one would have a moment of 24,913 in.-lbs, which with a section modulus of 0.731 would mean a stress of 34,015 psi. It would probably break.

However, change the length to a shorter distance like 6 in., and the multiplier goes to less than 3 and the stresses fall to less than 2000 psi. The shorter length is one of the mitigating methods to deal with vibration.

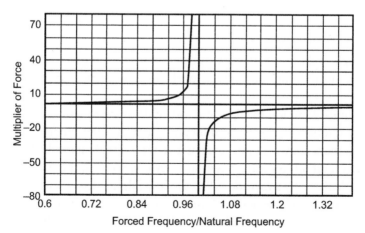

FIGURE 10.1 Natural versus forced frequency

If, however, the force multiplier goes to infinity, wouldn't that mean the system in the previous example would destroy itself as the equipment moved from start to running speed or from running speed to stop as, in this case, it went through the natural frequency as it was moving? We know that doesn't happen, but maybe we don't know why.

Every system has some damping, whether it is inherent or added. That damping reduces the effect of the multiplier depending on how close the damping ratio to what is called the critical damping ratio, C_c.

Once again, to get this ratio one has to do a great deal of math to arrive at the simpler solution. The two related equations can be written in dimensionless ratio form and are given here as a reference. Many scholars think they are the most important relationships in the field of mechanical vibration. Those equations are:

$$x_o = \frac{\frac{P_o}{k}}{\sqrt{\left(1 - \frac{w^2}{w_n^2}\right)^2 + \left(2\frac{c}{c_c}\left(\frac{w}{w_n}\right)\right)^2}} \qquad (10.7)$$

$$\tan \varphi = \frac{2\frac{c}{c_c}\left(\frac{w}{w_n}\right)}{1 - \left(\frac{w^2}{w_n^2}\right)} \qquad (10.8)$$

The $\tan \varphi$ is to determine the phase angle, and the x_o formula is used to determine the amplitude of the force.

It is convenient that when the ratio of frequencies is 1, Eq. (10.7) reduces to a simple equation of

$$\frac{1}{2\frac{c}{c_c}}$$

This simple relationship shows the power of the damping factor. Figure 10.2 shows the relationship of the ratio of damping to critical damping and its effect at that nasty ratio of 1, or resonance.

It is quite common to add some sort of damping to reduce the possible effect of resonance. It may be difficult to estimate the inherent damping ratio, but there is always some, even if it is just internal molecular friction. If one has a method to measure two successive vibrations, the damping ratio can be calculated by a method called the logarithmic decrement. It involves the natural log of the ratio of those oscillations. However, that is an empirical method and does not necessarily fit into the design. It is also most effective on an underdamped system, which is the most common.

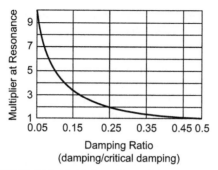

FIGURE 10.2 Multiplier effect of damping ratio

The Appendix contains natural frequency and spring constant formulas for the simple systems to calculate the natural frequencies, the k stiffness factors, and the rotational spring constants and uniform beams, as well as rings and plates. Many of the more common elements and the simpler ones are covered. As the elements get complex the computational effort gets massive. There is an estimating procedure called the Raleigh–Ritz method, after its developers, but it is beyond the scope of this book.

Suffice it to say that given the power of the various computer programs, they are the best way to determine the natural frequencies. As indicated, the methods here have been limited to those of a one-degree-of-freedom system. The compound systems and the Raleigh–Ritz methodology can be described as starting with a string, which would have an infinite number of degrees of freedom, and adding masses and parameters to it to get to an accurate approximation.

VIBRATION SEVERITY

With pipes, pumps, compressors, valves, and turbines, there is the question of how bad the vibration is. Also, in addition to the question of natural frequency, there is the question of severity. The more severe the vibration, the more it has to be watched. There are several standards that include charts and descriptions of what is severe for a particular type of equipment. They include, but are not limited to, those shown in Table 10.1.

These charts about the severity of vibration give indications about what one should do. Basically, they provide a severity level for vibration that, if exceeded, require some action to be taken. That action may be setting up monitoring or doing some analysis, which may lead to remedial action. The

Table 10.1 Standards that Include Vibration Requirements

Standard	General Topic
API 610	Pumps
API 612	Steam turbines
API 613	Gear units
API 617	Centrifugal compressors
API 618	Reciprocating compressors
API 674	Positive displacement pumps—reciprocating
API 541	Motors
ISO 2954	Rotating or reciprocating machinery

fundamental supposition is that below a certain level, the vibration is tolerable. This of course is dependent on many factors in the field.

Making use of the following relationship may be useful in the field:

$$\frac{iM}{Z} < \frac{S_D}{SF}$$

where S_D is an endurance limit. In Chapter 6 we discussed the high end of the cycles and the stress reduction factor f being on the order of 0.095 times the allowable stress range. This may not apply directly.

As another anomaly in pipe stress analysis where stress range is applied, it is important to note that a vast portion of the world utilizes the stress amplitude rather than the range. This often creates a "language problem" between the purists and the down-and-dirty pipe guys. The argument continues and may be exacerbated here because this formula includes the stress intensification factor i, which, as you know, is built with stress range.

Basically, the amplitude is twice the range, with certain exceptions. The choice of the figure to use for the endurance limit obviously has an effect on the static type for results of the analysis that the equation mentioned creates. It should be determined with care. The result may reject the situation by being too constrictive (i.e., assuming too low an endurance). On the other hand, if the stress intensity factor drives the decision it might cause the acceptance of a situation that actually has half the life.

In any event the M represents the moment created by the vibration. In the simple example we had of the valve on a short stub, that moment was 350 in.-lbs. Assuming that the calculation was made for the 6-in. length and we use the multiplier 3, the moment then becomes 1050 in.-lbs. For this purpose we set the stress intensity factor to 1, which is the lowest allowable. The Z or section modulus of an S80 pipe is 0.731; therefore the result is

1436. Now assuming an endurance limit of 10^9, which translates to a little over five years at 3600 RPM, and given an allowable range of 20,000 psi, we effectively have a safety factor of 1.8.

The question then becomes: Is that enough? Well, the allowable stress range has some margin in it, assuming it was code allowable. So it may be high enough. If the plant is a petrochemical plant the shutdown cycle may be less than five years, so there would be a natural opportunity to check and evaluate. However, "enough" is rather like beauty in the eyes of the beholder. It is also dependent on every assumption that went into the design and the analysis. One could add some damping to get the value of the multiplier lower. One could possibly shorten the pipe some more. Once we have a handle on the size of the situation it becomes easier to make a knowledgeable decision.

There is much more to mechanical vibration, including layout vibration analysis by instrument. However, one now can assert a level of command to make the decision as to whether more expertise is needed, including using more precise software analysis.

Flow-Induced Vibration

Earlier in the chapter we discussed the two types of flow-induced vibration—internal and external. We will not discuss much regarding external flow-induced vibration, due to two factors. First, unsupported pipe of any length has some degree of flexibility, which will give it a reasonable resistance to destruction by vibration. Second, it is reasonable to assume that in most cases it can be determined that the vibration is occurring in a "pipe beam" that is clamped on both ends. This translates to a spring constant k of anywhere from two to eight times higher than other forms of beams. That yields, for a given condition, a higher natural frequency.

The vortex shedding frequency is related to the Strouhal number. It is a dimensionless ratio that relates the wind velocity and the pipe diameter to the forcing frequency. When we hear electric lines "sing" in the wind we are hearing that vortex frequency. The wires are quite small in comparison to pipe and so they make a higher-frequency sound than pipe. The Strouhal number was developed by the Czech physicist Vincenc Stouhal in 1878; the expression is

$$f = S_{\text{strouhal}}\left(\frac{V}{D_o}\right) \tag{10.9}$$

where

   f is the frequency

   V is the wind velocity

   D_o is the pipe diameter

This is not the scientific form, because essentially the form here is to determine the value of the Strouhal number by measuring the vortex shedding frequency and the other two variables. Fortunately, the number most often falls into a narrow range from 0.18 to 0.22. It is basically dependent on the Reynolds number.

The Strouhal figure in the Appendix shows some measured Strouhal numbers over a range of Reynolds numbers. An examination of the graph in the figure shows that the use of the value 0.2 for the number is an appropriate approximation. It is possible to develop a more detailed formula from the graph. However, for most engineering purposes the 0.2 estimate is sufficient.

Note that the Strouhal number itself is dimensionless and therefore works in either the USC or SI system. For our purposes we will assume an 8 NPS (200 DN) pipe and a wind velocity of 14 m/sec (46 ft/sec). By using 0.2 and remembering to keep the units consistent we get a frequency of 12.8 hertz or cps for either system of units. And that is at a nominal speed of 71 km/h, which is a nearly tropical storm–force wind.

It is easy to surmise that the natural frequency would be much higher, on the order of twice as much as the vortex frequency for a 30-m pipe. That length might be considered long for unsupported pipe.

This is not to say that wind is not to be considered in piping design. Wind also creates forces and moments as the wind pressure creates a force on the pipe, which was considered in the flexibility-type analysis of the piping.

This then leaves the internal flow-induced vibration in piping to discuss. One of the more common places flow-induced vibration is encountered in piping is in the installation of safety relief valves. These are quite common in pressure systems to protect against runaway pressure excursions. In fact, they are required in ASME codes.

The situation in a safety valve is a simple one. The method of what happens to the fluid is dependent on the type of fluid to be relieved. If the fluid is steam it might be relieved to the atmosphere. In the case of other fluids, which might be harmful or toxic, some capture device might be included. That method not withstanding, the valve basically sits on a stub branch of some length. There is a requirement that there be no impediment between the entrance/opening in the header and the stop in the valve.

This creates what is in effect a tube open at one end and closed at the other. The tube can be modeled as an organ pipe with the same open–close characteristics. Depending on several properties, this can set up a vortex shedding–type situation, where if the conditions are such that the vortex shedding frequency and the resonate frequency of that tube are close together, pressure oscillations occur. These pressure excursions can exceed the set pressure of the valve and cause a partial opening. The partial opening causes the pressure to drop and the valve to close. See Figure 10.3.

This sets up what is called chatter. The net result is that fluid is lost, which costs money. The variation can cause other damage to the system. In short, it is an undesirable result. There was considerable research done on the problem to determine possible resolutions. One of those was the spatial location of the valves themselves. Setting them too close to other disturbances in the flow could cause a problem. This could happen shortly after a change in direction, like an elbow, or with the merging of lines, like in branch- or lateral-type merges. These are functions of the layout of the system.

One of the major elements of safety valve sizing is the amount of relief a particular valve must offer. In large high-capacity lines, at some point it becomes quite cost prohibitive to put in a valve that has enough relieving capacity to relieve the line completely. It is also not completely sensible to do such a thing.

A pressure excursion would be a time-dependent phenomenon. The monitoring instrumentation would signal an alarm and corrective action would begin as soon as possible. The rise might be stopped quickly or the pressure might continue to climb. Naturally, in the worst case, the entire

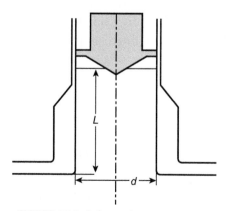

FIGURE 10.3 Safety valve arrangement

line capacity must be relieved. It is often wise for such lines to have multiple safety valves of which the total relieving capacity is the line's capacity.

There is flow-induced vibration that occurs because of the conditions in the safety setup. One of the things that the research shows is that if the vortex shedding frequency is too close to the natural frequency, a solution is to enlarge the opening and in effect change that shedding frequency. This enlargement accomplishes two things that work in concert: the larger opening changes the shedding frequency, and the larger opening quite often requires some taper down to the valve's opening. This in effect changes the natural frequency in a favorable direction.

To calculate this we first need to calculate the speed of sound in the fluid. For this example we use steam as the fluid. Many of the calculations are done for steam power plants. Given that, the speed of sound in a gas can be calculated using the following formula:

$$c = \sqrt{\frac{kRT}{M}} \tag{10.10}$$

where

c is the speed of sound; c is the universal symbol for speed (e.g., $E = Mc^2$ and other well-known formulas)

k is the adiabatic index, which is the ratio of the specific heats (C_p/C_v) of the gas (see the steam k factor chart in the Appendix)

T is the absolute temperature, Kelvin for SI or Rankine for USC

M is the molar mass

R is the universal gas constant

As an example, referring to the chart, one can read a value for k at 900°F and 2000 psia of 1.290. The ratio is a dimensionless number and the ASME chart is only in USC units, but it is a simple matter to convert the inputs from SI units to USC units, or there are probably similar charts for this available.

This index can be approximated by the ratio of the specific heats, and that has been further refined to only need the C_p specific heat at constant pressure. The formula is as follows:

$$\gamma = \frac{c_p}{c_p - 1.986}$$

A few words of caution regarding these numbers is needed. The most relevant ones for this type of calculation are in BTU/lb-mass/mol, or kg/mol. There are many charts that give the specific heats in lb-mass or kilograms rather than the molar basis. One needs to be careful, as the charts in the Appendix are on a mol basis.

Let's continue with the 900°F, 2000 psia values, and calculate the velocity of sound, which in this case would be

$$c = \sqrt{\frac{1.290(1545.35)(900 + 460)}{0.559}} = 2202 \text{ ft/sec}$$

The 0.559 factor in the denominator is because it needs to be in mass, and that is the molar weight of 18 for steam divided by the acceleration of gravity, or 32.2 ft/sec. If one is working with steam exclusively, he or she could substitute the factor 2763.83 in the numerator for the 1545.35, which is the universal gas constant when using pressure in lbs/ft^2, which of course we are, even though the pressure was stated in psia.

Say the header pipe has a flowing velocity of 366 ft/sec. This of course can be computed in many ways from the size of the pipe ID in ft^2 and the amount of flow in pounds. We know the Mach number, which is velocity divided by the speed of sound, at these conditions:

$$\frac{366}{2202} = 0.1662$$

This Mach number is an abscissa on an empirical chart with a family of curves for L over d and an imaginary Strouhal number, which indicates the problem region where excess vibration has occurred during the research. See Figure 10.4.

Careful analysis of the figure shows that the imaginary Strouhal number is based on the L and d dimensions, and a frequency defined as

$$f = \frac{c}{4L}$$

We have already calculated c as 2202 ft/sec. If we establish L at 18 in., and we convert that to feet by dividing by 12, then the frequency is

$$\frac{2202}{4 \times 1.5} = 367 \text{ cps}$$

The imaginary Strouhal number is defined as

$$S = \frac{fd}{v}$$

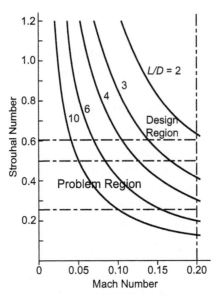

FIGURE 10.4 Safety valve design and problem regions. The problem region is where the vibration is likely to occur. The design region is where it is unlikely to occur

where d is in feet and v is the steam velocity in the pipe, comparable to the wind over the pipe we discussed, which we established as 366 ft/sec in this example.

We will start with a hole of 7.7 in. to match the 18-in. L. So making that calculation we find the imaginary Strouhal number to be

$$\frac{367\left(\frac{7.5}{12}\right)}{366} = 0.635$$

By examining Figure 10.4 we find that when that calculation gets over 0.6 for any conceivable L/d, it is in the design region.

The calculations for the L and d dimensions or for the velocity of the flowing steam are not shown in this example. They are very specific to a particular process and are quite often given to the designer, whose only means of manipulating the considerations is through adjusting the size of the opening in the header, which is limited by the size of the header. Some adjustment might be made in the L dimension by changing the height of the attaching fitting between the valve and the header.

There are other considerations depending on which stage of the design or process an analyst is in. He or she may have to ensure the reinforcement. This was covered in the discussion on pressure design in Chapter 6. There is a concern for the moments that are created if for any reason the valve has a safety trip. This is discussed along with the occasional load in Chapter 11.

As mentioned previously, there are other situations where flow-induced vibrations are met. This chapter has been an introduction to vibration that is far more extensive than that in the current codes. To go further is to attempt to become a vibration analyst. Readers are left with this thought. For complex issues, it is far better to use the extensive software available. The intention is to leave readers with an ability to handle simple problems in the field and to understand enough to deal with the complexities those problems introduce.

Occasional Loads Calculations

Contents

The ASME codes, especially the aboveground codes, recognize that sustained loads such as weight, pressure, and the like, are not the only loads of that type that can occur on a piping system. They call these occasional loads. As far as pressure and temperature are concerned, they are internal to the pipe. Those are generally taken care of by establishing the design pressure and temperatures in such a manner that they will be included in the design process.

Both the B31.3 and B31.1 codes allow certain variations with loads when they meet specific time and other short-term limits. See the codes for specific limitation details. However, there are other occasional loads that can operate on the pipe. These basically come from the environment and include wind, earthquake, snow, and ice.

These occasional loads are also short-term loads. For that reason when the stresses caused by these loads are calculated they are often allowed to exceed the prescribed stresses allowed for the normal or sustained loads. If one were prescient and could predict both the frequency and the duration of such loads they could be included in the variations noted above. However, we do not generally have the gift of prescience, so we use the methods described below.

Buried pipe codes deal with earthquakes, but wind and other elements only operate on the aboveground facilities, which are generally within a building.

The important thing about occasional loads is that they occur with a varying degree of frequency depending on the geographical location of the piping system. It is also true that they do not, usually, operate for a sustained length of time. Even if to the humans involved they may seem an eternity, they usually have a short duration with respect to the sustained loads. In the

Piping and Pipeline Calculations Manual
ISBN 978-0-12-416747-6
http://dx.doi.org/10.1016/B978-0-12-416747-6.00011-5

United States we have geographical maps locating occasional loads by region. There are sources of such maps for other regions of the world.

Figures 11.1 through 11.7 from the American Society of Civil Engineers (ASCE) are shown for reference for the general U.S. geographical maps. ASCE also provides more specific maps for coastal regions, which show more detail. This is true in all cases of wind, ice, snow, and seismic

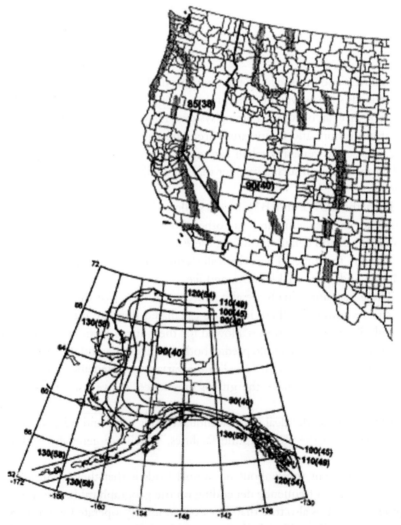

FIGURE 11.1 Basic wind speed for western United States (Source: *From ASCE; used with permission.*)

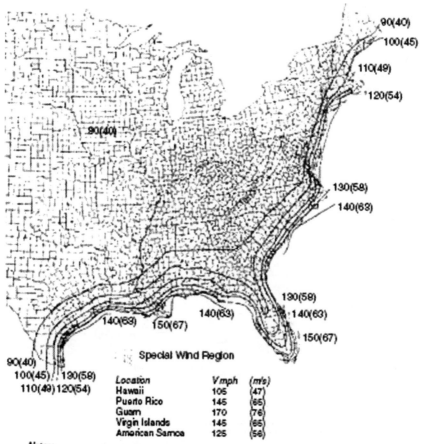

FIGURE 11.2 Basic wind speeds for eastern United States *(Source: From ASCE; used with permission.)*

acceleration. These maps give the basic values to use in the particular calculation procedure needed for computing the load on each variable. This is of course assuming that a particular occasional load actually occurs in the region one is designing for. It is important to remember that the operating word for these charts is basic. Anyone living in a region where the chart says this is the basic load long enough will know that it is

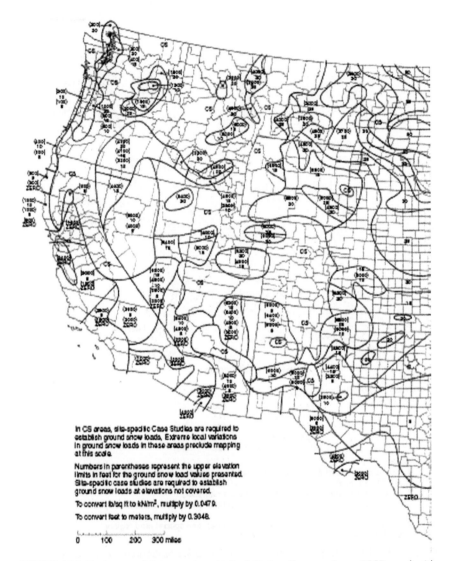

FIGURE 11.3 Basic snowfall in western United States (Source: *From ASCE; used with permission.*)

exceeded on occasion. It is simply not predictable what the maximum load for any region will be in the future. We can but hope our design efforts are enough.

The U.S. seismic zones are particularly useful since the new ASME B31-E procedure for piping uses a cutoff figure for acceleration

FIGURE 11.4 Basic snowfall in eastern United States (Source: *From ASCE; used with permission.)*

in determining how to design piping for seismic activity. This is a simplified procedure based on the experience that the detailed seismic design requires in some instances that is not essential in ordinary piping.

When a designer is dealing with these sorts of loads it is always the jurisdiction that determines the extent of analysis that is required within a particular territory. For instance, it is a well-known fact that California has more specific requirements than other regions for earthquake design of any type of structure. It is also true that there are regions of very high winds in, for instance, mountainous areas. So a word to the wise for an analyst or

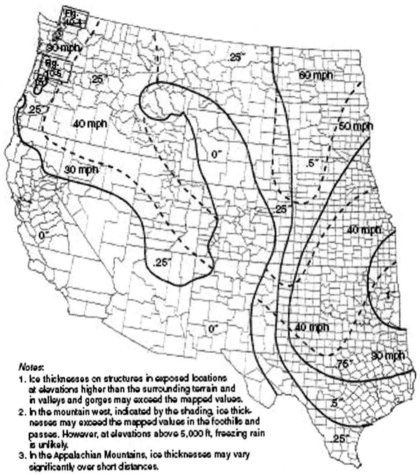

Notes:
1. Ice thicknesses on structures in exposed locations at elevations higher than the surrounding terrain and in valleys and gorges may exceed the mapped values.
2. In the mountain west, indicated by the shading, ice thicknesses may exceed the mapped values in the foothills and passes. However, at elevations above 5,000 ft, freezing rain is unlikely.
3. In the Appalachian Mountains, ice thicknesses may vary significantly over short distances.

FIGURE 11.5 Base ice load in western United States *(Source: From ASCE; used with permission.)*

engineer is to check the jurisdictional requirements before proceeding with any design or construction.

EARTHQUAKE OCCASIONAL LOADS

For earthquake requirements, ASME has developed the B31-E code for piping; therefore, it is the procedure used here. In addition, MSS has produced a standard practice, SP-127, in which bracing with seismic forces is one of the main considerations. SP-127 was written before B31-E was

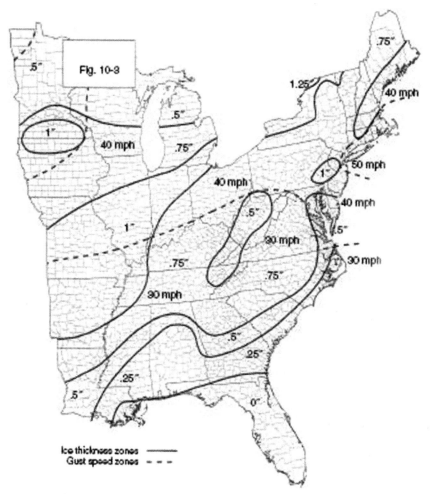

FIGURE 11.6 Basic ice load in eastern United States *(Source: From ASCE; used with permission.)*

published, so the two organizations are currently working to get the nomenclature and approaches consistent.

The B31-E approach establishes that the design goal for seismic earthquakes must first be determined on the basis of the desired outcome after a seismic event. This is based on whether the piping can be defined as critical or noncritical. By definition, noncritical piping has only to meet the requirements of retaining its position after the event. Critical piping, on the other hand, must meet one or both of the following requirements:

1. Leak tightness—that is, prevention of leakage to the environment

2. Operability—that is, the ability to deliver control (e.g., automatic shutoff during or after the event)

These parameters are to be determined in a project's specifications. Once determined, the designer is guided to one of two design methodologies by means of a chart based on two criteria: the criticality and the size of the piping. In the standard, those two are designed by rule or analysis. A designer always has the option to substitute design by analysis for design by rule. He or she is also allowed to use a more rigorous and detailed method at any time. This allowance of more rigorous methods is inherent in the ASME codes.

The chart also breaks the decision into two levels. The first level is where the seismic acceleration is less than or equal to 0.3 g (g is the universal sign for acceleration due to gravity) and the second is where it is more than 0.3 g. Figure 11.7 and Table 11.1 show the regions where that break occurs. It may be that one would want more specific accelerations once the necessary analysis is determined. These can be obtained from ASCE, as they offer charts of geographical regions in finer detail than shown in Figure 11.7.

To summarize, there are three cases where explicit seismic analysis is not required. There are two cases where design by rule is required. There are three cases where design by analysis is required. This is shown in Table 11.2.

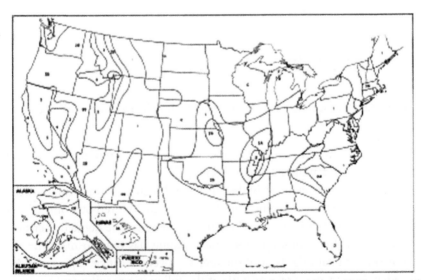

FIGURE 11.7 Seismic zones for the United States (Source: *From ASCE; used with permission.*)

Table 11.1 Seismic Acceleration per Seismic Zone

Zone	1	2A	2B	3	4
Acceleration fraction of g	0.075	0.15	0.20	0.30	0.40

The first step is to determine the type of system and the pipe size. If the pipe system is one of the three system types that require no explicit action for seismic consideration, the analyst has no more seismic thinking to do. If the system is one of the two system types requiring design by rule, there is more. First, the decision can be made to go directly to the analysis procedure in B31-E or choose a more rigorous one. The design by rule procedure is done to modify the suggested pipe support spacing as tabulated in B31.1. This particular tabulation was chosen because the other ASME code books do not publish a starting pipe span, and there is no reason to create a new or different one since this tabulation has stood the test of time.

As one might expect, there are two suggested lengths. One is for standard pipe filled with water, and the second is for steam gas or air service. The one with gas is longer than the one with water because of the weight difference.

It goes without saying that an analyst needs to be reminded that he or she should make adjustments in those span lengths as the weight of the contents of the pipe being analyzed varies. This is why the pipe dimension tables given in the Appendix only provide the volume of the pipe and not the weight; as the contents vary one has to make weight adjustments in any case. Careful readers will understand that the weight is that of air and will not be far off from the second consideration in the table. Remember to read the table notes as to the specific conditions under which the suggestions were developed. All of those content weights vary with temperature and pressure. It is incumbent on analysts to analyze the most difficult situations possible.

Table 11.2 Seismic B31-E Design Case Summary

Piping Type	Acceleration (a)	Design Type	Pipe Size NPS (DN)
Noncritical	$a \leq 0.3$ g	No explicit	All
Critical	$a > 0.3$ g	No explicit	Size $\leq$ 4 (100)
Noncritical	$a > 0.3$ g	Design by rule	Size $\geq$ 4 (100)
Critical	$a \leq 0.3$ g	Design by rule	Size $\leq$ 4 (100)
Critical	$a \leq 0.3$ g	Design by analysis	Size $\geq$ 4 (100)
Critical	$a > 0.3$ g	Design by analysis	All

Be that as it may, the dimensions given in Table 11.3 are what B31-E considers the base span length L_{max}. From that base span length a formula was developed to determine the adjusted span length for the particular seismic acceleration to be used in the project. The formulas for the adjustments are

$$L_{max} = \text{the smaller of } 1.94 \times \frac{L_T}{a^{0.25}} \text{ and } 0.0123 \times L_T \times \sqrt{\frac{S_\gamma}{a}} \text{ for USC}$$

$$L_{max} = \text{the smaller of } 1.94 \times \frac{L_T}{a^{0.25}} \text{ and } 0.148 \times L_T \times \sqrt{\frac{S_\gamma}{a}} \text{ for SI}$$

where

a is the peak spectral acceleration, the largest in any direction, g

L_{max} is the maximum permitted pipe span between lateral seismic restraints, ft or m

L_T is the reference span

S_γ is the material yield stress at operating temperature, psi or MPa

As an exercise consider a system where the pipe is 4 NPS (100 DN) filled with water in a noncritical system where the gravity (a) is 0.3. From Table 11.3 the base span in meters is 4.2, and S_y at operating temperature is 241 MPa.

$$L_{max} = 1.94 \frac{4.3}{0.3^{0.25}} = 11.2 \text{ or } L_{max} = 0.148 \times 4.2 \times \sqrt{\frac{241}{0.3}} = 18.03$$

Clearly, the 11.2-m calculation is smaller, and so that is the maximum span between seismic anchors.

Table 11.3 Reference Spans L_T for B31-E

Pipe Size NPS	Pipe Size DN	L_T (ft)	L_T (m)
1	25	7	2.1
2	50	10	3.0
3	80	12	3.7
4	100	14	4.3
6	150	17	5.2
8	200	19	5.8
12	300	23	7.0
16	400	27	8.2
20	500	30	9.1
24	600	32	9.8

Note: These values are from Table 121.5 of the B31.1 code for water-filled pipe.

There is also the cautionary requirement to reduce the calculated span lengths by a factor of 1.7 for threaded, brazed, and soldered pipe. This is a "should" caution, not a "shall" requirement, so an analyst has discretion as to the use. If straight runs become more than three times the adjusted span length, another "should" is brought to the analyst's attention. There is no relaxation of the requirement to fully consider and provide support or bracing for pipe that has heavy in-line components. This is considered when making the full-gravity support calculations, but there might be need for some extra lateral restraint. By including these cautions, the designer has fully analyzed the noncritical piping system by making the adjustments to the span lengths.

It is desirable if in this analysis seismic and nonseismic supports are considered simultaneously so that the support system has as little redundancy as possible. The use of one of the pipe stress analysis software packages would probably accomplish this desirable result quickly. However, in the first run of the stress software you choose the supports, and so on, that have to be placed at some location along the pipe; therefore a little prethought about the span adjustments before building the model in the software is strongly suggested.

B31-E in its approach to the analysis offers equations that must be equal to or less than certain stresses. In this case they are stresses, not stress ranges. The formulas are:

$$\frac{PD}{4t} + 0.75i\frac{M_{sustained} + M_{seismic}}{Z} \leq min(2.4S, 1.5S_\gamma, 60 \text{ ksi}(408 \text{ MPa}))$$

and

$$\frac{F_{SAM}}{A} \leq S_\gamma$$

where

A is the pipe cross-sectional area, less the corrosion/erosion allowance

D is the pipe OD

i is the stress intensification factor from the applicable ASME code and $0.75i \geq 1$

$M_{seismic}$ is the elastically calculated resultant moment amplitude due to seismic load

$M_{sustained}$ is the elastically calculated resultant moment due to sustained loads concurrent with the seismic load

P is the system operating pressure

S is the ASME B31 allowable stress at the normal operating temperature; for ASME B31.4 use $0.80\ S_\gamma$; for B31.8 use FTS where F and T are the location factor and temperature derating factor, respectively

S_γ is the specified minimum yield stress at the normal operating temperature

t is the wall thickness, deducting only the corrosion/erosion allowance

Z is the section modulus of the pipe with only the corrosion/erosion allowance deducted

Some comments are needed about the deduction of only the corrosion/ erosion allowance. This is somewhat different than two common uses of those terms in ASME. In pressure design it is traditional to deduct all mechanical allowances including the manufacturer's tolerance. This is based on the idea that it is a safety concern because, absent any other knowledge, the pipe may have that smallest wall. In ASME one is allowed to use the actual measured wall in lieu of deducting the manufacturer's tolerance. However, one is required to use the corrosion or erosion allowance and other mechanical allowances. This is with the concern that they actually will be there sometime in the life of the system.

It is also a tradition in compiling things like the stress intensification factors and moments of inertia or a section modulus that the nominal wall is used. This is true of all the pipe size charts I have examined. Naturally, the pipe chart constructor does not know the various corrosion-type allowances to be utilized for the pipe. However, it is deemed important to use the charts during a seismic analysis.

As a general comment, once the values of the various factors are established, the computation is straightforward. Therefore, there will be no example because it is the establishment of the values that is important. It is obvious that B31-E was written with ASME codes in mind. There is no reason, if appropriate values of the variables are established for some other reason than compliance, that this method could not be used.

However, this begs the question of how the values can be established. The most vexing might be the seismic moments. And this leads to the conclusion that the use of some other method might be advisable. This is especially true if that method is included in the piping stress analysis software that is used for the job. It has been suggested throughout the book that some partial hand calculations may be appropriate for a quick field-type calculation. So one of those is presented here.

Most ASME codes defer to the ASCE methods as a way that meets their requirements. ASCE specifically points to one set of equations to determine the forces on ASME pressure piping. This quick and far-from-complete analysis uses some of those techniques. It should be noted that this is a book on piping calculations, not the loads that ASCE imposes. This chapter recognizes that most of the occasional loads discussed are thoroughly handled in other regional or country-specific civil engineering sources. While we hope that none of the occasional loads visit our site, it is incumbent on an engineer to recognize the possibility and prepare for it in some manner.

The simplest way to get a magnitude for the seismic moments is to take advantage of the fact that ASCE says that the required horizontal seismic force can have a maximum amount of

$$F = 1.6(\text{spectral acceleration}) \times I_P W_P \text{ in USC}$$

$$F = 1.07(\text{spectral acceleration}) \times I_P W_P \text{ in SI}$$

where

I_P is the importance factor; unless unusual conditions exist, it would be 1.5—to make it compatible with SI it is squared

W_P is the operating weight

Assume we have a pipe section that is between two anchors; the computation of the weight has been discussed previously. For this exercise, assume that the pipe is 30 ft (9.1 m) and weighs 26 lbs/ft (38.7 kg/m). Assume for our purpose that the spectral acceleration is 0.35 (as a percent of g). Recall that Figure 11.7 is a zone map that gives nominal accelerations. In a complete analysis one might refer to the far more detailed charts. The maximum force would then be

$$F = 1.6 \times 0.35 \times 1.5 \times 30 \times 26 = 655 \text{ lbs in USC}$$

$$F = 1.07 \times 0.35 \times 1.5 \times 9.1 \times 38.7 = 297.3 \text{ kg in SI}$$

The remaining calculation would be to convert the force into a moment by using one of the beam equations. Here again, an analyst has the choice of which divisor to use between the end conditions. Since the weight is distributed along the length it is best to use the uniform load assumption. The force calculated uses the distributed weight, so the calculated values are the ones to use. For this exercise we will use the compromise value of 10 (3.08 in SI), which is the average in USC units of 8 for simply supported and 12 for fixed supports. Different conditions would cause the analyst to use different approaches.

$$M_{seismic} = \frac{437(20)}{10} = 874 \text{ ft-lbs in USC}$$

$$M_{seismic} = \frac{198(9.1)}{3.08} = 585 \text{ kg/m in SI}$$

Note that the 3.08 is an empirical divisor, as I do not have the SI version of Roark and Young, but I know a force by any other name has just as much push.

It is also important to recall that the load calculated in this manner is the maximum load that ASCE would require. This simply means that when this load is coupled with the sustained load and meets the criteria of the two previous formulas, there is little concern. It does not mean that the system failed, for it may by using far more complex calculation methods. Calculating the load with that more rigorous method might allow the system to pass.

It is also necessary to remind readers that the example calculation was only part of the total calculation. It was also for only one section of the system. When one gets into the analysis phase one has taken on a task that entails much labor.

ASCE has limits on displacements. However, B31-E offers some simple limits, and the first limit is a total diametrical sag of 0.5 in. (12 mm). If the designer chooses to multiply the load calculated for the seismic activity by a factor of 2 to allow for dynamic impact, that gap or displacement can take on the value of $0.1\ D$ or 3 in. (50 mm).

Since B31-E is also for retrofit, there are maintenance and equipment investigation requirements. These are details that have no calculation requirements so we are now at the end of the discussion of earthquakes. Readers have not been turned into experts, but have been given tools and guidance to work in that direction.

ICE AND SNOW OCCASIONAL LOADS

The next two subjects regarding occasional loads may or may not be a concern to a piping designer. They not only may not happen or be expected to happen in the particular region, but the type of piping may not be subject to concern. Many of the pipes exposed to weather in a code-covered piping system operate at temperatures where ice or snow do not accumulate. One might think that the hotter pipes do not have icing problems; however, ice can accumulate on supports, braces, and guy-wires, and these might add to the load on the pipe, which could overload the pipe in some manner. Since it possible for pipe to be subject to these phenomena, they are presented here.

Ice storms occur in most regions of the United States with the exception of a relatively large area in the west, which includes the Rocky Mountain regions, California, and the regions to the south. In the eastern half of the country, this is limited to Florida and the southern tip of Louisiana (see Figures 11.5 and 11.6). In the rest of the United States, there is a 50-year uniform ice thickness that occurs. That uniform ice thickness is the basis of the ASCE method of calculating the ice load.

For purposes of this discussion the base assumptions are as follows. All of the concerns are about pipe, so the D in the formula is the pipe or the pipe plus the insulation diameter, and all supports are assumed to be circular in shape. There are different D's for different structural shapes. The importance factor is 1.25, which will cover the majority of the categories of piping that are covered by ASME. Lastly, there will be no topographic factor to consider—that is, it will be 1.

Given the preceding, the following method can be used to calculate the design for ice thickness:

$$t_d = 2t \times 1.25 f_z(1)$$

The 1 represents the topographic factor. The procedure for calculating a topographic factor can be found in ASCE, Chapter 6. Also, t is the nominal ice thickness for the region from either Figure 11.5 or Figure 11.6, and f_z is the height factor calculated by the following formulas:

$$f_z = \left(\frac{\text{actual height}}{33}\right)^{0.10}$$

for heights from 0 to 900 ft; above 900, $f_z = 1.4$ in. in USC.

$$f_z = \left(\frac{\text{actual height}}{10}\right)^{0.10}$$

for heights from 0 to 275 m; above 275, $f_z = 1.4$ m in SI.

Assume we choose a region that has nominal ice of 0.75 in. (0.019 m). The pipe is 10 NPS (250 DN) at 50 ft (15.2 m). We calculate

$$f_z = \left(\frac{50}{33}\right)^{0.10} = 1.04 \text{ in USC}$$

$$f_z = \left(\frac{15.2}{10}\right)^{0.10} = 1.04 \text{ in SI}$$

So

$$t_d = 2 \times 0.75 \times 1.25 \times 1.04 \times 1 = 1.95 \text{ in. in USC}$$

$$t_d = 2 \times 0.19 \times 1.25 \times 1.04 \times 1 = 0.0494 \text{ m in SI}$$

Now that we have this design thickness, the formula for the cross-sectional area of the ice is

$$A_i = \pi t_d (D + t_d)$$

Given the 10 NPS (250 DN) pipe, we have a D of 10.75 in. (0.273 m), so the A_i is

$$A_i = 3.14 \times 1.95(10.75 + 1.95) = 77.76 \text{ in.}^2 \text{ in USC}$$

$$A_i = 3.14 \times 0.0494(0.273 + 0.0494) = 0.05 \text{ m}^2 \text{ in SI}$$

The density of ice shall be not less than 56 lbs/ft^3 or 900 kg/m^3. Readers are cautioned to convert the cross-sectional area from inches squared to feet squared before computing the volume, and the subsequent density by at least the minimum. In this instance, the area is already computed in meters.

After computing the final weight the designer can then compute a load case based on the sustained load, both absent the ice load and plus the ice load. Then compare it to the allowable stress range, remembering that the allowable range for an occasional load is higher by some factor. This makes sense considering that it is rare for occasional loads to occur simultaneously. It is not a stretch of the imagination to say that if someone sees these events occurring simultaneously, he or she is probably watching a movie called *Armageddon Day*, or some such thing.

The next concern is snow loads for the United Sates. Figures 11.3 and 11.4 give a picture of what is called the ground snow load in lbs/ft^2 (4.88 kg/m^2). This is the starting point for determining the snow load on a surface.

ASCE noticed that on curved surfaces where the slope exceeds 70°, the snow probably won't stay put. This knowledge allows a designer/analyst to consider any snow load in that area to be 0. It also allows a designer to use the horizontal projection of the surface as the width of the area. This is a fairly simple multiplication to determine the projected width of a pipe. From basic geometry we can set up a little rule to determine that width. See Figure 11.8.

The geometrical formula to determine the length of the chord and therefore the horizontal projection is

$$L = 2R \sin\frac{\alpha}{2}$$

For the 70° slope, the α in the figure is 140°, which makes

$$\frac{\alpha}{2} = 70°$$

if one makes the circle a unit circle (i.e., if the diameter (2R) is equal to 1), then the multiplier on any diameter is 0.93. To get the horizontal projection for the snow load one only needs the pipe diameter times 0.93.

Once again, decisions about the pipe may have to be made at some level of pipe temperature at which the snow would melt and drip off rather than accumulate on the pipe. There is a C_t factor to adjust for temperature. For

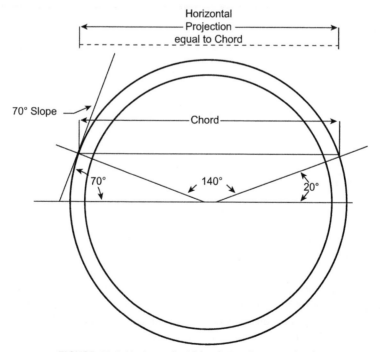

FIGURE 11.8 Horizontal width of pipe for snow loads

instance, if the temperature is 50°F (10°C) on the surface of a structure, there is an allowance of a multiplier of 0.85. Some pipes have heat tracers that might eliminate entirely the possibility of collecting snow. If the pipe can be expected to be below 50°F and above freezing that multiplier would be 1. If the pipe is kept near or below freezing the multiplier would range from 1.1 to 1.2; for all other structures the factor would be 1.

There is a factor C_e for exposure if the pipe is exposed to winds. Usually this is for what is known as exposure B or C; B is for urban areas with groups of buildings or other structures to keep the wind from getting a straight blow, and C is for rural areas that have fields and other sparse areas that do not impede the wind. Both of those have a factor of 0.9. As usual there is an importance factor I, which for most piping is 1.1. But if the material is toxic or hazardous, it is 1.2.

Now we have all the variables and can construct the snow load formula. That formula is as follows:

$$P_{\text{pipesnow}} = 0.7\,CeCtCsIP_g$$

Example

Let's do a sample exercise. Take a pipe that is 14 NPS (350 DN). The factors are $C_e = 0.9$, $C_t = 1.1$, $C_s = 1$, and $I = 1.2$, and the ground snow load = 20 lbs/ft^2 (97.6 kg/m^2).

$$P = 0.7 \times 0.9 \times 1.1 \times 1 \times 1.2 \times 20 = 16.6\frac{\text{lbs}}{\text{ft}^2} \text{ in USC}$$

$$P = 0.7 \times 0.9 \times 1.1 \times 1 \times 1.2 \times 97.6 = 81.2\frac{\text{kg}}{\text{m}^2} \text{ in SI}$$

The pipe is 14 in. in diameter, so the horizontal projection is

$$\frac{14 \times 0.93}{12} = 1.085 \text{ ft in USC}$$

The DN pipe is 0.356 m in diameter, so the horizontal projection is 356 · 0.93 = 0.331 m.

In either, the remaining calculation is to multiply the load from the horizontal projection by the appropriate length to get the load in lbs per square of kg per square meter. Then one has the uniform load in either pounds or kilograms to apply to the moment equation for loads and combine it with a case for snow load plus sustained load to get the occasional snow case.

Again, this calculation needs to be repeated for each section of pipe being analyzed. Then one calculates the snow load plus the sustained load case and that occasional load is complete.

WIND OCCASIONAL LOADS

There have been references to wind loads throughout the previous discussions. For example, the discussion on snow loads had an exposure multiplier based on exposure categories that come from the ASCE wind load section. Each type might have a different value for the multiplier based on what the wind might do to that particular type of load, but the category comes from the wind criteria.

There are two separate types of wind load problems. Wind is considered to vary according to the elevation at which it is being measured. The wind figures in Figures 11.1 and 11.2 are taken as 3-second gust wind speeds in miles per hour or, for SI, meters/sec at 33-ft elevation. All other wind heights require some adjustment for wind speed. This adjustment is applied in the calculation of the wind pressure force. It is this variation in wind speed that causes a problem in computing loads on vertical risers in that the elevation varies, unlike a horizontal run where the elevation is stable. So in essence there are two different methods.

The first step is the same for either method (horizontal or vertical pipe runs): to calculate the velocity pressure q_z. That formula is given in ASCE as

$$q_z = 0.00256 K_z K_{zt} K_d V^2 I \text{ lbs/ft}^2 \text{ in USC}$$

$$q_z = 0.613 K_z K_{zt} K_d V^2 I \text{ in N/m}^2 \text{ and } V \text{ in m/s in SI}$$

where

K_z is the height factor.

K_d is the directionality factor that is used when combining loads, as is done in the piping case. For round structures, such as pipe, that factor is 0.95.

K_{zt} is the topographic factor; as previously, we will make this 1. The factor depends on a combination of the height of the hill or escarpment and the distance to the structure. The slope should be less than 0.2 and is not applicable over 1. That is generally the case for piping.

V is the velocity for the area picked from graphs such as Figure 11.1 or 11.2 or local climatological data.

I is the importance factor, which for most piping facilities can be taken as 1.15.

K_Z can be computed by the following formulas. For the other factors where the typical use in piping is given here, it is recommended to consult the ASCE procedures for the details of there values.

$$K_z = 2.01 \left(\frac{\text{height}}{900} \right)^{0.210}$$

This is good up to 900 ft and is based on exposure C in USC. The following formula is good up to 275 m and is based on exposure C in SI:

$$K_z = 2.01 \left(\frac{\text{height}}{274.3} \right)^{0.210}$$

Example

As an exercise, assume an area, say Puerto Rico, where the V from Figure 11.2 is 145 mph (65 mm/sec) and the height of the piping is 59 ft (15.2 m). Calculate

$$K_z = 2.01 \left(\frac{59}{900} \right)^{0.210} = 1.09$$

$$q_z = 0.00256 \times 1.09 \times 0.95 \times 1 \times 1.15 \times 145^2 = 64.09 \text{ psf in USC}$$

Note that K_z will be the same because the variable is a ratio:

$$q_z = 0.613 \times 1.09 \times 0.95 \times 1 \times 1.15 \times 65^2 = 3084 \text{ Pa in SI}$$

That wind pressure can be converted into a force. The first consideration would be whether it is a flexible or rigid structure. Here we might get into a little interdisciplinary garble. In piping we talk about flexibility in terms of stress produced because of thermal expansion. In wind the flexible/rigid boundary is based on whether the natural frequency is greater than 1; if it is, it is considered rigid. It would seem that under this definition most pipe would be classed as rigid. In that case, a gust effect factor of 0.85 is accepted. It is possible that it could be different with a more detailed analysis.

In addition, there is a force coefficient factor of 0.7 that is applied to a structure for rounds such as pipe. It applies to situations where the diameter of the pipe times the square root of the wind pressure coefficient is greater than 2.5. It is highly unlikely that the wind pressure for an occasional load will be less than 6.25 in psf in USC or 28 Pa in SI. This makes 0.7 a reasonable choice. The gust factor and the force coefficient combined would make the convert-to-force multiplier 0.6.

Given the 64.09-psf velocity pressure calculated in the preceding exercise, the force would be 39.54 psf and the metric force would be 1850 Pa. The horizontal moments can then be calculated by figuring the ft^2 or m^2 of pipe for the section in question, using the appropriate beam formula, and combining that with the sustained moment. Depending on the situation the sustained moment might be at a different angle than the wind load. For instance the sustained load might be down from the weight and the wind load might be horizontal. This would require the designer to consider resolution of the moments into one moment. Thus that occasional load case is closed.

As mentioned previously the case of a section of vertical riser is somewhat different. The procedure recommended in this book is as follows. Calculate the velocity pressure at the high end of the riser using the techniques just applied. Calculate the velocity pressure at the low end of the riser in the same manner. Convert those two pressures into forces as described before. Make the reasonable assumption that these represent the extremes of a gradually increasing load on the appropriate area.

The moment load will not be the simple uniform load moment calculation. The best way to do this is by superposition. Calculate one load as the uniform load based on the low-end loading, and then calculate the second as a triangular load uniformly increasing from the low-end to the high-end load. Combine them by superposition and you have the risers' moment load to combine with the sustained load for the occasional load case on that section of the piping system.

REACTIONS

This covers the sources of external occasional loads. There can be occasional loads that occur within the piping system. The most prominent of these are called discharge reactions. These can occur during a safety-relieving operation or a letdown that is a planned discharge. One of the more complex of these is a discharge when a safety release valve releases.

As noted earlier, a safety release valve sits on the pipeline it is protecting. Its inlet is open to the flow in the pipe and senses the pressure. We have discussed the effects of that flow when the system is working as it should. Under certain conditions, those effects and the size of the inlet chamber can cause flow-induced vibrations. These vibrations can cause "chatter" in the valve, which can be disruptive. However, the occasional load situation is the situation where the system has gone awry and the pressure has reached a level that requires relief. That can cause a discharge, which will create a thrust and most likely a relatively high moment of a short duration on the piping.

Like all occasional loads, it can and possibly will happen. Therefore, a designer has to provide for that eventuality. The problem then is to determine what that load will be. It is a function of the relieving capacity of the safety valve, the pressure, and the fluid. And it is hopefully of a short duration until the anomaly that caused the overpressure and safety trip to occur is corrected.

B31.1 has a nonmandatory appendix, Appendix II, for the installation of safety valves. In that appendix they outline a procedure for computing the moment that is generated when such a discharge occurs. There are several configurations that might be used. Figure 11.9 is a typical safety valve installation showing a common configuration. It will be used as a reference in the following calculations required to compute those forces.

Using Figure 11.9 we can set an exercise. Note that this exercise will be limited to calculating the reaction moments on the pipe only. In a real situation the project design will set many of the parameters. By choosing a specific safety valve many of the dimensions are set by the valve design. This includes the discharge capacity of that valve. For high-capacity lines it is common to choose more than one valve and place those valves in a line with a progressively increasing set pressure.

This is based on the practical assumption that the anomaly might be corrected and the pressure rise in the line abated before there is a need to discharge the line's entire capacity. By setting this increment, the damage that may occur when one valve is discharged is considered to be less if there

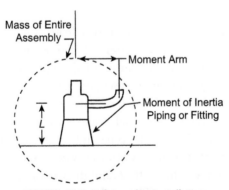

FIGURE 11.9 Safety valve installation

is a partial discharge of the line. If one valve is chosen to relieve the entire line capacity, there may be disastrous effects from that occasional load, so incremental discharges are preferred.

The following list contains the givens in this exercise. The demonstration will be in USC units only, as the actual specification utilized for the demonstration was in that system. It is of a hot reheat line in a power-generation steam system.

- The steam temperature is 1095°F
- The design pressure is 905 psig, the valve set pressure is 920 psi, and the absolute pressure is 920 psi
- h is the enthalpy of steam at this temperature and pressure, which is 1561.125 Btu/lb gravity 32.2 lbm/ft^2
- J is a conversion factor to convert heat in Btus to mechanical energy, 778.16 ft-lbs
- The equation constants are $a = 823$ and $b = 4.33$; they represent steam $\geq$ 90 percent quality and from 19 to 1000 psia
- A is the discharge opening of the elbow, 78.85 in.^{2}
- W is the mass flow given for the valve as 621,000 lbm/hr and equals

$$\frac{621,000}{3600} = 172.5 \text{ lbm/sec}$$

The pressure at the discharge for such an open-vented system as shown in Figure 11.9 is calculated by the formula

$$P_d = \frac{W}{A} \frac{(b-1)}{b} \sqrt{\frac{2(h-a)}{g(2b-1)}}$$

$$P_d = \frac{172.5}{78.85} \frac{(4.33-1)}{4.33} \sqrt{\frac{2(1561.125-823)}{32.2(2 \times 4.33 - 1)}} = 114.8 \text{ psi}$$

In a like manner the velocity at that discharge point can be computed as

$$V_d = \sqrt{\frac{2g(h-a)}{(2b-1)}}$$

So the velocity is

$$\sqrt{\frac{2 \times 32.2(1561.125-823)}{(2 \times 4.33 - 1)}} = 2197.5 \text{ fps}$$

Given those two values at the elbow one can calculate a reaction force based on the givens. That force can be computed by the formula

$$F_r = \frac{W \times 1.11}{g} V + (P_d - P_{atm})A$$

In this exercise we assume P_{atm} to be 14.7:

$$F_r = \frac{172.5 \times 1.11}{32.2} 2197.5 - (114.8 - 14.7)78.85 = 20{,}966 \text{ lbs}$$

Before this force can be applied to any moment arm, one must consider a dynamic amplification of the load due to the opening time of the valve and the period of the valve assembly's reaction to the dynamic factors involved.

By making the assumption that the valve pipe arrangement is a one-degree-of-freedom system, it allows the designer to assume a single ramp from no load to static, making the calculations considerably less complex and suitable for engineering situations.

In that case, the first step is to calculate the period T of the assembly system. The formula suggested for that requires a little more data.

First, Young's modulus E at the design temperature in this example is $E = 20.7 \cdot 10^6$. L is the distance from the header pipe to the centerline of the outlet piping. This is obviously a function of the valve size and the method of establishing the attachment to that header. For our exercise that distance is $L = 22$ in. As a reminder, distance is a combination of the distance from the attachment point of the valve and the valve's distance to the outlet centerline, which is basically fixed by the valve choice. The other element in the combination is the fitting or attachment to the header pipe, which is somewhat under the designer's control. It is used to bring the system into a reasonable configuration regarding flow–induced vibration.

The next element that is included is the weight. This includes the valve, installation fitting, any flanges, and other elements that might be included. For this exercise the weight is $W = 1132$ lbs.

Lastly, the moment of inertia of the inlet piping has to be determined. This most likely will not be a straight piece of pipe, but some configuration that has a larger OD at the bottom than at the top. Quite likely, in the modification of the inlet for the flow–induced vibration, this may also be the case for the ID. In that case, B31.3 suggests that the average ODs and IDs may be used to calculate a working moment of inertia where the formula for that ID would be

$$I_{working} = \frac{\pi}{64}\left(OD_{avg}^4 - ID_{avg}^4\right)$$

In this exercise, $I_{working} = 370$ in.^{4}. The period formula is

$$T = 0.1846\sqrt{\frac{WL^3}{EI_{working}}} = 0.1846\sqrt{\frac{1132 \times 22^3}{20.7 \times 10^6 \times 370}} = 7.91 \times 10^{-3}$$

We must know the opening time of the valve, which is commonly 0.04 sec, but could be any other number. From that opening time we compute a ratio, as follows:

$$\frac{\text{open time}}{T} = \frac{0.04}{7.91 \times 10^{-3}} = 5.06$$

With that ratio we can determine the dynamic load factor (DLF), which in these cases varies from 1 to 2. It can be computed as part of the three-section graph that is provided in the B31.1 appendix. For computational and programming simplicity it can be broken into four straight-line segments.

- Segment 1 would be when the ratio is 0.4, or less than 1. In those cases the DLF is 2.
- Segment 2 is when the ratio is above 0.4 and less than 1. In those cases the DLF can be computed by the straight-line equation DLF = 2.467 − 1.167 · ratio.
- Segment 3 is when the ratio is above 1 and less than 9.5. In these cases the DLF can be computed by the straight-line equation DLF = 1.2665 − 0.0165 · ratio.
- Segment 4 is when the ratio is greater than 9.5, where the DLF is 1.11.

In the exercise we computed the DLF to be 5.06, so we choose segment 3 and calculate DLF = 1.2665 − 0.0165 · 5.06 = 1.18. That is used to make the previously calculated force higher by multiplying, and the force used to calculate moments on the run pipe is 20,966 · 1.18 = 24,740 lbs.

The final calculation is certainly not complete at this point. It depends on several factors (e.g., the type of elbow), if any, used to deflect the force upwards. In that case there would be a moment arm based on the distance from the centerline of the valve and piping installation that would be multiplied by the force. The stress that moment would create on the run pipe would depend on the stress intensification factor of the attached fitting or paraphernalia. Some valves and rupture discs create a force in a straight line coincident with the centerline of the installation piping. In that case the moment would be computed as a single force on the run pipe being considered along with the forces involved. If there is an elaborate config-uration of discharge the forces may have to be resolved into their x, y, or z axis before computing the net moments. They are all variations of the procedure described in the exercise.

It should also be noted that if there is a series of safety valves, a designer has to consider the effect if for some unfortunate reason all of the valves discharge in one event. That may be a catastrophic event and might include considerations beyond the discharge forces and moments. The amount of safety relief is basically covered in the boiler code rules, which are discussed in Chapter 13, on fabrication. The fundamental rule is that the capacity of the valves must be equal to or exceed the line capacity.

This concludes the discussion of the major known types of occasional loads. The methods described and explained here are certainly not the only ways to calculate these loads. They do, however, present a proven and workable methodology.

It should be pointed out that MSS SP-127, among others, has tables and calculation procedures that are often simpler than those given in this book. This does not imply that they are wrong anymore than it implies that the ASME codes are wrong because they are simplified. In order to simplify anything one either has to ignore variables that can be considered insig-nificant or must assume worst cases to be on the conservative side.

This is an engineer's constant battle. On the one hand, simplification and standardization make life simpler; but on the other hand, it can produce a situation that is more than necessary for the specific occasion or problem. One is reminded of the bridges and cathedrals that have stood the test of time for centuries even though the builders did not know the strength of the design materials or of the structure in the ways one can now. For those readers who have the interest, it is very informative to read The Roman author Vitruvius's *Ten Books on Architecture* (read engineering). This book was written in the first century BC. Many of the structures still stand in somewhat ancient form. Then consider modern bridges collapsing in a few short years and/or a building collapsing because of some unplanned occurrence or even a mistake in the calculations.

CHAPTER 12

Slug Flow and Fluid Transients Calculations

Contents

OVERVIEW

We discussed steady-state flow in Chapter 4. Certainly there are differences between laminar and turbulent flow. However, fluid transient flow is flow that varies. To those who are already familiar with fluid transients, it is a truism to say that the steady-state flow is a special case of transient flow. In many cases, it is the special case that is the desirable one.

Flow control with control valves and the like are a case of variable flow that is quite often desirable. This is a variation of the transient flow that is discussed in this chapter. The techniques discussed here are certainly applicable to that flow. They will not be discussed in detail, but readers can utilize the techniques and develop their approaches.

The major effort in this chapter will be directed toward the phenomenon most popularly known as water hammer. In spite of the name, water hammer can occur in any fluid flow, whether incompressible such as water, or compressible such as a gas. The severity of the hammer, as one might expect, is a partial function of the density of the fluid.

The reason for discussing water hammer is that a pressure spike accompanies its occurrence. This spike travels back and forth over the length of the pipe between stations such as a reservoir and a valve. Depending on many factors, that spike can be injurious to the piping or equipment. Any readers who have been involved in an operating piping system that has had some damage to the piping or equipment may have asked when this damage occurred. The vast majority of the answers would be that it was noticed right after a power outage or some other event that stopped the pumping or closed the safety or check valves. That event most likely would have created

Piping and Pipeline Calculations Manual
ISBN 978-0-12-416747-6
http://dx.doi.org/10.1016/B978-0-12-416747-6.00012-7

241

a water hammer situation. One might recall the disaster at a Russian dam in August 2009 causing millions of dollars (rubles) of damage and loss of life. The prime cause was water hammer. For that reason, it is incumbent on a designer/analyst to determine as best as possible the probable extent of a pressure spike.

The ASME codes do not, in general, address such flow changes and fluid dynamics with specific requirements. The usual approach is to point to such elements of fluid dynamics and require that they shall be taken into account. Fortunately, ASME has recognized this as a need and the Mechanical Design Committee has undertaken a project to develop a new code book, B31-D, which will address this issue. Some of the concepts that are being considered for that book will be discussed in this chapter. They will not necessarily be in the final published edition as the technology evolves and consensus is reached.

WATER HAMMER

Before working the calculations it is important to develop an understanding of what is happening that creates the hammer effect. Assume that there is a horizontal pipe of some length L between a reservoir and an open valve. The reservoir is creating a pressure head that causes flow along the pipe and through the valve. It is important to realize that the reservoir in many plant and commercial situations would be a pump or pumping station. Now, close the valve instantly.

A series of events happens within the pipe. At the valve stop the water wants to keep flowing and a pressure builds up that forms a pressure wave that begins to travel back down the pipe in the opposite direction of the flow. It does so until it arrives at the reservoir. This backward wave is then reflected back in the direction of the original flow toward the valve. Once it reaches the valve the process is repeated. It continues to repeat itself until internal friction or some other similar damping mechanism causes the wave to die down and the fluid becomes motionless (see Figure 12.1).

The problem for designers is to calculate the value of the pressure spike that is created by this pressure wave. As is true in many of the previous chapters, there are very rigorous computer programs that can calculate such results. They often include computational fluid dynamics and partial differential equation solving. Some examples are Boss Fluids and PipeNet, and many of the pipe stress programs deal with water hammer in their

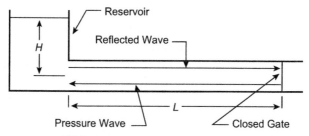

FIGURE 12.1 Water hammer diagram

programming modules. They are beyond the scope of this book, as again it is not intended to make readers fluent in operating such programs. The methods discussed in this book are basically hand or calculator-type and field-type calculations, intended to provide an understanding of the underlying principles that makes working with the programming experts possible.

One might ask, what is the magnitude of the pressure rise? One simple way to find out is to use a couple of formulas that have empirical constants that cover the elements discussed in the following text, to give readers an estimate of the order of magnitude of the potential problem. The formulas are as follows (the medium is water):

$$P_{total} = 0.070 \frac{VL}{t} + P_{initial} \text{ in USC}$$

$$P_{total} = 29,000 \frac{VL}{t} + P_{initial} \text{ in SI}$$

where
   P is the total or initial pressure, psi or Pa
   V is the velocity, fps or m/sec
   t is the time of valve closing, sec
   L is the length of pipe, ft or m
Take the situation of $P_{initial} = 50$ psi (344,738 Pa), $L = 50$ ft (15.2 m), $V = 5$ fps (1.524 m/sec), and $t = 0.04$ sec. We find that the total pressure is

$$P_{total} = 0.07 \frac{5 \times 50}{0.04} + 50 = 437.5 + 50 = 487.5 \text{ psi in USC}$$

$$P_{total} = 29,000 \frac{1.52 \times 15.2}{0.04} + 345000 = 3.36 MPa$$

This is a lot of pressure rise in a relatively short valve closure. However, at the beginning of the investigation we said close the valve instantly, so the 0.04-sec closing time is a reasonable substitute for instantaneous. It is relatively easy to see that water hammer pressure can be large. So how are the empirical constants developed?

One of the first considerations would be the speed of propagation of the wave. This speed is based on the bulk modulus of the fluid, Young's modulus, the thickness of the pipe wall, and the weight in pounds of the fluid. It is important to remember that we are working in hydraulics, so most of the USC units are in foot measurements rather than inch measurements.

Note that the two measurement systems use a different formulaic expression. This is not an uncommon thing among engineering disciplines, as the people who develop the approaches take different paths to the results. It is also true that as one develops the formulas it makes a difference which source one uses for the variables.

This amounts to a scientific Tower of Babel. I have tried throughout to keep the same symbol for the same reference. As an example, in the case of water hammer for the thickness of the pipe wall, the term t did not come up. This is the reference used in most piping, but the different sources use the symbols e and b. It is offered as an explanation as to why the definitions of the symbols for a particular formula are usually placed near that formula and there is no common set of symbols. That being said, the formulas are

$$a = \sqrt{\frac{\frac{K}{\rho}}{1 + \left(\frac{K}{E}\frac{D}{t}\right)c_1}} \text{ in SI}$$

$$a = \sqrt{\frac{\frac{k \times g \times 144}{\rho}}{1 + \left[\left(\frac{k}{E}\frac{D_i}{t}\right)c_1\right]}} \text{ in USC}$$

where
a is the velocity of wave propagation, m/s or fps
K is the bulk modulus of fluid, GPa in SI and psi in USC
D is the internal pipe diameter, m or ft
t is the pipe wall thickness, m or ft
E is the modulus of elasticity, Gpa or lbs/ft^2
ρ is the density, kg/m^3 or lbs/ft^3
g is the gravity, ft/sec^2

c_1 is the constant depending on how the pipe is anchored; here, it is assumed the pipe is between two anchors, and therefore c_1 is $1 - \mu^2$; for steel, it is 0.91.

By examination one can see that the formulas could be manipulated algebraically to come close to the same form. They include some assumptions about the pipe.

In the formulas there is an assumption that the wall thickness is relatively thin. This assumption is reasonable for most pipe. As the wall gets thicker, the C_1 formulae get more complicated. There are other differences as well. Actually, these formulas are algebraic devices that make simplifying assumptions to make the results calculable. The programs mentioned earlier in the chapter go into the most difficult (read tedious) math that is required to solve these transients more completely.

Example Calculations

Let us now assume we are given a 6 NPS (125 DN) S80 0.432-in. (0.0109-m) wall of steel pipe and see what the results of the formula are. With water at a bulk modulus of 2.2 GPa and a ρ of 998.2 kg/m^3, we can calculate the wave velocity as follows:

$$a = \sqrt{\dfrac{\frac{2.24e9}{998.2}}{1 + \frac{2.24}{204}\frac{.146}{.109}.91}} = 1408 \text{ m/s in SI}$$

$$a = \sqrt{\dfrac{\frac{3.25e5 \times 32.2 \times 144}{62.4}}{1 + \frac{3.25e5}{30e6}\frac{5.761}{.432}.91}} = 4620 \text{ fps in USC}$$

The next item to consider was hinted at in the order of magnitude calculations. The question is, what is instantaneous? Remember that the wave has to travel from the valve to the reservoir or reflection point and back. The wave has to travel $2L$ lengths to get to the valve. A wave speed has just been calculated. In the previous exercise we posited a 50-ft length; therefore, the wave has to travel 100 ft to get back to the valve. If the valve is closed that would be effective instantaneous closing. The speed was calculated in the exercise as 4620 fps. So the time involved to return is

$100/4620 = .021$

In this case the valve that has a 0.04-sec closing time would be only half-closed. However, the valve would be fully closed on the second wave as the wave repeats its cycle. This sets up the two major cases of closing. One is where the valve is effectively closed instantaneously. The other is when the valve doesn't close that fast. The second case is the more general one, especially in larger valves and motor-operated valves. Anyone who has been around when such a valve closes knows it is not in less than a second and sometimes not even in less than a minute. There is also the case of control valves where they never really close, but just adjust to control the rate or volume of the flow being delivered down the line.

For the instantaneous closure case it is a relatively simple thing to calculate the change in pressure. For discipline reasons it is easier to work in units of head pressure and then convert the answer to feet of water or meters of water. If you need the answer in psi or Pa, convert the calculated answer and then there is no need to make multiple conversions during any complicated calculations. Of course, it is easier to work in one system. However, we must remember that for the time being when working with ASME and other U.S. standards, the conversion to one system is not complete.

The formula for the increase in pressure for the instantaneous case is

$$\Delta head = -\frac{a}{g}\Delta V \text{ head in ft (m) of } H_2O \text{ and } V \text{ in fbs(m/sec)}$$

where

a is wave velocity

g is ft/sec^2 (m/sec^2)

For example, if we calculate $a = 4620$ fps (1408 mps) and we have a velocity of 10 fps (3.048 mps), the change in pressure is

$$a = \pm\frac{1408}{9.8}3.04 = 438 \text{ mps}$$

which converts to 4.29 Mpa. The actual calculation for the USC system is left to readers as a test of their calculating ability. It should come out around 622 psi. The word "around" is used because conversions are according to the accuracy of the converter and the calculator.

Unfortunately, that is not the common case as discussed. The problem for a designer, then, is to determine what the effect is when a valve doesn't close before the first wave returns. The technique is to approximate the movement by a series of stepwise movements and basically calculate them as instantaneous movements. In each iteration one uses the actual valve gate opening. Naturally, as the valve closes there is an effective area through the gate for each time increment. This effective area is a combination of a coefficient of discharge (C_d) and an effective gate area. Primarily this is based on test data from the valve manufacture and can be found in the published literature. For demonstration purposes in this book, Table 12.1 shows the C_dA_g factor for the example valve (gate) used in the calculation exercise for this approximation method.

The approximation method as shown is quite tedious. When one uses a program to calculate this, the power of the computer can break the steps into very small increments of time. The largest possible increment of time is the segment describing the round trip of the wave:

$$\frac{2L}{a}$$

The smallest is whatever the program chooses. The benefit of the smaller increments is that, depending on the size of an increment, one can calculate the pressure wave amplitude for distances along the pipe intermediate to the full length between the gate and the reflection point, which is the effective reservoir.

There is a set of formulas that are used in the process of building the analysis of the gate closing for closure times longer than

$$\frac{2L}{a}$$

This is commonly called slow gate closing. The set of formulas is the same for each system; when one uses the appropriate units of measure these formulas are

$$B = \frac{(C_dA_g)}{A}\sqrt{2g}$$

Table 12.1 Gate Closure Time Relations

Time (sec)	C_dA_g (ft^2)	C_dA_g (m^2)
0	0.418	0.0388
0.5	0.377	0.035
1	0.293	0.0272
1.5	0.209	0.0194
2	0.123	0.0113
2.5	0.042	0.0041

Note: This table is for 3-ft gate converted from a table for a 10-ft gate. The metric is a straight area conversion from the square foot portion of the table. The medium was water and takes all of the factors from the original development.

where B is a factor used in subsequent equations. To calculate V in the valve at that point in time, use the following formula:

$$V = \frac{B}{2}\sqrt{\left(\frac{aB}{g}\right)^2 + 4\left(H_o + \frac{aV_o}{g} + 2f\right)} - \frac{aB^2}{2g}$$

where

V_o is the original velocity

H_o is the original head in the system

Once V is calculated, calculate F by the following equation:

$$F = (-1)\frac{a}{g}(V - V_o) + f$$

where

F is the pressure equivalent to the instantaneous calculation of ΔH

Now calculate f by the following equation:

$$f = -F_{\left(t - \frac{2L}{a}\right)}$$

This equation is the pressure of the reflected wave taken at time

$$\frac{2L}{a}$$

back from the current time step.

It is usually best to make the calculations in table form so that one can keep track and so that the movement back and forth between time steps is a little easier to calculate. The typical way to set up the table is to pick your time steps. Pick the valve $C_d A_g$ that corresponds and make the calculations in the order of the formula.

For this exercise we will work through the first time step after the 0 time and show the calculations. The entire set of results will be shown in a table. In that way readers can run some calculations themselves and then have a method to check their results.

In the example, the following data are used:

- A 3-ft (0.91–m) diameter is chosen and the wall thickness is 0.0315 ft (0.00952 m); this constitutes an area for the gate of 7 ft^2 (0.65 m^2)
- Young's modulus is 30 e6 psi (207 Gpa); the bulk modulus is 4.68 e7 lbs/ft^2 (2.2 GPa)
- The density is 998 kg/m^3; Poisson's ratio is 0.3
- The original head is 500 ft (152.4 m) in H$_2$O

- The original velocity is 10 fps (3.048 mps)
- The respective g for gravity is 32.2 ft/sec^2 (9.8 m/sec^2)
- The length of pipe is 900 ft (152.4 m)
- The wave speeds are 4730 fps (1430 mps); these were calculated per the proper equation shown in the previous discussion
- The reflection times are 0.38 seconds for USC and the metric method

Checking in Table 12.1, the valve posited has a closing time of 3 sec. So the choice was made to perform the stepwise calculations in 0.5-sec time to ensure that it wasn't instantaneous. The first time step after the 0 time step is 0.5 and the C_dA_g factors for that step are 0.377 for USC units and 0.035 for metric units. The first calculation step is to compute B using the following formulas:

$$B = \frac{(C_dA_g)}{A}\sqrt{2g} = \frac{0.377}{7} \times 8.029 = 0.432 \text{ in USC}$$

$$B = \frac{(C_dA_g)}{A}\sqrt{2g} = \frac{0.035}{0.65} \times 4.42 = 0.238 \text{ in SI}$$

These values would go in the appropriate B column.

Having calculated B at that step, one can then calculate the velocity associated with that time period. For this, use the following formula:

$$V = \frac{0.432}{2}\sqrt{\left(\frac{4730 \times 0.432}{32.2}\right) + 4\left(500 + \frac{4730 \times 10}{32.2}\right) + 2 \times 0}$$

$$-\frac{4730 \times 0.432^2}{2 \times 32.2} = 9.86 \text{ in USC}$$

$$V = \frac{0.238}{2}\sqrt{\left(\frac{1430 \times 0.238}{9.8}\right) + 4\left(152.4 + \frac{1430 \times 3.048}{9.8}\right) + 2 \times 0}$$

$$-\frac{1430 \times 0.238^2}{2 \times 9.8} = 3 \text{ in SI}$$

These numbers are placed in the V column of the table. The next step is to calculate the pressure from the formula given before for F. It is important to remember F is from the step removed by reflection time.

$$F = (-1)\frac{4730}{32.2}(9.86 - 10) + 0 = 20.78 \text{ ft in USC}$$

$$F = (-1)\frac{1430}{9.8}(3 - 3.048) + 0 = 6.64 \text{ m in SI}$$

Note that in most cases when the velocity is slowing down, as it should in a closing situation, it establishes the need for the negative because it is a positive pressure as expected.

The reflective pressure is the negative of the previous step as indicated by the formula, so the only thing remaining is to compute the total pressure at that time step. By the fundamental equations of water hammer, it is the sum of F and f as calculated. Tables 12.2 and 12.3 are samples of the type of table that is recommended.

A table for a much larger valve is included in Table 12.2. The gate had the same relative closure characteristic curves, so the B factor, in spite of the different wave velocities and sizes, came to almost the same value. But the total pressure pattern, while similar, is slower to develop. A graph showing this pressure versus time relationship is shown in Figure 12.2.

Table 12.2 Stepwise Water Hammer in USC units

Time (sec)	C_dA_g (ft²)	Factor B	Velocity V (mps)	F (m H₂O)	f (m H₂O)	Total (F + f)
0	0.418	0.48	10	0	0	0
0.5	0.377	0.432	9.86	20.78	0	20.78
1	0.293	0.336	8.77	160.18	−20.78	139.40
1.5	0.209	0.24	7.12	220.79	−160.82	60.61
2	0.123	0.141	4.45	232.82	−220.79	12.03
2.5	0.041	0.048	1.71	180.78	−232.82	−52.04
3	0	0	0	18.99	−180.78	−161.78

Time (sec)	C_dA_g (ft²)	Factor B	Velocity V (mps)	F (m H₂O)	f (m H₂O)	Total (F + f)
0	4.7	0.48	10.733	0	0	0
1	4.29	0.432	10.16	53.02	0	53.02
2	3.29	0.336	8.77	182.04	0	182.04
3	2.35	0.24	6.99	348.37	−53.02	295.4
4	1.41	0.144	4.5	527.86	−182.04	345.82
5	0.48	0.048	1.51	676.81	−348.37	328.44
6	0	0	0	651.59	−527.86	123.73

Note: For extra measure this includes a chart from computations for a 10-ft gate with a 3000-ft pipe and a wave velocity of 3000 fps. There is a closure time of 6 sec.

Note: With the bigger valve the fall in total pressure comes much slower, but if one runs the calculations the vacillating total pressures go negative. Also note that the B factors remained the same because the two gates had the same relative C_dA_g in spite of the size difference.

Table 12.3 Stepwise Calculation in SI units

Time (sec)	$C_d A_g$ (ft²)	Factor B	Velocity V (fps)	F (ft H₂O)	f (ft H₂O)	Total (F + f)
0	0.0388		3.048	0	0	0
0.5	0.035	0.238	3	6.65	0	6.65
1	0.0272	0.185	2.67	55.43	−6.65	48.79
1.5	0.0194	0.132	2.16	122.5	−55.43	67.11
2	0.0113	0.076	1.318	196.93	−122.5	74.38
2.5	0.004	0.0272	0.459	255.15	−196.93	58.22
3	0	0	0	247.82	−255.15	−7.32

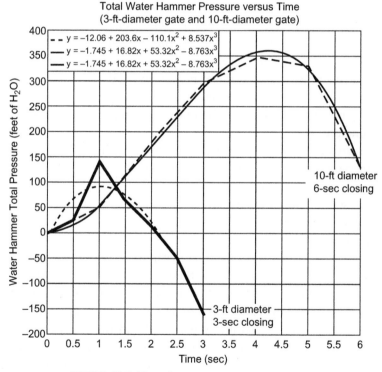

Total Water Hammer Pressure versus Time
(3-ft-diameter gate and 10-ft-diameter gate)

$y = -12.06 + 203.6x - 110.1x^2 + 8.537x^3$
$y = -1.745 + 16.82x + 53.32x^2 - 8.763x^3$
$y = -1.745 + 16.82x + 53.32x^2 - 8.763x^3$

10-ft diameter
6-sec closing

3-ft diameter
3-sec closing

FIGURE 12.2 Water hammer pressure versus time

The figure shows the relationship of the water hammer pressure as it rises and falls in a similar manner. Sometimes such a graph is useful in predicting the water hammer at different time periods than the one calculated by these methods. As in all graphs, the curve becomes smoother if one has more points, which in a way defeats the purpose of graphing. However, in the graph shown a very good curve fit can be developed with the big gate curve, while the smaller curve completely misses the peak point. This is one of the problems of regression analysis. The data can lead to an erroneous conclusion.

These calculations did not consider any friction from the pipe or conduit. As was noted earlier it is friction or some other similar mechanism that finally causes the pressure waves to die out. Remember that as fluids travel along a pipe with friction, that friction dissipates the pressure and thus the velocity until finally, barring any other events, the fluid settles into quiescence and is still. The techniques for calculating that kind of dissipation are beyond the scope of this book. They lie in the realm of hydraulic engineering.

Also note that we have discussed the source of the pressure head as being a reservoir. This is, of course, a simplification. The major sources of developing heads in a fluid are pumps, turbines, and other such equipment. They in fact act as the reservoir. Again, that adds complications to the calculations. For centrifugal pumps and turbines it is a relatively acceptable assumption to consider the fluid source as a steady source. That may not be the case with reciprocating pumps. In any case, it is an approximation.

One can see that this is a calculation-intensive method. There are several other methods that are not discussed in this book:

1. Characteristics method
2. Rigid-water column theory
3. Graphical method
4. Implicit method
5. Finite-element computational fluid dynamics (CFD) method

The plethora of methods implies that there is no one good way. Many folks seem to like the characteristics method. It came about because there is not a general solution to all the partial differential equations that are involved. The methods discussed above basically simplify by eliminating many of the details that in the more critical cases need to be considered; more critical cases means complex piping situations rather than simple point-to-point sets. The characteristics method basically transforms the original partial differential equations into total differential equations by using eigenvalue multipliers. They work on the "characteristic" lines developed and are valid within the restrictions inherent in the process. Programs to solve by this method have been developed.

Once a program is developed it eliminates much of the drudgery of the hand methods. That elimination and resulting speed of computation are the benefit. It is probable that at some point in time the inputs are more important than the method. As the famous, but often ignored, saying goes, "Garbage in, garbage out." We make our calculations based on constants at some temperature, some density, some pressure, and so forth. Computers do not eliminate the necessity of understanding the underlying science in making the best possible inputs.

This discussion was based on closing or sudden stopping. Essentially that is the most damaging case. There are similar methods that can be utilized for openings also. Opening or starting can cause the problems in the system that an unexpected stop does. The little calculations done so far indicate the magnitude of the pressures that can arise, and therefore the emphasis is put on that.

It should be noted here that the highest pressure can be assumed to occur in the instantaneous closing situation. One of the devices used to prevent backflow of fluids is a check valve. Depending on the layout of the system, the check valve is usually there to protect the equipment from some damage. However, it could in turn cause a damaging closure when it acts too instantaneously. This possibility should be checked. It is not uncommon to equip a check valve with some sort of dashpot or bypass to obviate the possibility of damaging something upstream of the check valve.

SLUG FLOW

Slug flow is a problem similar to but different from the closing or opening problems that were discussed in the water hammer section. It is basically a two-phase flow situation. We discuss gas entrainment as bubbles in a more general sense in the "Other Transients" section that follows.

In certain cases those bubbles decide to become much more substantial than little gas bubbles. They occur fairly often in gas and oil well extraction. Often the gas contains liquids and other things. In oil, the fluid often contains gases. In steam and other gaseous transport you often can get condensation and create two-phase flow.

This flow then acts as "slugs" of liquid alternating with slugs of gas. Once they develop, they tend to remain a consistent size as they flow through the pipe. As they go through elbows, equipment, and other changes, the slugs tend to create havoc to that portion of the system. This may be the origin of the term "slug." The gas is usually less dense than the liquid at these points and can alternately hit or slug whatever it is going through. This "slugging" can create vibration, which under the worst conditions could approach the resonance of the system. We know that vibration at the resonant frequency is not a good thing unless of course you are making a TV commercial and want the glass to shatter.

If one has ever turned a bottle partially filled with a liquid upside down rather quickly, you can feel the bottle shake as the liquid and gas spurt in jerky motions as they struggle in taking turns to get out, since they both can't go through at the same time. A word to the wise would be don't pour out anything of value as a thirst quencher.

It is important to gain an understanding of the characteristics of what is happening. Here is the similarity to the water hammer discussion. The velocity of the slug can be approximated by the following formula:

$$V_s = \sqrt{\frac{2L_v P}{\rho L s}}$$

where

V_s is the velocity of the slug
L_v is the length of the vapor portion
L_s is the length of the slug portion
ρ is the mass density
P is the pressure, usually a delta P

This value can then be substituted in the equations for the ΔP to gain a feeling for how much damage might be caused. It is next to impossible to know the lengths of the two portions. Some suggest calculating a short set and then a long set and using your judgment to decide what to do. Many of the pipe stress programs include some method of including the dynamic forces in the stress calculations so you can determine the increase in stress due to this dynamic force that is generated. Of course, one can also take steps like putting in a separating function in the line where this is expected.

OTHER TRANSIENTS

Water hammer and slug flow are not the only concerns. For instance, fluids can have air entrainment. Air bubbles can be entrained in a pipeline. One of their effects is to reduce the velocity of a pressure wave to a great extent. It is a function of the percent of gas volume in the liquid volume, and the bulk modulus K of the combination as well as the density of the combination. The slowing of the wave velocity affects the timing of the reflection cycle as well as the efficiency of the system. A similar effect results if there is a solid of some sort mixed in the fluid. The effect of both can be estimated by using the following formulas.

In either, use the appropriate variable for whatever foreign material is in the fluid.

$$K_{combined} = \frac{K_{liquid}}{1 + \left(\frac{Vol_F}{Vol}\right)\left(\frac{K_{liquid}}{K_F} - 1\right)}$$

$$\rho_{combined} = \rho_F(\% \text{ by volume}_F) + \rho_{liquid}(\% \text{ by volume}_{liquid})$$

$$a_{mix} = \sqrt{\frac{K_{combined}}{\rho_{combined}}}$$

where

K is the bulk modulus

ρ is the density

F is the foreign material, gas or solid

liquid is for the major fluid

combined denotes the mixture's properties

These formulas work in either system of units when the variables are consistent with the system. Since this is one of those calculator-only exercises, there seems to be no need to demonstrate the math skills.

As the entrained gas in the mixture varies from 0 to one percent, the wave velocity can be reduced by as much as five times. This is not an inconsiderable problem. If one is experiencing this phenomenon it would pay to investigate the possibility of air entrainment. Naturally, it should be avoided if possible.

Another problem that is associated with gases or vapors is cavitation. This occurs when the pressure in the system drops below the vapor pressure of the fluid. At that time the fluid can vaporize. This happens when the fluid goes through a low-pressure portion of the piping system. The vapor can then return to the fluid state when it enters a pressure state that is above the vapor pressure. One who is familiar with fluids can understand that this phase change from liquid to vapor and back to liquid can cause problems, such as noise, vibration, and damage.

This problem most often occurs on the suction side of pumps. It can also occur in propellers in ships, mixers in tanks, and so on. There are two basic types of cavitation. They are inertial and noninertial. Noninertial cavitation is when the bubbles oscillate in response to some input energy. This is often associated with acoustic cleaning devices and will not be discussed here. Inertial cavitation is where the bubbles collapse rapidly and cause a shock wave. This is the type that can affect piping through pumps, valves, and other components. We discuss it with regard to pumps.

Every pump manufacturer provides a figure for the pump that is the net positive suction head required. This figure is usually supplied as that of fresh water being pumped at 68°F (20°C). It is also related to the capacity and speed of the pump. A designer must then be assured that such a head is available in the piping layout. That is, it is necessary to determine that the net positive suction head available to the pump for the capacity and/or speed at which it is going to be operated is above the comparable net positive suction head required by the pump. That can be calculated as follows:

$$NPSH_{available} = H_{ss} - H_{fs} - \text{vapor pressure}$$

where

H_{ss} is the static suction head, which is defined as the distance from the free surface of the source plus the absolute pressure at that free surface, ft or m

H_{fs} is the friction head of the piping intake system, ft or m

Vapor pressure is in ft or m

This $NPSH_{available}$ must be higher than the requirement by the pump manufacturer. It is usually a good idea to have a margin above that required. Some experts have suggested 10 ft (3 m), but it is in fact a case-by-case policy.

For instance, if the source is a tank of fluid, will the fluid be replaced at a rate that would keep the static head above a certain level? There are many other things to consider—for example, will the temperature of the fluid be controlled so the vapor pressure cannot rise above a certain level? These are questions that must be asked.

If there is a reason, check the $NPSH_{available}$ for an existing situation. One of the premises of this book is that it will be useful in the field. This can be accomplished by utilizing the gauges at the pump suction flange as follows:

$$NPSH_{available} = H_{atm} + H_g + H_v - \text{vapor pressure}$$

All heads are converted to consistent units. The new pressures are as follows:
- H_{atm} is the atmospheric pressure
- H_g is the gauge pressure at the suction flange connection
- H_v is the velocity pressure

Vapor pressures of water and other selected materials are given in the Appendix.

Fluid characteristics do change as the fluid flows through the pipelines. It is certain that sometimes we want the velocity to change. This, of course, will result in a change in pressure. The Bernoulli equation holds throughout the piping system. Sometimes we want to change the pressure, but this will change the velocity. The problem is that we want to control much of that change to keep the ill effects, such as water hammer or cavitation, to a minimum.

Fluid transients occur in many ways, but we discussed in this chapter the ways that are deemed to be the most important to piping engineers and designers.

CHAPTER 13

Fabrication and Examination Elements Calculations

Contents

OVERVIEW

There are not a lot of calculations for fabrication. However, in this chapter we cover some miscellaneous calculations that are loosely connected to the fabrication portion of a pipeline project. Any pipeline project is much like a three-legged stool. The design is the leg of the stool that is the most calculation oriented, and it has been discussed in the bulk of this book to this point. The other two legs are fabrication and examination. It is a truism to point out that the ultimate success of a project is dependent on all three legs.

HYDROTEST

One of the things that clearly requires a calculation is the hydrotest. This is the final test of the system after it is installed, and may occur as spools or other sections are fabricated in shops. Most people understand that the hydrotest is not to test the design but to test the fabrication. The intent is to subject the assembly being tested to something close to the kinds of stresses that will occur during operation. This may include several considerations.

One of the first considerations is that the process itself may be such that exposing the piping and tested material to water, which is the usual medium for a hydrotest, is detrimental to the operation. For example, a large thin-

Piping and Pipeline Calculations Manual
ISBN 978-0-12-416747-6
http://dx.doi.org/10.1016/B978-0-12-416747-6.00013-9

walled pipe used to convey some gases may not have enough strength to withstand the weight of water in the system. This would be especially true if the pipe were installed in its working position. The hangers and structure to support the pipe, as well as its ability to withstand sustained loads, might overstress the assembly or the supporting components.

Another reason may be that for any of several reasons water in the system may be very detrimental to the intended fluid service. There may not be sufficient means to ensure that the water vapor can be eliminated after testing and subsequent draining. In those cases most codes allow testing by method, for example, pneumatic testing. There is also a combined hydro-pneumatic test allowed.

Typically the requirement in all codes is to test the assembly at some level of pressure above the design pressure. However, the test should never test the material above yield strength in either the hoop or the longitudinal direction at the test temperature. There is a temperature adjustment required in ASME B31.3. This adjustment is not specified in all ASME B31 codes, but it is specified in some codes and standards. It requires additional calculations.

The basic temperature adjustment calculation can be expressed as it is in B31.3. There are other ways to express the same adjustment. The goal is to make the stresses developed in the test as close as possible to the same relative stress on the material as it will be in service. The test is looking for weaknesses in the fabrication or material. The relative strain compatibility achieved by the adjustment will ensure that any microflaws will be expanded in the same manner as will be experienced in operation.

The formula, which works in both USC and SI if appropriate units are used, is

$$P_T = 1.5PRr$$

where
   P_T is the test pressure at the test temperature
   P is the internal design pressure
   R_r has two meanings; for components that have no established ratings (e.g., pipe), it is

$$\frac{S_T}{S}$$

where
   S_T is the allowable stress at the test temperature
   S is the allowable stress at the design temperature

For components that have established ratings, R_r is the ratio of the established rating at the test temperature to the established rating at the design temperature. In no instances may the ratio used exceed 6.5. Also, the prohibition of exceeding yield strength must be followed.

Example

Let's run through an example using a straight pipe spool with the following properties:

- The design pressure is 765 psi, 53 bar
- The design temperature is 950°F (500°C); the test temperature is 68°F (20°C)
- The allowable stress design is 20,000 psi (137.9 MPa)
- The allowable stress temperature is 14,400 psi (99.3 MPa)
- Flange rating from Table 2-2.1 (SI) (test, 99.3 bar; design, 53 bar)
- Flange rating from Table F2-2.1 (USC) (test,1440 psi; design, 765 psi)

These are figures for material 304 H from B31.3 and B16.5; note that they are exact figures from the tables rather than converted figures.

This was done to eliminate the need to linearly interpolate for an intermediate temperature or pressure. Because there are two components, the check will have to be done for each component of the pipe that does not have an established rating. The flange has an established rating per B16.5 Table 2-2.1 for SI and Table F2-2.1 for USC. For the pipe, the USC calculations are:

$$\text{Pipe adjustment}: P_T = 1.5 \times 765 \times \frac{20{,}000}{14{,}400} = 1594 \text{ psi}$$

$$\text{Component adjustment}: P_T = 1.5 \times 765 \times \frac{1440}{765} = 2160 \text{ psi}$$

Neither fraction exceeds 6.5. At the test temperature the yield strength is 30,000 psi. The higher test stress is 2160 and the pipe is a standard-wall 8 NPS, so the 2160 pressure does not exceed the yield strength at the test temperature. The temperature-adjusted test should be run at 2160 psi.

The SI calculations are:

$$\text{Pipe adjustment}: P_T = 1.5 \times 53 \times \frac{137.9}{99.3} = 110.4 \text{ bar}$$

$$\text{Component adjustment}: P_T = 1.5 \times 53 \times \frac{99.3}{53} = 149.25 \text{ bar}$$

Once again, neither fraction exceeds 6.5. At the test temperature the yield is 206.8 MPa, and the higher test pressure is 149.25 bar. Since this is the same pipe 200 DN standard, the 149.25 test pressure does not exceed the yield. The temperature-adjusted test should be run at 149.25 bar.

The ASME codes allow a preliminary pneumatic test at some low level to find any gross leaks before the final hydrotest. This is a good idea. B31.3 sets this level at 170 kPa (25 psi). This is not the alternative pneumatic leak test that may be run. That test pressure is set at 1.1 times the design pressure rather than the 1.5 used in the hydrostatic test. The reasons for this are that air, which is the usual medium for the pneumatic test, is a compressible gas, and so as the pressure increases in a pneumatic test, the stored energy also increases and it becomes quite dangerous (explosive is a more accurate description). If one has been around when there is a structural failure in a hydrostatic test procedure, he or she is aware this is not a totally safe failure. It is, however, not an explosion. Certainly, if the temperature is low enough to make the failure a brittle fracture due to the metal being at or near its null ductility point, failure will cause shrapnel. That is always a warning in ASME codes, and steps should be taken to avoid that event. They include watching the temperature of the test environment and the fluid, as well as being aware of the region of that point metalurgically.

PNEUMATIC TESTING

If a hydro failure comes from some weakness in the attachment weld of the cover or some other weak point, a projectile would be the result of that cover giving way. That projectile could travel a certain distance. There will be a booming noise and danger attendant with it, but it is not a true explosion.

This is not so in the case of pneumatic failure, which is most certainly an explosion. In any dictionary the definition of *explosion* will include the idea of rapid expansion. In the case of pneumatic failure it is the rapid expansion of a highly compressed gas, whereas water is defined as basically incompressible, so its expansion is very minor compared to gas. One just has to remember the old formula $PV = RT$ and think that in a pneumatic failure for all practical purposes the volume and pressure instantly reverse. The pressure goes to zero and the volume has to rise instantly.

There is no temperature adjustment for the pneumatic test. However, for reference we will explore one way to understand the amount of energy that the explosion of a pneumatic test might engender. One formula to calculate the stored energy can be written as follows:

$$\text{Stored energy} = \left(\frac{k}{k-1}\right) PV \left[1 - \left(\frac{P_a}{P}\right)^{\left(k-\frac{1}{k}\right)}\right] 144 \text{ in USC}$$

$$\text{Stored energy} = \left(\frac{k}{k-1}\right) PV \left[1 - \left(\frac{P_a}{P}\right)^{\left(k-\frac{1}{k}\right)}\right] 1000 \text{ in SI}$$

where
> k is the ratio of specific heats of the gas; for air it is common to use 1.4
> P is the pressure, psi or kPa
> P_a is the absolute pressure of atmosphere; it is common to use 15 psi or 103 in this book
> V is the volume, ft^3 or m^3

The constants at the end convert to a usable energy—ft-lbs (N-meters). The usable energy gives readers an explosive comparison to the understood explosive TNT by using the heat of combustion of TNT in ft-lbs and N-meters.

For exercise purposes use the pipe spool in the previous hydrostatic example, and use a 20-ft (6-m) length. The volume in feet will be 6.95 ft^3 (0.2 m^3) for that spool. The test pressure will be 842 psi (1.1 × design of 785) and 58.3 bars (5830 kPa).

Make the calculation:

$$\text{Stored energy} = (3.5)842 \times 6.95[0.936]144 = 2.76 \times 10^6 \text{ ft} - \text{lbs}$$

$$\text{Stored energy} = (3.5)5830 \times 0.2[0.936]1000 = 3.82 \times 10^6 \text{ N} - \text{meters}$$

The heat of combustion of TNT is
- 5.706 × 10^6 in ft-lbs
- 6.86 × 10^6 in N-meters

Using either system, one finds that the spool piece is equal to 0.56 lbs (0.25 kg) of TNT. Not being an explosives expert but understanding that this is one big firecracker explosion, it is easy to see the danger involved.

One can estimate the volume of any similar assembly to be tested at that pressure and again estimate the amount of TNT by using the ratio of that estimated volume to the volume we just calculated, or one can estimate based on pressure ratios. The graphs in the Appendix give a multiplier by which one can multiply the PV to get an estimate of the explosive power for a pressure range.

One reads the pressure in the appropriate units and then selects the multiplier. That multiplier times the pressure times the volume in psi ft^3 for USC units or kPa m^3 for SI units and the approximate pounds of TNT are the answer. The result will be in lbs or Kg depending on which system you use. This might be handy for fieldwork.

It is also relatively obvious from the graphs that the low pressures that the ASME codes allow for preliminary tests for leaks are considerably less explosive, as the amount of compression is not linear but more or less follows a power curve.

DISSIMILAR METAL WELDS

As stated earlier, this chapter is about calculations that are loosely connected to fabrication rather than precisely a fabrication concern. The discussion now moves to those sorts of calculations that occur in piping systems where pipe of different materials has to be welded directly together. When that happens, a differential expansion of the two sections of pipe causes a strain difference. We use strain in this discussion because it is more easily calculated from the radius of the pipe. It is then related to the allowable stress range of the joint. This supposes that an allowable stress range S_A as discussed in Chapter 7 has been developed for that section of the system. It further assumes that the flexibility stress analysis of the system has produced an actual stress S_E for the loads that are on the system. It then becomes a matter of calculating that stress due to differential thermal expansion between the two pipes.

That load is independent of the supposed calculations for S_A and S_E. It must be checked using the following equation and algebraic manipulation. Certain simplifying assumptions make the manipulation simple and the computation relatively easy. The basic assumption equation is

$$\sigma = 0.5E\Delta T\Delta a < S_A - S_E$$

Assume E is the same for both materials. This is usually within 10 percent and therefore a reasonable engineering judgment. It is also safe to assume that each pipe has the same wall thickness at the point of the weld. Once these assumptions are made, the rearrangement can be accomplished and the working equation becomes

$$\Delta T\Delta a < \frac{2(S_A - S_E)}{E} \text{ in USC}$$

$$\Delta T\Delta a < \frac{50.6(S_A - S_E)}{E} \text{ in SI}$$

where

ΔT is the temperature difference from installation to operating, °F or °C
a is the mean coefficient of thermal expansion of the materials, in./in. or mm/mm
Δa is the difference between the two a's
E is the common Young's modulus; it is conservative to use the larger one

As an example problem we need to determine some parameters. Notice from the way the assumptions were made and the formulas derived that the only factors involved in solving the problem are the materials and the Δ temperature. It was established at the outset that the S_A and S_E were given at this point in the calculation procedure.

Assume three thermal materials—304H, P22, and TP310, or their European or other standard equivalents. The three materials are assumed because there often needs to be a transaction material when the co-efficients are disparate. For demonstration purposes such materials were chosen. It will be noted here that rather than choose a's from an SI chart for the Δ temperature, the coefficients were mathematically converted to mm/mm from in./in. charts available in ASME. The data are shown in Table 13.1.

The first step is to calculate the acceptable level of strain using the right half of the rearranged inequality with the appropriate data. Taking the most conservative set of data to find the weak point,

$$\text{Allowable} = \frac{2(26,875 - 8000)}{30.6 \times 10^6} = 0.00123 \text{ in USC}$$

Table 13.1 Data for Differential Expansion

Factor	Pipe 1 (P22)	Pipe 2 (TP304H)	Pipe 3 (TP310)	Comment
S_A	26,875 psi (185.3 MPa)	28,400 psi (195.8 MPa)	27,700 psi (190.9 MPa)	Given
S_E	8000 psi (55.2 MPa)	8000 psi (55.2 MPa)	8000 psi (55.2 MPa)	Given
E	30.6 e6 psi	28.3 e6 psi	28.3 e6 psi	B31.3 Table C-6
E_{SI}	210.9 GPa	195.1 GPa	195.1 GPa	Math conversion
a	7.97 e^{-6} in./in.	10.29 e^{-6} in./in.	9.18 e^{-6} in./in.	B31.3 Table C-3
A_{SI}	3.64 e4 mm/mm	4.74 e4 mm/mm	4.2 e4 mm/mm	Per °C, math conversion
ΔT	1000°F (555.56°C)	1000°F (555.56°C)	1000°F (555.56°C)	Given

The next step would be to determine what the actual thermal strain is at the point of the weld. That strain would use the left side of the inequality:

$$\Delta T \Delta a = 1000\left(10.29E^{-6} - 7.97E^{-6}\right) = 0.00232 \text{ in USC}$$

Clearly, the actual strain in this two-piece similar weld is greater than that allowed for the allowable stress range. This would call for a piece with an intermediate thermal coefficient to be placed between the two pipes to reduce the actual strain. That possibility was anticipated, so the properties of TP310 pipe are also in Table 13.1. Interposing a pup piece between the two would add two welds and both would have to be checked. However, one can tell by examination that it is logical to first check the thermal strain on the weld that is between the two materials that have the greatest difference in their thermal coefficients. If that weld passes, the second weld most likely does not have to be checked. In this exercise the second weld will be checked in the SI system, because in reality one could also have a different allowable strain due to the change in S_A as indicated in the table. First, we calculate between the P22 and TP310, which is the greatest disparity:

$$\Delta T \Delta a = 1000\left(9.18E^{-6} - 7.97E^{-6}\right) = 0.00232 \text{ in USC}$$

This is clearly less than the allowable strain of 0.00232.

For SI the allowable calculation between the TP310 and TP304H is

$$\text{Allowable} = \frac{50.6(190.9 - 55.2)}{195.1} = 0.0354 \text{ in SI}$$

The differential thermal expansion is equal to

$$\Delta T \Delta a = 555.56\left(4.74E^{-4} - 4.24E^{-4}\right) = 0.0278 \text{ in SI}$$

This is clearly under the allowable strain. The actual calculations between the P22 and the TP310 material will not be carried out. However, readers may perform them as an exercise.

The welding of different materials may require different weld procedures. That would depend on project-specific factors such as jurisdiction, code, or even the way the weld procedure was qualified.

Thermal strains between different materials in different geometrical configurations may be computed in a similar manner to the techniques previously demonstrated. The actual calculations are simple. Determining how to apply them might be different.

Metal versus Fluid Temperatures

Another calculation that may be handy from time to time is determining the pipe metal temperature through insulation. Sometimes it may also be useful to determine the metal temperature of an uninsulated pipe by calculating that pipe's or the cylinders' temperature at the center of the wall.

It is important once again to caution that these are not exhaustive heat transfer equations as might be needed in a design office. However, the equations are accurate and within the theories of conduction in heat transfer. But for high accuracy, there is a need to calculate boundary layers for the fluid flowing on the inside and fluid flowing around the outside of the pipe. If the pipe is outdoors, there also may be a need to calculate heat absorption from the sun or other external elements, such as wind, rain, and snow. The sun requirement may be important for aboveground sections of pipelines. These methods are based on one-dimensional steady-state conduction only.

The discussions in this book are based on a unit length of pipe. This eliminates the length factor in the heat calculations, as it becomes one unit in the formulas given here. It also assumes that the materials in the layers of the pipe have a known thermal conductivity, k, which is valid for the temperature range in which the work is being performed. Thermal conductivity tends to fall with temperature. It also tends over fairly large temperature ranges to be linear. This lends itself to using some sort of average or the k for the closest temperature range.

Heat flow has three primary elements: the ΔT across which one is making the set of calculations, the thickness of that calculation, and the thermal conductivity of the material through which the heat is flowing. These elements allow certain assumptions that make reasonable estimates possible.

The flowing fluid has a temperature. In most cases the design temperature is chosen. Many codes require that the pipe metal temperature be used as the temperature of the fluid. One of the main reasons for the use of insulation in hot pipes is to make that requirement as close as possible in practice as well as in calculation. The main purpose of insulation is to slow as much as practical the flow of heat. As that heat flows out through uninsulated pipe it begins to reduce the temperature of the fluid, which may be detrimental to the process for which the piping is transporting the fluid. It certainly would require excess energy to maintain the fluid temperature.

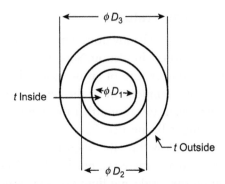

FIGURE 13.1 Pipe with one layer of insulation

However, there are two temperatures in the piping system: the internal fluid temperature and the external, say air, temperature. In both of these cases there is a small boundary layer between the metal and the surrounding fluid or air that can have an impact on the heat flow. We will not address these boundary layers; however, readers are made aware that they exist because as in all calculations they may become important in marginal cases between workable solutions and risky solutions.

The thickness of the pipe wall is known from the other design calculations that are made. One of the reasons for temperature calculations at material boundaries is to help determine the appropriate insulation thickness. That can be done by the procedure shown in the method that follows or by other more sophisticated methods. Even these additional methods often require repeated calculation rather than direct solving. However, they do lend themselves to use of a spreadsheet or other software calculations, but they can be done with a modern hand calculator as well.

In the example we will assume a pipe that has one layer of insulation around it. There can be as many layers as required. Figure 13.1 is a representation of the sample.

The first step is to utilize the two assumed known or given temperatures to calculate an overall heat flow. Those temperatures are called t_i for the fluid and t_o for the outer temperature.

The heat flow calculation uses the following formula:

$$q = \frac{2.729(t_i - t_o)}{\frac{1}{k_{m1}} \log \frac{D_2}{D_1} + \frac{1}{k_{m1}} \log \frac{D_3}{D_2} + repeatasmanyaslayers} \text{ in USC}$$

$$q = \frac{106(t_i - t_o)}{\frac{1}{k_{m1}} \log \frac{D_2}{D_1} + \frac{1}{k_{m1}} \log \frac{D_3}{D_2} + repeatasmanyaslayers} \text{ in SI}$$

where

q is the total heat flow

k_{m1} is the k thermal conductivity for the numbered layer using the appropriate D

D is the diameter where 1 is the innermost diameter, 2 is the next out, 3 the next after that, etc.

t_x is the temperature where $x = i$ means inside and o means outside

As an exercise, make the following assumptions:

- A 10 NPS (250 DN) standard-wall pipe
- A layer of silica 2 in. (50 mm) thick
- The k of pipe is 24 Btu/ft/deg (43 $W/m/K$)
- The k of calcium silicate is 0.525 in USC and 0.0763 in SI
- Inside and outside temperatures are, respectively, 425 and 95°F (491.3 and −08.1°K)

The calculation then becomes, in USC units,

$$q = \frac{2.729(425 - 95)}{\frac{1}{21} \log \frac{10.75}{0.365} + \frac{1}{0.525} \log \frac{14.75}{10.75}} = 3422.4 \text{ Btu/hr/ft}$$

We want to know the temperature of the outside of the steel pipe, and that formula is

$$(t_i - t_o) = \frac{q}{2.729k_m} \log \left(\frac{D_2}{D_1}\right)$$

which makes the preceding calculation compute to the following:

$$(t_i - t_o) = \frac{3422.4}{2.729 \times 21} \log \left(\frac{10.75}{10.020}\right) = 1.8$$

This makes the temperature of the outside of the steel 423.2°F.

A very similar calculation is made when working in SI units. The logarithms are ratios, and when that is the case, the numerical logarithms for the same pipe will have the same numerical value. The q, or heat flow, calculation becomes

$$q = \frac{106.(218.33 - 35)}{\frac{1}{43} \log \frac{10.75}{0.365} + \frac{1}{0.0763} \log \frac{14.75}{10.75}} = 10,976.3$$

The temperature at the outside of the steel pipe then becomes

$$(t_i - t_o) = \frac{10{,}796.3}{106 \times 43} \log\left(\frac{10.75}{10.02}\right) = 0.07$$

This makes the temperature outside the steel pipe 218.08°C (491.23°K).

In the case where the pipe is not insulated, a two-layer technique can be used by making the assumption that (1) there are two pipes encircling each other; (2) the OD of the inside layer of pipe is at a position one-half the wall thickness from the pipe's ID; and (3) the second layer inside is the same diameter as the OD of the first layer of the pipe. This would make an approximation of the metal temperature.

As in all such things one could make an infinite number of layers and calculations and plot the temperature through the pipe wall. This is a very crude calculus. If one has a calculus-type calculator, then it can be set up as a calculus problem. But once again the question becomes: For what purpose? As the old saying goes, these calculations are not rocket science where a minute error midcourse will cause one to miss the moon by a wide margin. This is not wise in space travel, but in pipe a small miss in temperature may not be so devastating.

Flanges have already been discussed in earlier chapters, but in those discussions the flanges were made to a standard such as ASME B16.5. There are other standards. All standards are specific to a particular set of dimensions and, in the case of most flange standards, to a specific pressure temperature rating. When making the temperature adjustment for hydrostatic testing use the appropriate ratings, such as a flange rating, that were used in an exercise earlier in this chapter.

Flange standards rate flanges for static pressure and temperature. It was discussed in Chapter 6 how one might handle such problems as moments and rigidity to ensure that loads not anticipated by the standards can be handled. One special flange that was mentioned was a swivel flange. The major use for swivel flanges is as their name implies—the flange needs to swivel during final assembly. This quite frequently occurs when the final assembly is underwater as in offshore piping.

In final assembly the last two flanges must align within the tolerance of the bolt holes in order to insert the bolts. Visualize a situation where the mating flange is attached to a piece of equipment. The flange on the pipe is attached to a long string of pipe where rotating the pipe assembly is not possible or practical because of the shape of the pipe behind the final flange. It then becomes important to be able to rotate the flange ring to align the bolt holes.

ASME B16.5 has a standard flange that essentially does this. It is the lapped flange in that standard's terminology. The lapped flange is available in the highest class of the standard. However, a lapped flange requires a stub end with a flare that is basically a flared–end pup piece welded to the pipe. As such, it has little or not enough ability to withstand any moment loads. Fatigue tests on the different flange styles by A.R.C. Markl and others have shown that the lapped–joint flange has a fatigue life 10 percent of that of a comparable weld–neck flange. The environments in which swivel flanges are utilized are not conducive to shorter fatigue life because replacement would be expensive and difficult.

A swivel flange can be described as a heavy–duty lapped–joint flange. This is a special design. The basic flange is a weld–neck flange where the ring is not integral as in an ordinary flange and the hub is modified to have a retainer on the back of the ring to keep it from sliding off. The design is a modified version of the ASME Appendix 2 flange method. This is discussed in Chapter 6, but is best handled by a proprietary program or flange designer. Figure 13.2 is a sketch of a typical swivel ring flange.

Two issues in that design are where to cut the ring and the clearance needed to allow it to swivel freely in the environment where it is to be installed. Another issue would be to ensure that the retaining ring in the front of the hub can take the bolt thrust and that the retaining device in the back will hold during preinstallation.

As mentioned in Chapter 6, the method of calculating flanges involves many factors.

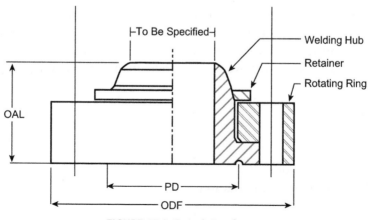

FIGURE 13.2 Swivel ring flange

Rectangular Tanks

There is one more calculation that is loosely connected to fabrication. Occasionally a field person has to design a holding tank for some liquid. ASME Section VIII has some detailed rules and examples of how to design vessels that are not circular. These rules specifically apply to code-stamped or certified vessels. However, there are times when that is not required. There is a relatively simple way to determine if a rectangular-shaped vessel/tank made from plate is adequate for holding a full amount of some liquid, be it water or another liquid of some specific gravity.

The procedure given here will help to design such a tank. The tank is posited to be 80 in. long × 120 in. wide × 60 in. high (2 m × 3 m × 1.52 m in SI units). For purposes of this exercise the fluid is water that has a specific gravity of 1. The first calculation will be to determine the equivalent pressure exerted on the wall by the increasing depth of water. The formula for that is

$$P_e = 0.4336HSg \text{ in USC}(9807HSg \text{ in SI})$$

where

P_e is the equivalent pressure, psi or Pa

H is the height of water, ft or m

Sg is the specific gravity of fluid, 1 for water

Using that we get that P_e for this exercise is 2.16 psi or 14,710 Pa.

Making the assertion that all sides of the plate will have the same thickness we only need to investigate the longest side since it will have the largest moment and deflection from the fluid. It is determined that the top or open side will have some sort of stiffener welded to it (say an angle commensurate with the plate). This then allows us to use the calculated factors with such a stiffener so that those factors can be modeled as supported on all four sides of the plate.

To calculate the factors we need to compute the ratio of the plate:

$$\frac{a}{b}$$

where

a is the height

b is the longest side

In this case that ratio turns out to be

$$\frac{60}{120} = 2$$

Since this is a ratio and dimensionless, it is the same for both systems of measurement.

There is a need to find a β and a γ to calculate the moment and the deflection. These factors often can be found in a standard like the Roark formula for stress and strain, and it becomes a search to find the correct case. For that reason we are offering formulas to calculate them for this specific case. Readers are warned that these formulas are only guaranteed to apply to a tank such as is being designed in this exercise. They are as follows:

$$\beta = 0.2636 \times 0.646^{ratio} \times ratio^{1.515}$$

$$\gamma = 0.2493 \times 0.9355^{\frac{1}{ratio}} \times ratio^{-0.481}$$

Using these equations, or a chart, we get the following values: $\beta = 0.32$ and $\gamma = 0.056$.

Establish the maximum stress value for the chosen plate of 20,000 psi for USC. The formula for the thickness that produces that stress is, in USC units,

$$t^2 = \frac{\beta p_e b^2}{stress} = \frac{0.32 \times 2.16 \times 120^2}{20,000} = 0.497$$

$$t = \sqrt{0.497} = 0.705$$

In SI units it is

$$t^2 = \frac{\beta p_e b^2}{stress} = \frac{0.32 \times 14907 \times 3^2}{1.38e8} = 3.07e - 4$$

$$t = \sqrt{3.07e - 4} = 0.01752 \text{ m or } 17.52 \text{ mm}$$

The next calculation is to determine the deflection, which is maximum at midcenter:

$$\Delta_{max} = \frac{\gamma p_e b^4}{E t^3} = \frac{0.056 \times 2.16 \times 120^4}{30e6 \times 0.707^3} = 2.36 \text{ in.}$$

This might be a bit more deflection than is desired in a 10-ft tank although it is ≈ 2 percent. Changing the thickness to 1 in. would reduce the deflection to less than an inch, as follows:

$$\Delta_{max} = \frac{\gamma p_e b^4}{E t^3} = \frac{0.056 \times 14907 \times 3^4}{207 GPa \times 0.01752^3} = 0.06074 \text{ m or } 60.74 \text{ mm}$$

Again this is in the 2 percent range. This time one must be careful when changing to the equivalent of 1 in., which was suggested for the USC units; in meters that change would be 0.0254 m.

As an alternative one could add a second stiffener of approximately 57 percent or 34.2 in., which is not quite 1 m in SI, around the tank. This would stiffen it and reduce the deflection. However, the calculation then moves itself from the simplified method offered here to a complex method, which is beyond the intended scope of offering a simplified way of designing a tank.

CORROSION ASSESSMENT

There are two handy ways to make decisions on existing pipe or pipelines. It is worth learning the basic methodology for them. The first has to do with corrosion. It is standard practice to add a corrosion allowance when determining the thickness of the wall to use in any service. It is not quite so standard to think about on which side of the pipe the corrosion allowance should be added. Most often the corrosion allowance is considered to be on the inside of the pipe. However, corrosion can attack the outside of a pipe, albeit possibly at a slower rate than that of the fluid, but the wear can come from the outside in rather than the inside out. The amount of corrosion allowance is determined by assessing in some manner how the fluid reacts with the material, and adding an allowance of material based on planned life and severity of attack.

Corrosion on the inside is relatively hard to see during the pipe's lifetime. There are inspection means like ultrasonic measurements of the thickness of existing pipe to determine if it is still at least the proper thickness for the pressure temperature of the service. There are, in pipelines mainly, "pigging devices" that travel through the pipe, recording in some manner the condition of the pipeline.

It is common to coat pipe in pipelines to slow the attack. This coating is applied both inside and outside. In buried pipe there is often some version of cathodic protection added to the outside of the pipe to prevent the electrical corrosion coming from the "battery effect." Be that as it may, checking for corrosion is a continuous battle in the piping world.

Its process has recently been given the name "fitness for service." ASME and API jointly published a book in 2007 with that exact title, *Fitness for Service*. It includes much of API 579, which was published as a standalone earlier. ASME published B31-G, which is about corrosion alone. As the piping systems of the world become older or as processes change in pressure and/or temperature, there is a need to determine if the system is in fact still suitable.

B31-G has a relatively simple way to determine what to do with corrosion after it is found. The basic assumption is that some corrosion is acceptable or is at a level that doesn't yet affect the operation. At some point the corrosion has deteriorated the pipeline to such an extent that the section in question needs repair. What is that point?

First one needs to determine the corrosion spot and its extent. As some may already know, corrosion is not even erosion all over the service, but is quite local and quite irregular. The first step then is to locate the local erosion and determine its size. Figure 13.3 defines the parameters used to determine the action. Once those dimensions are known one can perform some calculations to determine the appropriate action.

If the maximum depth d of the corrosion should not penetrate more than 80 percent of the nominal wall, the wall should not include any thickness that is added for external loads. That is to say, it should be the wall for pressure containment, not other loads.

It is necessary to calculate a factor B by the following formula. Note that B31-G does not recognize the SI system, so the empirical development of the factors and formulas may not be accurate in native SI dimensions. It is recommended that the dimensions be converted to USC from SI (when one is working in that manner) and the decision made that way.

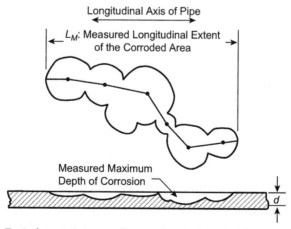

FIGURE 13.3 Typical corrosion spot (Source: From ASME, B31-G, Figure 2.2; used with permission.)

$$B = \sqrt{\left(\frac{\frac{d}{t}}{1.1\left(\frac{d}{t}\right) - 0.15}\right)^2 - 1} \text{ in USC}$$

Then one calculates the maximum L for that corrosion site by use of the following formula:

$$L = 1.12B\sqrt{Dt}$$

where
   d is the maximum depth of the corrosion, as shown in Figure 13.3
   L is the maximum longitudinal extent, L_M in Figure 13.3
   D is the nominal OD of the pipe
   t is the nominal thickness with the limitation discussed earlier
Assume a pipe with a 12.75-in. OD, a 0.375-in. nominal wall, and a 0.07-in. d. The B factor for that situation would be

$$B = \sqrt{\left(\frac{\frac{0.070}{0.375}}{1.1\left(\frac{0.070}{0.375}\right) - 0.15}\right)^2 - 1} = 3.222$$

and $L_{maximum}$ would be

$$L = 1.12 \times 3.22\sqrt{12.75 \times 0.375} = 9.79$$

The maximum calculated B would be 4; when it is above 4 one should use 4 in the L equation. One should also remember at all times that the depth of d needs to be less than 20 percent of the thickness of the nominal wall.

When those conditions are met and the measured L is less than the maximum calculated, no other steps are needed. In this case the depth was close to the maximum 20 percent and hopefully there were records that showed the rate of the corrosion based on the maximum depth from the previous time. From that rate the diligent investigator could make a decision about the frequency of subsequent inspection or possibly opt to repair at this time.

There is an additional option available. The assumption is that the line has been operating at some maximum allowable operating pressure (MAOP). That pressure could be reduced to a safer level. B31-G gives some guidance on this option.

Once again there is a factor to calculate, factor A. This can also be used when the length L exceeds the maximum L calculated. In fact, most times when one calculates the new P' or new MAOP, the procedure when the dimension is less than the maximum L will give a slightly higher pressure. It is not a method to increase the MAOP. The slightly higher pressure just gives a margin of safety measure that exists with the current condition.

First, calculate A using the following formula:

$$A = 0.893\left(\frac{L_{measured}}{\sqrt{Dt}}\right)$$

The A for an L longer than the 9.79 calculated above, using 11 as the longer measured value, is

$$A = 0.893\left(\frac{11}{\sqrt{12.75 \times 0.375}}\right) = 4.49$$

Then there are two formulas to calculate the P' or new MAOP (or, as mentioned, the margin with the current MAOP). The first is for the situation where the calculated A is less than 4. This is basically the margin calculation formula, as it is not allowed to be used to increase MAOP. That formula is

$$P' = 1.1P\left[\frac{1 - \frac{2}{3}\left(\frac{d}{t}\right)}{1 - \frac{2}{3}\left(\frac{d}{t\sqrt{A^2+1}}\right)}\right]$$

Since the purpose is to determine the new lower MAOP, the calculation of P' by this formula for A less than 4 is left for readers.

When the calculated A is greater than 4 the formula becomes much simpler to apply:

$$P' = 1.1P\left(1 - \frac{d}{t}\right)$$

In both formulas, P is the MAOP. For our exercise we need to establish P, as it was not involved in the L calculations. Set it at 1000 psi, and the new P' is

$$P' = 1.1 \times 1000\left(1 - \frac{0.070}{0.375}\right) = 894.66 \text{ psi}$$

An operator now has the option of running at that pressure, say 890, or repairing. This is an operational decision based on economics.

The foregoing is a simplified way that one can make this determination. As mentioned, there are more sophisticated ways to make those decisions in ASME/API FF-1.

PIPE DENTING OR FLATTENING

Another concern that can arise in the field is the denting or flattening of a pipe. The immediate question that comes to mind is: What is the amount of damage? ASME has a procedure in its B31.8 code that helps to answer that question. There are several steps to calculate the damage, and Figure 13.4 shows the basic forms of the dents.

There are five basic formulas to complete the strain estimate for pipe denting; they are:

$$\varepsilon_1 = 0.5t\left(\frac{1}{R_0} - \frac{1}{R_1}\right) \text{ for bending strain in the circumferential direction}$$

$$\varepsilon_2 = \frac{0.5t}{R_2} \text{ for bending strain in the longitudinal direction}$$

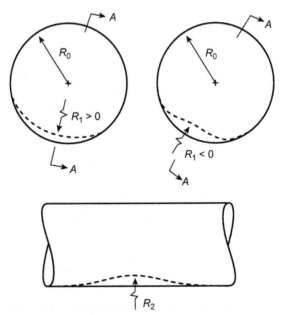

FIGURE 13.4 Dent nomenclature for pipe (Source: From ASME, B31.8, Figure R-1; used with permission.)

$$\varepsilon_3 = 0.5 \left(\frac{d}{L}\right)^2 \text{ for extensional strain in the longitudinal direction}$$

These are combined to calculate the strain on the inside and outside of the pipe:

$$\varepsilon_{inside} = \left[\varepsilon_1^2 - \varepsilon_1(\varepsilon_2 + \varepsilon_3) + (\varepsilon_2 + \varepsilon_3)^2\right]^{0.5}$$

$$\varepsilon_{outside} = \left[\varepsilon_1^2 + \varepsilon_1(-\varepsilon_2 + \varepsilon_3) + (-\varepsilon_2 + \varepsilon_3)^2\right]^{0.5}$$

Once the total strains are calculated, the respective stresses can be calculated from the traditional relationship between Young's modulus and strain. Then the decision can be made as to the extent of the damage.

Note that in some cases there needs to be a sign convention with respect to R_1 depending on whether the dent is a flattening of the pipe, as on the left of Figure 13.4, or an actual inversion, as on the right. The radii change direction. These formulas are set up so that in the left view R_1 is positive and in the right view R_1 is negative.

The mathematical calculations are straightforward and basically hand calculations with little chance for making a mistake, so the actual math will not be computed here. Note when making the calculations that d is the depth of the longitudinal dent and L is the length.

It is unfortunate that the equations were developed in USC; the conversion to metric is once again recommended after making the calculations. In fact, when SI units are used, the answer is the same numerically, which means the results require a multiplier factor to get the strains in units that are meaningful in SI.

Note that while it is relatively easy to measure L in the longitudinal direction and to measure the depth, measuring the radius can be somewhat tedious. Therefore, a little technique to convert depth and L into a radius is offered. This assumes that the radius is smooth and continuous and that the measured depth is at the low point in the arc.

$$R = \frac{L^2 + 4d^2}{8d}$$

This formula was developed in working with circles and the drop that is required from fitting a certain diameter opening on the top of a circle. It has worked for thousands of such fittings and offers a good approximation of the radius of a dent.

BENDING NON STANDARD RADII

On occasion one wants to change the direction of a pipe in some amount other than 90°. This is often accompanied by a radius other than the traditional long short and 3D radii that are found in the B16 and other standards. This quite often happens in pipeline construction. One has at least two things to consider, especially if they are doing the bending in the field or so called cold bending. The pipeline codes have established minimum radii for such occurrences and those are dependent on the size of the pipe. There is usually a factor involving the strain the bend would put on the pipe. It is also dependent on any need for a tangent length at the ends of the bends. This can be translated into a horizontal distance and a vertical distance to be covered.

Eventually the question is what length of straight pipe is needed to accomplish the desired results. Fortunately there are three rather simple formulas to calculate this having determined the pipe size and thus the minimum radius and the angle of change desired. Given the radius R, the angle required A, and the desired tangent lengths the formulas are:

$$Verticaldis\!\int tance = R(1 - cosA) + T * \sin A$$

$$Horizontaldistance = R \sin A + T(1 + \cos A)$$

$$Lengthstraightpipe = 2T + R\frac{A}{57.296}$$

Note the division in the length assumes that A is in degrees rather than radians. Since the calculation is simple algebra no example is offered.

There are many more things that are calculable and those are found as a chart or through graphs or some other method of determining values rather than giving a calculation procedure for them. These charts and graphs can be found in the Appendix.

The number of calculations that one runs into working in a plant is innumerable. There are handbooks and manuals for many of them. In this book, I have tried to bring together the calculations that have come across my desk over the years. The Bibliography provides many reference sources for further investigation.

In the next chapter, we move to valves, which help move and control fluid flow through pipes.

CHAPTER 14

Valves and Flow Control Calculations

Contents

OVERVIEW

It is difficult to tell whether valves or fittings are the most ubiquitous separate pieces of pipe in a large piping system. Certainly, valves will most likely have greater value than fittings, which gain their numbers in the smaller pieces. And most certainly, the majority of the dollars will be in the equipment—pumps, turbines, boilers, vessels, and so forth—along with the thousands of meters or feet of pipe.

Even so, valves are generally recognized to represent upwards of 5 percent of the total cost. If one includes the cost of the controls that move the valves, which control the process, their percentage of the total cost can get much higher. Valves are simply an important part of the piping system.

It is difficult to determine the actual number of different valves that may exist. If one uses the popular website *EngineeringToolbox.com* and simply enters the term "valves" in the search box, at least 1560 entries pop up for that specialized area. This website is a good reference on engineering subjects in general.

Within that site one is further apprised of the multiple various valve standards. There are at least 15 national valve standard societies. Included in those are five American standards, as well as Chinese, British, DIN, JIS, and ISO standards specific to valves. This is to give readers the idea of the size of the valve universe.

Piping and Pipeline Calculations Manual
ISBN 978-0-12-416747-6
http://dx.doi.org/10.1016/B978-0-12-416747-6.00014-0
281

In spite of their number and variety, valves fall into one of four basic functions with a fifth category for the variety of functions that might fall outside the basic four:

1. On–off service to allow or stop fluid flow
2. Control service to change the amount of or to throttle the service
3. Prevention of reverse flow
4. Pressure control
5. Special service for any of a number of other uses

These are fundamental and in one form or another have existed for centuries. The major changes over the years have been in such things as the method of operating the valve, the materials, and the shapes of control devices such as gates.

Even so, there still are only four methods to cause the valve to perform one of the functions just mentioned. Those four methods can be described in the following ways:

1. Interject a plate in the conduit; this can be a flat plate, a cylinder with a hole rotated to open or close, or a spherical surface such as a ball with a hole. Examples include a gate, plug, or ball valve.
2. Move a plug into the opening. Usually this plug or disc is tapered. The best example might be a cork in a bottle, which could be considered a bodiless valve. The more familiar valves in this category include globe, angle, Y, and needle valves.
3. Rotate a disc on a shaft on a casing and insert it into the conduit. The most common of this type are the damper and butterfly valves.
4. Close a flexible material, such as one does in crimping a hose to stop the water flow while moving a lawn sprinkler. These valves are the diaphragm and the logically named pinch valve.

These then are the basic flow control elements in a valve. There are again innumerable ways to fine-tune the design of that element and the mechanism used to move the element into or out of its required position.

If readers are beginning to suspect that we need a book to cover the variety of valves and issues within this subject, they are correct. That, however, is not the purpose of this chapter. We are addressing pipelines and piping. It is significant that the major codes and standards do not necessarily go into detail about valves. Valves are an integral and important part of pressure technology and therefore are in the books. However, the B31 codes in general only have requirements regarding "listed" and "unlisted" valves, which are descriptions of the valves that may be used in their respective systems. Since they are pressure codes they have requirements for pressure-temperature ratings through those listed and unlisted valves. As mentioned before, there are pressure relief requirements including capacities.

Most requirements now have an added requirement for stem retention. This is a requirement that the valve stem cannot be removed from the valve assembly while the valve is under pressure. This nonremoval requirement is neither accidental nor purposeful. It has been added in the past few years because a stem was forcefully ejected from a valve in an accident, causing much damage to a facility. It was discovered in the investigation that there was no codified requirement for this protection. Further, it was discovered that many valves currently in use had no such protection. Presumably this type of accident had never happened before. This is an example of learning the hard way.

In the pipeline-specific books there is usually an additional requirement for the location of the valves. This is a function of how long the committees believe it is prudent for a section of pipe to be before there is a means to stop the flow. The flow must be stopped in case of an accident between valves, since the capacity of the line before flow is stopped has a great effect on any damage that might occur because of the volume in the unstopped line.

The standards that the various codes refer to by and large establish the pressure-temperature ratings of the valves. For instance, ASME B16.34 develops flanged valve ratings (actually flange ratings) in concert with the ASME B16.5 flange standard. Much of that standard is related to the pressure requirements and little is related to the actual flow design. There are several sections and even tables on minimum wall thicknesses and discussions about bolt strength, all of which apply to the actual design of the valve body.

CLOSURE TESTS

There is an ASME section that concerns closure tests. This is the leak test. It is primarily important for on–off and check valves, which stop flow. B16 references MSS SP-61 and API 598, which both establish acceptable leak rates.

The 2009 version of the MSS standard practices is quite similar to the API standard and is discussed here. The first requirement is that the test is conducted at 1.1 times the 100°F (38°C) pressure rating of the valve. This is the highest rating and the rating at which the test is most likely to be run. It is a liquid test. There is an option for a gas test at 80 psi (5.6 bars). There are provisions within the standards for the valve manufacturer, who will presumably be testing valves in quantity and therefore might want more automatic equipment. These are not of concern for the purposes of this book. We assume that one might need to conduct a field test to see if the valve in question has a rate that is reasonably close to the rate at which the test was passed.

In addition, it is assumed that the available measurement tool is drops for the liquid tests and bubbles for the gas test. The bubbles for the gas test assume that the tested surface is covered with a liquid and therefore the bubbles can be seen and counted. It could be called the "inner tube test" for those who remember inflating an inner tube or tire and running it around in a tub of water.

The other test feature is that the number of drops or bubbles is per unit of either NPS or DN size. These drops are based on a specific drop size, as are the bubbles. That drop size is based on spherical drops of 0.5-cm ($3/16$-in.) diameter.

- The volume of a sphere is expressed as

$$V = \frac{\pi D^3}{6}$$

- The volume of the 0.5-cm sphere is

$$V = \frac{3.14 \times 0.5^3}{6} = 0.065 \text{ cm}^3 \text{ in SI}$$

- The volume of the 0.187 is equal to 0.00345 in.^{3}, which converts to 0.056 cm^3 in USC.

The specified number of drops per minute is 2.66 per NPS and 0.11 per DN. While the conversion from 1 NPS as 2.66 and 25 DN as 0.11 is accurate to two decimal places, the math of the little spheres is more proximate, which is why the term *approximate diameter* is used in standard practice.

The same logic applies to the size of the approximate diameter of 0.42 cm and $5/32$ for the sphere used to size the bubble count, which is 1180 bubbles per minute per unit of NPS and 50 bubbles per minute per unit of DN (although not quite 50 when one divides 1180 by 25). It goes without saying that one would hope that a 2 NPS or 50 DN valve would be allowed more bubbles or drops.

The question one might ask is: Why is there such a disparity in the drop/bubble count between liquid and gas? The gas has a lower pressure; presumably the allowable leak orifice is the same size. Think back to the discussion on fluid flow. The density of the fluid has a great deal to do with the mount of flow. Gas is considerably less dense than a liquid and so forth. SP-61 has a note briefly discussing how the bubble rate is determined in some situations from a needle valve with a specific diameter and a gas that has a specific density. This question and this discussion allow us to switch our focus to some of the things that one needs to know about valves that essentially do not come from codes—that is, the flow through the openings and, in the case of control, the partial openings of valves. There is unavoidably a pressure drop during these events, and how that affects piping design and valve choice is the point of the following discussion and calculations.

For starters, we must pay homage to Bernoulli and his balance equation. First, let me point out that this equation and principle came to light in 1738 when Daniel Bernoulli published his book *Hydrodynamica*—the mathematics were simpler then. Since then there have been many forms developed for special situations and conditions. Most, if not all of the fluid flow mathematics can be traced back to this principle.

To state it in terms relating to valves, it can simply be said that the energy conditions of the fluid on the upstream side of the valve are equal to the energy equations on the downstream side when all the changes in energy form have been taken into account. We will work with more specific forms of the equation that make the calculations simpler, but they can all be traced in some way back to the basic equation, which, stated in algebraic language, is

$$Z_1 + \frac{144P_1}{\rho_1} + \frac{v_1^2}{2g} = Z_2 + \frac{144P_2}{\rho_2} + \frac{v_2^2}{2g} + h_L$$

The 144 shows that this is in USC units. It is a conversion from psi to psf, which is what the other elements' basic measurements are in. The fundamental SI equation is the same when the units are made compatible and a conversion factor is required. The symbols have the following meaning:

Z is the potential energy or elevation above a reference level, ft or m

P is the pressure, lbs/in.^{2} or Pa

ρ is the weight density, lbs/ft^3 or kg/m^3

v is the velocity, ft/sec or m/sec

g is the gravity constant, 32.2 ft/sec^2 (9.8 m/sec^2)

h_L is the head loss, ft or m of fluid

There will be no long calculation procedures with this equation. Knowledgeable readers may know the equation in its more derivative form or even some higher-order form.

One of the things that the valve and piping industries have done is develop a simple way to calculate the nominal pressure drop through a valve. The key to the first calculation is that when working with a piping system, the pressure loss of running through a pipe is a function of many things, among which are the pipe's diameter and the pipe's friction loss at various flow levels. This information, once gathered and coupled with the length of pipe, would garner a pressure drop along that length.

Early reasoning said that if we added a certain number of valves of specific designs in that length it would certainly be easy if we had the h_L from that valve expressed in an equivalent length of the same or matching pipe. This would also apply to fittings placed in the flow. It would then be simple to add up those losses and get a total loss through the system, and thus to size the pumping equipment.

Most, if not all, of the resistance coefficients (K) are confirmed by testing the valve and fitting portions of the piping system. The Darcy formula is one of the theoretical approaches discussed in Chapter 4 on fluid flow. To refresh one's memory, it is repeated here:

$$h_L = \frac{f L v^2}{D 2g}$$

This is expressed in feet or meters of head. Valve testing has shown that the head loss through a valve can be expressed as

$$h_L = \frac{K v^2}{2g}$$

It then becomes an equality in h_L that can be expressed as

$$\frac{K v^2}{2g} = h_L = \frac{f L v^2}{D 2g}$$

By algebraic manipulation the formula evolves to one for valves (or fittings) that can be expressed as

$$K = \frac{f L}{D}$$

which can be written in terms of length as

$$L = \frac{K D}{f}$$

This gives a simple way to calculate the equivalent length of valves or fittings in a piping system that one can use to determine the total head loss in a given system.

The K factors are given by many manufacturers for their products. The diameter is in feet in the USC system, as is the friction factor. The SI system may not be as readily available, and certainly, as the computer world has digitized the information into its programs, these things can be calculated more readily with the use of some of the other formulas. Many tables offer the data in terms of S40 pipe.

Each of the valve types would inherently have a different K factor, as they offer considerably different resistance to the flow. It is generally agreed that the least resistance to flow would come from a full-bore open-gate valve. The resistance would move upward toward a globe valve, which is considered the most resistant.

There is an empirical relationship that is handy for use in the field. It requires only three pieces of information: friction factor, diameter, and relationship of the valve type to the gate valve. Those data are developed here. The formula that describes the K factor for a gate valve is as follows:

$$f = 0.238(NPS)^{-0.238} \text{ in USC}$$

where the K factor is considered to be $8 \times f$, and

$$f = 0.0354(DN)^{-0.124} \text{ in SI}$$

where the K factor is considered to be $7.71 \times f$.

If one studies the previous formula for L it becomes apparent that the L in feet or meters can be calculated by multiplying by 8 for USC and by 7.71 for SI, since the friction factors would cancel out.

Naturally, the diameter should be in feet or meters, and to be very accurate, it should be the internal diameter of the pipe. As noted, these data were based on clean S40 pipe. There is a rough conversion factor where

$$K_{new} = K_{base} = \left(\frac{d_{new}}{d_{base}}\right)^4$$

Table 14.1 shows the multiplier for various valve K's. This is based on a large sample of head losses, and the multiplier and its standard deviation are listed per valve type. Designers can use higher or lower than the average depending on how critical the situation is. It goes without saying that the

Table 14.1 K Multiplier per Valve Type

Gate	Butterfly	Y Valve	Swing Check	Angle	Globe
1	5.42	10.16	11.27	24	47.61
0	1.48	0.61	0.61	1.54	3.24

Note: The top row of numbers is the average multiplier. The second row is the standard deviation for that average. The procedure is to develop *L* for a gate valve of the specific size by any means available using the formulas in this chapter for the specific valve size, and then multiply *L* by the multiplier shown in this table.

K coefficient will actually vary from manufacturer to manufacturer. If one has access to specific data for the valve in question, it is always best to use that data rather than the average data. That then sizes the pressure drop for a given line.

INCOMPRESSIBLE FLOW

These calculations are for fully open valves and apply to general sizing of the shutoff valves, which are normally open when operating or shut when there is no flow and therefore no pressure drop across a closed valve. That is the intent of closing valves in any case.

Control valves are used to control the flow through the line as the process parameters call for more or less flow at a given time. As the flow is changed, the pressure drops and velocity changes. Assuming a horizontal line where there is no change in elevation, that factor in the Bernoulli equation can be ignored. Typically the flow change is accomplished by changing the area of the opening where the fluid can flow. It is assumed that there is a reasonably constant pressure as well as velocity on the upstream side. The variables on the downstream side then become some combination of the increase in head loss and change in velocity and pressure to create the necessary energy balance.

In doing this, the calculation is generally changed from feet of fluid as a measure of the pressure to using a flow coefficient in the control valve expressions for capacity and control. In USC units this is commonly called the C_v flow coefficient, and is defined as the flow of 60°F water in U.S. gal/min at 1-psi pressure drop of valve opening. The SI system uses a similar coefficient. However, it is noted as k_v, and its definition is the flow of water within a temperature range of 5–30°C in m^3/hr at a pressure drop of 1 bar. There is a difference in the two systems as far as the numerical value is concerned. Consequently, there is a standard conversion from one to the other, which takes the following form:

$$C_v = 1.16 k_v \quad \text{and} \quad k_v = 0.862 C_v$$

Many valve manufacturers make the conversion automatically in the literature and in sizing equations. But first, using one form of the C_v equation, C_v can be expressed as follows:

$$C_v = \frac{q}{a} \sqrt{\frac{S_{gravity\,F}}{P_1 - P_2}} \text{ in USC}$$

where

q is the flow, gal/min or m^3/h

S_g is the specific gravity at flowing conditions

$P_{1\&2}$ are the upstream and downstream pressure, respectively, psia or kPa

a is the equal factor, which is 1 for USC and 0.0865 for SI

Should for any reason a reducer be on either side of the valve, another factor would have to be included with the factor as a multiplier.

Now assume that one has to select a valve that will work to control a flow region that is normal at 100 gpm and has a high flow of 200 and a low flow of 50 with $_{za}$ normal pressure drop of 5 psi.

The best way to determine control valve sizes is by using the manufacturers' catalog data. Figure 14.1 shows the C_v ratings of various-size valves to help in the selection.

It is a problem not so much of determination but of where to start. In designing, about all one gets is pressure, size, etc. But to really know what has to happen, one has to figure out what the opening will be and the flow through that opening.

$$\text{Area} = \pi(D)(\sin \theta \times D) = 3.14 \times 24 \times 0.174 \times 24 = 314.2$$

There is a similar issue in determining the amount of flow through an orifice, which is basically what a valve is. For instance, if one takes a globe valve and lifts it out of its seat, there is an increasing annular opening area. A similar concern comes from the opening and closing of a butterfly valve as it is rotated from fully open to fully closed. This is essentially a geometry/trigonometry problem. Assume a butterfly valve that has an opening of 24 in. (609.6 mm) when fully open and a disc shaft combination that has a thickness of 3 in. (76.2 mm). The open area for the fluid to flow through is

Standard Trim
Models 41315, 41415, 41515, 41615 and 41915

Flow characteristic : Linear

Valve size (inches)	Valve size (mm)	ANSI class and equivalent PN	Orifice diameter (inches)	Orifice diameter (mm)	Travel (inches)	Travel (mm)	Percent of travel 10 / F_L 0.94	20 / 0.94	30 / 0.93	40 / 0.93	50 / 0.92	60 / 0.92	70 / 0.91	80 / 0.91	90 / 0.90	100 / 0.90
							Rated C_v									
2	50	900 - 1500	1.84	46.7	0.8	20.3	1.4	2.7	4.2	6	8	10	12.5	14	15.5	16
2	50	150 - 600	2.50	63.5	1.5	38.1	2	4.9	8.3	13	19	25	30	35	38	40
3x2	80x50	150-1500					2.7	5.1	7.9	11	15	19	23	26	29	30
4x2	100x50						4	9	15	24	35	47	57	65	71	75
3	80	150 - 1500	3.50	88.9	2.0	50.8	5	10	16	22	30	38	46	52	58	60
4x3	100x80						8	19	31	50	73	96	118	135	147	155
6x3	150x80						9	16	25	35	48	60	72	83	91	95
4	100	150 - 1500	4.38	111.3	2.0	50.8	12	29	48	77	113	149	182	209	228	240
6x4	150x100						7	15	28	41	58	74	94	117	144	165
8x4	200x100						20	52	92	148	204	260	308	348	376	400
6	150	150 - 1500	5.12	130.0	0.8[1]	20.3[1]	17	37	71	104	145	187	237	295	361	415
8x6	200x150				2.0	50.8	32	83	147	237	326	416	493	557	602	640
10x6	250x150						20	46	87	128	179	230	291	362	444	510
8	200	150 - 1500	6.50	165.1	1.5	38.1	50	130	230	370	510	650	770	870	940	1000
10x8	250x200				2.5	63.5	31	69	131	193	270	347	439	547	670	770
12x8	300x200						70	182	322	518	714	910	1078	1218	1316	1400
10	250	150 - 1500	8.00	203.2	1.5 / 3.0	38.1 / 76.2	51	128	211	320	448	576	730	922	1114	1280
12 / 16x12	300 / 400x300	150 - 1500	9.75	247.7	2.0 / 3.75	50.8 / 95.25	104	268	464	744	1024	1304	1544	1720	1880	2000
16	400	150 - 1500	13.00	330.2	2.5 / 4.0 / 5.0	63.5 / 101.6 / 127	130	335	580	930	1280	1630	1930	2150	2350	2500

Notes: 1. Travel of 1.5 inches (38.1mm) for 41405. 2. Ex. 3x2 size = valve with 3" body x standard 2" trim.

FIGURE 14.1 Typical catalog page for a control valve

Table 14.2 Open Area of 24-in. (609.6-mm) Butterfly at Various Steps of Degrees Closed

Degrees Closed (q)	Open Area (in.²)	Open Area (mm²)
0	380	245,161
10	373.8	241,161
20	297.7	192,064
30	226.2	145,936
40	161.6	104,258
50	105.8	68,258
60	60.6	39,097
70	27.3	17,613
80	6.9	4,452
90	0	0

the area of the 24-inch opening minus the area of the disc rectangle. Or to put it mathematically,

$$A_{forflow} = D^2 \times 0.7854 - D \times \text{Thickness}_{shaftdisc} = 24^2 \times 0.7854 - 24 \times 3$$
$$= 380.4 \text{ in.}^2$$

As the disc closes the open area becomes less of a round disc at say $10°$ closed, which creates a shadow equal to the perpendicular area of a circle on a $10°$ slant to the axis of the valves (see Table 14.2). In our exercise that open area becomes

$$\text{Area}_{open} = 24^2 \times 0.7854 - \pi 24 \times (\sin 10) \times 24 \times 0.7854 = 373.8 \text{ in.}^2$$

One can plot the pressure drop through the valve based on the changing areas. A procedure for calculating the C_v and thus the flow quantity through the valve is outlined in the following text. One will note that it is somewhat tedious and serves, among other things, as a reminder to work with the valve vendor of choice to get the same kind of information. Presumably they have gone through this or the experimental work to settle these issues. A butterfly was chosen in this book as it is a somewhat simpler set of calculations. The procedure is the same for any valve.

A method has been given to calculate the resistance coefficient for butterfly valves by relating it to the gate valve. Recall that the gate valve K could be calculated for a 24 NPS gate valve by multiplying the ID in feet by 8 when using a 24 NPS S40, the ID of which is

$$\frac{22.62}{12} = 1.88 \text{ ft}$$

That would make the K factor

$$8 \times 1.88 = 15 \, \text{ft}$$

This K factor is for a gate valve. Refer to Table 14.1; the multiplier for a butterfly valve is 5.42. This ignores the possibility of adjustment because the range of multipliers is larger by the standard deviation and the concern one might have in being certain. The K factor for the butterfly is then 81. The question becomes: What does this mean with respect to C_v?

There is a conversion factor to get to C_v from K. That factor is an equation:

$$C_v = \frac{29.9d^2}{\sqrt{K}} = \frac{29.9 \times 1.88^2}{\sqrt{81}} = 12$$

This formula for C_v requires a flow to calculate the C_v factor. This is the reason this circuitous route was taken. It also shows why it is much better to get the information from the manufacturer of the valve—there is a higher degree of reliability based on actual tests rather than general theory. However, this is one of those on-the-spot ways.

Once one has the C_v in hand, the C_v formula can be manipulated algebraically to determine an estimated flow:

$$q = \frac{C_v}{\sqrt{\frac{sg}{\Delta P}}}$$

This requires a change from feet of water, which is what the K is, to psi, or, to establish our inclusion of SI, the flow q instead of gal/min would be in m^3/hr. The pressure in kPa and the ever-present conversion factor would be

$$q = \frac{C_v}{11.7\sqrt{\frac{sg}{\Delta P}}} \quad \text{in SI}$$

However, this would not be the K_v that is used in Europe. The conversion from K to C_v would have to be different by the 1.16 factor previously mentioned.

One thing readers should be garnering from this is that the systems, as we saw in Chapter 12 on pipe flow, in this general area are not quite as cross-compatible as some of the others between the USC and SI systems.

At any rate the flow would be calculable and the key in the USC system is the C_v or K factor. It can be asserted that for control valves the C_v methodology is much less conversion-heavy.

COMPRESSIBLE FLOW

It should be pointed out that when we move into compressible flow there is a caution that needs to be repeated. It has to do with the fact that most of the calculations so far are based on the Darcy formulas and their derivatives. Those cautions or limitations to keep the computation in the simpler area are simple:

1. If the pressure drop is no more than 10 percent of the inlet pressure and the specific volume is based on the known upstream or downstream conditions, the accuracy of the incompressible flow equations is quite acceptable.
2. If the pressure drop remains between 10 and 40 percent, the methods have reasonable accuracy if one uses the average of upstream and downstream specific volumes.
3. If the velocity approaches the sonic velocity, there comes a time when the velocity or mass flow changes no more; this is at the sonic velocity of the pressure wave in the fluid. In compressible flow terminology a pressure drop greater than the pressure drop that produces this condition creates what is known as unretarded flow. This is based on the fact that it can increase no more. Conversely, flow at pressures below this is retarded flow in the sense that it has not reached the maximum.

The condition is more readily determined by calculating the critical downstream pressure. Lowering the pressure below that occurring downstream achieves no increase in flow. As downstream pressure is increased above that pressure the flow decreases.

This is based on the weight of flow per unit of time being at its maximum at the critical pressure. This is calculable using the relationship that deals with the exponent n that is based on specific heats of gauges at constant pressure and volume, all of which revolve around the gas laws of Boyle and Charles. It varies with various gases, but it is common to set that relationship equal to 1.4. If one uses 1.4, the critical pressure can be expressed as

$$p_{critical} = p_1 \left(\frac{2}{n+1} \right)^{\frac{n}{(n-1)}} = p_1 \left(\frac{2}{1.4+1} \right)^{\frac{1.4}{(1.4-1)}} = p_1(0.53)$$

Whenever the downstream pressure is less than 0.53 upstream, it is unretarded flow. Whenever it is higher than that, it is retarded flow.

For those who work with steam there are some deviations and empirical formulas that may be useful. First, dry-saturated steam has an n of 1.135, and superheated steam has an exponent of 1.30. The 1.4 just noted is for a diatomic gas. Other gases have exponents and some are listed in the heat capacities tables in the Appendix. It should be noted in the flow discussions that the term G is flow in lbs/sec.

Some empirical formulas for dry-saturated steam are:

1. For retarded flow, developed by Rankine:

$$G = 0.0292a_t\sqrt{p_2(p_1 - p_2)}$$

2. For unretarded flow, developed by Rankine:

$$G = \frac{a_t p_1}{70}$$

3. For unretarded flow, developed by Grashof:

$$G = \frac{a_t p_1^{0.97}}{60}$$

It is interesting to note that the difference in results is very small and probably within experimental error.

It is stated that the formulas are workable in either inch or foot units; however, if one changes the Rankine divisor to 700, the answer comes out reasonably close in kg to the answer in lbs, both per second. However, the same increase of a factor of 10 in. in the Grashof equation does not yield an answer nearly as close. Neither of the metric conversions should be taken as gospel since the original data were not available and the test was on only one set of area pressure parameters.

Note that the formulas were on dry-saturated steam. There are conversions to superheated steam and wet-saturated steam. If one wants to convert to superheated steam, divide the result for dry-saturated steam by the quantity $(1 + 0.00065\Delta t)$ where the Δt is the degrees of superheat. To convert to wet-saturated steam, divide by $\sqrt{\text{dryness fraction}}$, which is the percentage of steam by weight of the total weight of the mixture.

4. The equation for air or other gas can be expressed in USC as

$$C_o = \frac{q\left(SG\sqrt{T + 460}\right)}{660p_1}$$

where
   q is the free gas, ft^3/hr
   SG is the specific gravity with respect to air, at 14.7 psi and 60°F

T is the flowing temperature, °F

p_1 is the inlet pressure, psia

This particular expression does not hold beyond that critical pressure.

As one might expect, there are vastly more complex equations to calculate the flow through the conduit. The conduit can be a valve orifice and so forth. These equations may become necessary in the pursuit of some answers. They in fact look somewhat more imposing than they really are, especially if one takes the time to set them up in a spreadsheet environment, which is relatively easy these days. I have done this on a relatively simple Texas Instruments 30x calculator from time to time.

For that reason the more fundamental flow equation in terms of G is

$$G = a_t \sqrt{\frac{2gk}{k-1}\left(\frac{p_1}{v_1}\right)\left(\left[\frac{p_2}{p_1}\right]^{\frac{2}{k}} + \left[\frac{p_2}{p_1}\right]^{\frac{(k+1)}{k}}\right)}$$

The terms are in USC units, where units must be in feet:

G is the mass flow, lbs/sec

a_t is the area at the throat, ft^2, which is somewhat enigmatic, but related to the *vena contracta*, which is probably not in the actual physical location of the opening, be it an orifice, valve, or whatever

g is the gravitational constant, ft/sec^2

1 and 2 represent upstream and downstream, respectively

p is the pressure, psfa

k is the constant previously labeled n. It is not mathematically exactly the same k as the actual ratio of the specific heats

$$\frac{C_p}{C}$$

which is the assumption of a perfect gas, unless one is working with pipeline metering or some other process where minute errors in flow are economically unacceptable.

v is the specific volume, ft^3/lb

Note that while temperature is not in the equation one needs to know it to get the accurate specific volume.

The previous formula goes wild at the critical pressure, essentially trying to take the square root of a negative number. While that may work in imaginary math, it doesn't work in the real world. The critical pressure forces one to classify the equations, and in the case of retarded flow of diatomic gases ($k = 1.4$), the equation reduces to the following:

1. For retarded flow:

$$G = \frac{15.03a_tp_1}{\sqrt{RT}} \sqrt{\left(\frac{p_2}{p_1}\right)^{1.43} - \left(\frac{p_2}{p_1}\right)^{1.71}}$$

Note that the derivation of the 15.03 and the square root of RT relates to the gas laws where $PV = RT$. The same is true for k's other than 1.4 where the numerical exponents change. The constants in the unretarded flow constants have the same deeper meaning. Both are left as an exercise for the ambitious. The 1.4 formulas are simple shortcuts for most gases. Note that steam is not necessarily diatomic.

2. For unretarded flow:

$$G = \frac{0.532a_tp_1}{\sqrt{T_1}}$$

where

R is the gas constant for the particular gas. As calculated for this constant, it is a function of the universal gas constant divided by the molecular weight of the gas. That gas constant is 1545.35. If one takes the molecular weight of oxygen, which is 32, one gets an oxygen gas constant of 48.3.

T is the absolute temperature, or 460 plus the thermometer temperature, °F. The simpler versions are less mathematically complex, and for many gases in the retarded zone, quite acceptable.

One thing readers will note as they work through the variety of formulas is that they do not always give the same precise answer. The business of valves is still as much art as it is science, since many of the approaches are experimentally developed. The flow results are empirical and small production differences may make large differences in results.

OTHER VALVE ISSUES

We have not talked about the actuation of the valves. This is again one of the experimental sciences. There are several types of actuators from manual to highly sophisticated electrical controls. The means of actuation can be air cylinders, hydraulic cylinders, and/or motor-operated gear drive actuators.

The speed of actuation in some instances is very important; as was discussed in Chapter 12 on water hammer, it can cause problems in the entire system. Yet there are times when it is extremely important to close the valve quickly for safety reasons, or, as in the case of safety relief valves, to open the valve quickly.

All of these issues involve the torque or force to accomplish the goal, which is a function inherent in the design. It involves the flowing medium and its relative viscosity as well as the velocity. It also is a function of the type of valve and it orientation.

For instance, It is obvious that opening a valve against the flow is more difficult than opening it with the flow. We all experience that sort of thing when opening a door into the wind as opposed to opening it with the wind. It should also be obvious that valves of different types would require different torques to cause the stem to rotate so that the valve could be opened.

Again it is much more worthy to use the experimental results of the particular valve manufacturer than to use the generic analytical methods we present in this book. As a note, those who manufacture actuators are a source of good information. They are the experts as to both the amount of torque and the torque curves of their product.

Note that when the mode of actuation is extremely rapid it also involves the means of stopping the device and absorbing the shock of the sudden start and stop, which involves impact. In point of fact it is often more difficult to have a means of sudden deceleration to stop the rapid action than to get it started. In short, the issue is too complex to go into it in any more detail, as it is a field unto itself.

So far we have discussed two and to a certain extent three of the major actions of valves. The two are on–off and throttling or control. Check valves have been alluded to in the on–off discussion, because they are essentially on–off. The off action is when the flow reverses and should not be allowed to go past the checkpoint. The on action is when the flow is going in the correct direction. Check valves have the most important function of preventing back flow where such flow could cause severe damage. One might recall that they are often involved in sudden closure situations and can cause damage from such transients as water hammer

Another type of valve is the pressure-reducing valve, which hasn't been discussed in any detail. It has been pointed out that the pressure drops through a valve; in any case, the ΔP to 0 valve has yet to be developed. But the pressure-reducing valves of the blow down or extreme drop in pressure are highly specialized. As was noted in discussions about compressible flow, there is a limit to the amount of pressure drop one can achieve in any one step.

For instance, the essential pressure from an extremely high pressure, say 1000 psi or the SI equivalent, can only achieve a drop of 530 in one step and then it is at its maximum. The art in pressure reduction is to develop some multistage drops of pressure through the valve. This often takes the form of large rings set in layers that have multiple paths for the pressure to dissipate. That is a very specialized design.

Safety relief valves are a subset of pressure-reducing valves. They are subject to the same critical drop phenomenon. That is a factor in the relief capacity of the individual valve, and is one of the reasons that for large systems the way the relief capacity for the entire line is achieved is through the use of multiple valves.

Only highly specialized valves are left, some of which are extremely high-temperature valves. One variety is called slide valves, which are used in fluid catalytic cracking (FCC) units in refineries. These are essentially sliding-gate valves, although there is at least one manufacturer that makes them in another form. They handle hot gases in the 1400°F (760°C) range. The gases are full of catalyst particles that could be likened to sand flowing through the valve. The issue in the design and use of such valves is the clearances of the moving parts. This is an issue in any high-temperature valve. The abrasive protection and/or the materials of construction, as well as the heat transfer strength of the material, are certainly calculable, but beyond the scope of this book.

On the other end of the temperature spectrum are cryogenic valves, which operate in the negative range, say −400°F (−240°C). The issues are the same—materials, clearances, and brittleness in this case. Again, these are all calculable, but beyond the scope of this book.

One might conclude that there is a dearth of books on valves. This might be true. In a personal search for detailed books on valve design this author has found that one of the best English books on the subject was written in 1960s and is out of print. There is a book in an Asian language but this old man can't read the kanji. The vast majority of books available seem to be on the business of selecting the valve for the application. This is where most of the readers of this book would probably fall. Valve designers are grown more than they are taught.

One can spend much time on the subjects discussed in this book and maybe should; however, the intent is to expose readers to the types of calculations and some of the variability of those calculations to give them a sense for the general field.

There are details in the charts and information in the Appendix regarding all the topics discussed, and readers are urged to get familiar with these. With so many different disciplines being referenced, it is hoped that this will narrow the search. The Bibliography should also be helpful for those seeking more information.

This appendix should be considered a reference source. It contains a selection of charts, graphs, and other information about certain attributes that can be found in several sources, including helpful conversion charts. What is here was chosen for inclusion because this author has had occasion to use them often, so having them in just one place is much easier than carrying around several reference books. The reader is reminded that these are used for field-type reference as opposed to a more precise source; they should not be considered as replacements for other technical resources currently available.

Many of the charts have both U.S. customary units and SI units. In some cases the units are side by side, so one who is familiar with one set of units but is working in the other has a handy reference. I have found that it is useful to have the value in the customary units for comparison. In other cases, it is helpful to have separate charts for different units.

A rigorous attempt has been made to group the charts and figures into categories that have affinity. For instance, the charts that give the estimated weight of fittings are grouped together so that one can find related fittings in the same area. It is expected that this additional information will reduce the readers' efforts to find the necessary ancillary information, rather than forcing them to consult specialty references from specific sources and thus having to carry those books in the field. The major categories the charts are grouped into are the following

- Basic properties of piping and piping components (e.g., sizes, weights)
- Basic properties of fluids
- Dynamic properties of systems
- Pipe hangers
- Pressure area and other burst data
- Conversion factors

Note also that the Web, especially *www.EngineeringToolBox.com*, is a helpful and general source of data, as well as being useful for crosschecking the data and other related research in this book.

CONTENTS OF APPENDIX

NPS	DN	SCh	OD (in)	OD (mm)	Wall (in)	Wall (mm)	ID (in)	ID (mm)	Inside area (in²)	Inside area (mm²)	Metal area (in²)	Metal area (mm²)	Moment of inertia (in⁴)	Moment of inertia (cm⁴)	Section mod (in³)	Section mod (M³)
1/8	6	40/std	0.405	10.3	0.068	1.727	0.269	6.846	0.05683233	36.8099041	0.07199291	46.5131819	0.0010639	0.04428299	0.00525384	8.6095E-08
		80/xs	0.405	10.3	0.095	2.413	0.215	5.474	0.03630512	23.5342565	0.09252012	59.7888295	0.00121608	0.05061714	0.00600534	9.841E-08
1/4	8	40/std	0.54	13.7	0.088	2.235	0.364	9.23	0.10406236	66.9105037	0.12496028	80.5012223	0.00331304	0.13789916	0.01227052	2.0108E-07
		80/xs	0.54	13.7	0.119	3.023	0.302	7.654	0.07163162	46.0116505	0.15739102	101.400075	0.00376658	0.15677682	0.01395029	2.286E-07
3/8	10	40/std	0.675	17.145	0.091	2.311	0.493	12.523	0.19089068	123.17077	0.16695719	107.698365	0.0072924	0.30353246	0.0216071	3.5408E-07
		801/xs	0.675	17.145	0.126	3.2	0.423	10.745	0.14053084	90.6783766	0.21731704	140.190758	0.00862091	0.35882922	0.02554343	4.1858E-07
1/2	15	40/std	0.84	21.336	0.109	3.734	0.622	13.868	0.30385869	151.049246	0.25031955	206.484387	0.01709623	0.71159893	0.04070532	6.6704E-07
		160	0.84	21.336	0.188	4.75	0.464	11.836	0.16909348	110.02739	0.38508476	247.506244	0.02216958	0.92276764	0.0528472	8.6499E-07
		xxs	0.84	21.336	0.294	7.468	0.252	6.4	0.04987604	32.169984	0.5043022	325.363649	0.02424748	1.00925606	0.05773208	9.4606E-07
3/4	20	40/std	1.05	26.67	0.113	2.87	0.824	20.93	0.53326775	344.056172	0.33263575	214.59013	0.03704584	1.54196435	0.07056351	1.1563E-06
		80/xs	1.05	26.67	0.154	3.912	0.742	18.846	0.43241297	278.951866	0.43349053	279.694436	0.04479813	1.86463882	0.08532977	1.3983E-06
		160	1.05	26.67	0.219	5.537	0.612	15.596	0.29416686	191.036939	0.57173664	367.609363	0.05279345	2.19742922	0.10055895	1.6479E-06
		xxs	1.05	26.67	0.308	7.823	0.434	11.024	0.1479348	95.4485436	0.7179467	463.197758	0.05793939	2.41161926	0.11036074	1.8085E-06
1	25	40/std	1.315	33.401	00.133	3.378	1.049	26.645	0.86425495	557.599462	0.49387837	318.613827	0.08736543	3.63642369	0.13287518	2.1774E-06
		80/xs	1.315	33.401	0.179	4.547	0.957	24.307	0.7193078	464.038078	0.63882551	412.175212	0.10563569	4.3968891	0.16066265	2.6328E-06
		160	1.315	33.401	0.25	6.35	0.815	20.701	0.52168232	336.568562	0.836451	539.644727	0.12515709	5.20943111	0.19035299	3.1193E-06
		xxs	1.315	33.401	0.356	9.093	0.603	15.215	0.28557851	181.817135	1.07255481	694.396154	0.14032817	5.84089919	0.21342688	3.4974E-06
1 1/4	32	40/std	1.66	42.164	0.14	3.556	1.38	35.052	1.49571576	964.97598	0.66853248	431.310415	0.19475967	8.10650889	0.2346502	3.8452E-06
		80/xs	1.66	42.164	0.191	4.851	1.278	32.462	1.28278125	827.639946	0.88146699	568.646448	0.24185259	10.0666644	0.29138867	4.775E-06
		160	1.66	42.164	0.25	4.851	1.16	32.462	1.05683424	827.639946	1.107414	568.646448	0.28393018	11.8180655	0.34208455	5.6058E-06
		xxs	1.66	42.164	0.358	9.703	0.944	22.758	0.69989882	406.779923	1.46435001	989.506871	0.33384111	13.8955154	0.40221821	6.5912E-06
1 1/2	40	5S	1.9	48.26	0.065	2.108	1.77	44.044	2.46057966	1523.57699	0.37471434	305.641288	0.15795655	6.57464741	0.16627005	2.7247E-06
		10S	1.9	48.26	0.109	3.048	1.682	42.164	2.22193399	1396.28639	0.61330001	432.931883	0.24688253	10.276026	0.25987635	4.2586E-06
		40/std	1.9	48.26	0.145	3.683	1.61	40.894	2.03583534	1313.43953	0.79945866	515.778749	0.30997407	12.9920943	0.3262885	5.3469E-06
		80/xs	1.9	48.26	0.2	5.08	1.5	38.1	1.76715	1140.09449	1.068144	689.123783	0.39130736	16.287441	0.41190248	6.7499E-06
		160	1.9	48.26	0.281	7.187	1.338	33.886	1.40605764	901.844186	1.42923636	927.374091	0.48251191	20.0836607	0.50790727	8.3231 E-06
		xxs	1.9	48.26	0.4	10.16	1.1	27.94	0.950334	613.117483	1.88496	1216.10079	0.5679888	23.6414773	0.59788295	9.7975E-06
			1.9	48.26		13.35		21.56		365.080309		1464.13797	0.63987611	26.6336529	0.6735538	1.1038E-05
				48.26		15.875		16.51		214.084411		1615.13387				

(Continued)

—cont'd

NPS	DN	SCh	OD (in)	OD (mm)	Wall (in)	Wall (mm)	ID (in)	ID (mm)	Inside area (in²)	Inside area (mm²)	Metal area (in²)	Metal area (mm²)	Moment of inertia (in⁴)	Moment of inertia (cm⁴)	Section mod (in³)	Section mod (M³)
2	50	10s	2.375	60.325	0.109	2.769	2.157	54.787	3.65419052	2357.46871	0.77595635	500.684847	0.49932283	20.7833839	0.42048238	6.8905E-06
		40/std	2.375	60.325	0.154	3.912	2.067	52.501	3.35561286	2164.84122	1.07453401	693.31234	0.66591817	27.7176051	0.56077319	9.1894E-06
		80/xss	2.375	60.325	0.218	5.537	1.939	49.251	2.95288487	1905.11415	1.477262	953.039408	0.86814439	36.1348953	0.73106896	1.198E-05
		160	2.375	60.325	0.343	8.712	1.689	42.901	2.24052707	1445.5254	2.1896198	1412.62816	1.16262089	48.3919323	0.97904917	1.6044E-05
		xss	2.375	60.325	0.436	11.074	1.503	38.177	1.77422567	1144.70741	2.65592121	1713.44615	1.31163426	54.5943365	1.10453411	1.81E-05
21/2	65	10s	2.875	73.025	0.12	3.048	2.635	66.929	5.45320892	3518.19226	1.03861296	670.071537	0.9875082	41.1031919	0.68696222	1.1257E-05
		40/stf	2.875	73.025	0.203	5.156	2.469	62.713	4.78776777	3088.91566	1.70405411	1099.34814	1.52994697	63.681197	1.06431094	1.7441E-05
		80/xss	2.875	73.025	0.276	7.01	2.323	59.005	4.2382768	2734.44081	2.25354508	1453.823	1.92472933	80.1132783	1.33894214	2.1941 E-05
		160	2.875	73.025	0.375	9.525	2.125	53.975	3.54657188	2288.10631	2.94525	1900.15749	2.35334766	97.953719	1.63711141	2.6827E-05
		xss	2.875	73.025	0.552	14.021	1.771	44.983	2.46336076	1589.23356	4.02846111	2599.03024	2.87152976	119.522085	1.99785592	3.2735E-05
3	80	10s	3.5	88.9	0.12	3.048	3.26	82.804	8.34691704	5385.097	1.27423296	822.084136	1.82242596	75.8550907	1.04138626	1.7065E-05
		40/STF	3.5	88.9	0.216	5.486	3.068	77.928	7.39267489	4769.55606	2.22847511	1437.62508	3.01793196	125.615805	1.72453255	2.826E-05
		80/XS	3.5	88.9	0.3	7.62	2.9	73.66	6.605214	4261.41986	3.015936	1945.76127	3.89531904	162.13541	2.22589659	3.6476E-05
		160	3.5	88.9	0.437	11.1	2.626	66.7	5.41602101	3494.15821	4.20512899	2713.02293	5.03321125	209.498057	2.87612072	4.7131 E-05
		XXS	3.5	88.9	0.6	15.24	2.3	58.42	4.154766	2680.48883	5.466384	3526.6923	5.99404944	249.491159	3.42517111	5.6128E-05
4	100	10S	4.5	114.3	0.12	3.048	4.26	108.204	14.253125	9195.54615	1.65122496	1065.3043	3.96370131	164.981695	1.76164503	2.8868E-05
		40/STD	4.5	114.3	0.237	6.02	4.026	102.26	12.7302941	8213.01231	3.17405587	2047.83814	7.23445893	301.120897	3.21351508	5.269E-05
		80/XS	4.5	114.3	0.337	8.56	3.826	97.18	11.496902	7417.28021	4.40744803	2843.57023	9.61296375	400.121736	4.27242833	7.0013E-05
		120	4.5	114.3	0.437	11.1	3.626	92.1	10.3263418	6662.08481	5.57800819	3598.76563	11.6463166	484.756265	5.17614069	8.4822E-05
		160	4.5	114.3	0.531	13.487	3.438	87.326	9.28330548	5989.3271	6.62104452	4271.52335	13.274371	552.521002	5.89972044	9.6679E-05
		XXS	4.5	114.3	0.674	19.05	3.152	76.2	7.80303068	4560.37798	8.10131932	5700.47247	15.2875898	636.317492	6.79448437	0.00011134
5	125	10S	5.563	141.3	0.134	3.404	5.295	134.492	22.0202794	14206.3922	2.28547002	1474.66071	8.42752986	350.780255	3.02985075	4.965E-05
		40/STD	5.563	141.3	0.258	6.553	5.047	128.194	20.0058729	12907.0293	4.2998765	2774.02366	15.1160796	631.259856	5.45248234	8.935E-05
		80/XS	5.563	141.3	0.375	9.525	4.813	122.25	18.1937667	11737.8521	6.1119828	3943.20084	20.6759659	860.598623	7.43338698	0.00012181
		120	5.563	141.3	0.5	12.7	4.563	115.9	16.3527891	10550.129	7.9529604	5130.92395	25.7383251	1071.30991	9.25397748	0.00015164
		160	5.563	141.3	0.625	15.875	4.313	109.55	14.6099865	9425.74444	9.695763	6255.30848	30.0335685	1250.09143	10.7976159	0.00017694
		XXS	5.563	141.3	0.75	19.05	4.063	103.2	12.9653589	8364.6985	11.3403906	7316.35443	33.6434108	1400.3444	12.09542	0.00019821

6	150	10s	6.625	168.27	0.134	3.404	6.357	161.462	31.739152	20475.3603	2.73254483	1763.07806	14.4011161	599.41967	4.34750673	7.1243E-05
		40/std	6.625	168.27	0.28	7.112	6.065	154.046	28.8903303	18637.6756	5.58136656	3600.76273	28.1494107	1171.66686	8.49793531	0.00013926
		80/xs	6.625	168.27	0.432	10.973	5.761	146.324	26.0667356	16815.974	8.40496124	5422.46437	40.5010782	1685.78205	12.2267406	0.00020036
		120	6.625	168.27	0.562	14.275	5.501	139.72	23.7669902	15332.3262	10.7047067	6906.11213	49.6233462	2065.47948	14.9806328	0.00024549
		160	6.625	168.27	0.719	18.263	5.187	131.744	21.1311627	13631.7806	13.3405342	8606.65775	59.0432217	2457.56428	17.8243688	0.00029209
		xxs	2.375	168.27	0.864	21.94	0.647	124.39	0.32877551	12152.3937	4.10137137	10086.0446	1.55359357	64.6654426	1.30828932	2.1439E-05
8	200	10s	8.625	219.08	0.148	3.759	8.329	211.562	54.4849581	35153.3101	3.94143879	2542.78477	35.4235518	1474.43946	8.21415694	0.00013461
		40/stf	8.625	219.08	0.322	8.179	7.981	202.722	50.0271219	32276.9628	8.39927495	5419.13207	72.5078693	3018.0052	16.813419	0.00027552
		80/xs	8.625	219.08	0.5	12.7	7.625	193.68	45.6636469	29461.8796	12.76275	8234.21528	105.743371	4401.37114	24.520202	0.00040181
		120	8.625	219.08	0.718	18.237	7.189	182.606	40.5908237	26189.1247	17.8355732	11506.9701	140.571572	5851.03021	32.5963065	0.00053416
		160	8.625	219.08	0.906	23.01	6.813	173.06	36.4558875	23522.5443	21.9705094	14173.5505	165.930017	6906.52832	38.4765256	0.00063052
		xxs	8.625	219.08	0.875	22.22	6.875	174.64	37.1224219	23954.0164	21.303975	13742.0785	162.026356	6744.04569	37.5713289	0.00061568
10	250	10s	10.75	273.05	0.165	4.191	10.42	264.668	85.2759046	55016.6026	5.48688294	3539.9174	76.8835674	3200.13549	14.3039195	0.0002344
		40/std	10.75	273.05	0.365	9.271	10.02	254.508	78.8544742	50873.7525	11.9083133	7682.76743	160.775548	6691.98316	29.9117299	0.00049017
		80/XS	10.75	273.05	0.5	12.7	9.75	247.65	74.6620875	48168.9924	16.1007	10387.5276	212.004594	8824.29689	39.4427151	0.00064635
		100	10.75	273.05	0.718	18.237	9.314	236.576	68.1339181	43957.4272	22.6288694	14599.0927	286.205152	11912.7571	53.2474701	0.00087257
		120	10.75	273.05	0.873	21.412	9.004	230.226	63.6739614	41629.3503	27.0888261	16927.1697	332.997164	13860.3876	61.9529607	0.00101523
		160	10.75	273.05	1.125	28.575	8.5	215.9	56.74515	36609.701	34.0176375	21946.819	399.410279	16624.71	74.3088892	0.0012177
12	300	10S	12.75	323.85	0.18	4.572	12.39	314.706	120.568403	77785.9111	7.10818416	4585.91609	140.455537	5846.20046	22.032241	0.00036104
		40/STD	12.75	323.85	0.365	9.271	12.02	305.308	113.474906	73209.4705	14.2016813	9162.35673	272.602239	11346.5611	42.7611355	0.00070073
		80/XS	12.75	323.85	0.5	11.75	12.7	298.45	108.434288	69957.4649	19.2423	12414.3623	361.636844	15052.461	56.727348	0.00092959
		120	12.75	323.85	1	25.4	10.75	273.05	90.7627875	58556.52	36.9138	23815.3072	641.829063	26714.9409	100.679069	0.00164983
		140	12.75	323.85	1.125	28.575	10.5	266.7	86.59035	55864.6302	41.0862375	26507.197	700.730842	29166.6179	109.918563	0.00180124
		160	12.75	323.85	1.312	33.325	10.126	257.2	80.531677	51955.6551	47.1449105	30416.1721	781.326362	32521.2566	122.560998	0.00200841
14	350	10	14	355.6	0.25	6.35	13.5	342.9	143.13915	92347.654	10.79925	6967.24413	255.366031	10629.1361	36.4808616	0.00059781
		STD	14	355.6	0.375	9.525	13.25	336.55	137.886788	88959.0398	16.0516125	10355.8583	372.856002	15519.4376	53.2651431	0.00087286
		40	14	355.6	0.437	11.1	13.126	333.4	135.318039	87301.5768	18.6203606	12013.3213	428.717115	17844.5525	61.1245021	0.00100363
		XS	14	355.6	0.5	12.7	13	330.2	132.7326	85633.7642	21.2058	13681.1339	483.8805	20140.6258	69.1257857	0.00113277
		160	14	355.6	1.406	35.712	11.188	284.176	98.3095736	63425.7616	55.6288264	35889.1365	1116.93341	46490.2757	159.561916	0.00261475

Pipe Size	OD	5S	10S	20	30	STD	40.000	60	XS	80.000	100	120	140	160	XXS
NPS 1/8	0.405		0.049			0.068	0.068		0.095	0.095					
DN 3	10.3		1.245			1.727	1.727		2.413	2.413					
P/S ratio			0.23791			0.34789526	0.34789526		0.52847009	0.52847009					
weight kg/m			0.277			0.364	0.364		0.468	0.468					
volume cm3/m			47.9			36.800	36.800		23.500	23.500					
weight lbs/ft			0.19			0.250	0.250		0.321	0.321					
volume in3/ft			0.888			0.682	0.682		0.436	0.436					
NPS 1/4	0.54		0.065			0.088	0.088		0.119	0.119					
DN 6	13.7		1.651			2.235	2.235		3.023	3.023					
P/S ratio			0.23654			0.33573177	0.33573177		0.48718747	0.48718747					
weight kg/m			0.489			0.630	0.630		0.794	0.794					
volume cm3/m			84.6			66.900	66.900		46.000	46.000					
weight lbs/ft			0.335			0.432	0.432		0.545	0.545					
volume in3/ft			1.568			1.240	1.240		0.853	0.853					
NPS 3/8	0.675		0.065			0.091	0.091		0.126	0.126					
DN 10	17.145		1.651			2.311	2.311		3.200	3.200					
P/S ratio			0.18455			0.26909054	0.26909054		0.39551478	0.39551478					
weight kg/m			0.629			0.843	0.843		1.098	1.098					
volume cm3/m			150.5			123.200	123.200		90.700	90.700					
weight lbs/ft			0.431			0.578	0.578		0.753	0.753					
volume in3/ft			2.790			2.284	2.284		1.681	1.681					
NPS 1/2	0.84	0.065	0.083			0.109	0.109		0.147	0.147				0.188	0.294
DN 15	21.336	1.651	2.108			2.769	2.769		3.734	3.734				4.75	7.468
P/S ratio		0.071386059	0.09209			0.12277351	0.12277351		0.16938321	0.16938321				0.22224	0.373428
weight kg/m		0.799	0.997			1.265	1.265		1.617	1.617				1.938	2.247
volume cm3/m		255.4	230.2			196.000	196.000		151.000	151.000				110	32.2
weight lbs/ft		0.548	0.684			0.868	0.868		1.109	1.109				1.329	1.541
volume in3/ft		4.735	4.268			3.634	3.634		2.799	2.799				2.039	0.597
NPS 3/4	1.05	0.065	0.083			0.113	0.113		0.155	0.154				0.219	0.308
DN 20	26.67	1.651	2.108			2.870	2.870		3.910	3.910				5.563	7.823
P/S ratio		0.11462908	0.14889			0.20866553	0.20866553		0.29975465	0.2966107				0.45413	0.720176
weight kg/m		1.016	1.273			1.680	1.680		2.190	2.190				2.878	3.626
volume cm3/m		428.9	396			344.100	344.000		279.000	279.000				191	95.4
weight lbs/ft		0.697	0.873			1.152	1.152		1.502	1.502				1.974	2.487
volume in3/ft		7.951	7.341			6.379	6.377		5.172	5.172				3.541	1.769

NPS 1	1.315								
DN 25	33.401								
P/S ratio		0.090473984	0.15672	0.19479446	0.19479446	0.2720882	0.2720882	0.40451	0.647077
weight kg/m		1.289	2.086	2.494	2.949	3.227	3.227	4.225	5.436
volume cm3/m		711.5	609.7	557.600	557.600	464.000	464.000	336.6	181.8
weight lbs/ft		0.884	1.431	1.711	2.023	2.213	2.213	2.898	3.729
volume in3/ft		13.190	11.303	10.337	10.337	8.602	8.602	6.240	3.370
NPS 1 1/2	1.9	0.065	0.109	0.145	0.145	0.200	0.200	0.281	0.4
DN40	48.26	1.651	2.769	3.683	3.683	5.080	5.080	7.137	10.16
P/S ratio		0.061735436	0.1058	0.14335391	0.14335391	0.20359896	0.20359896	0.29951	0.459532
weight kg/m		1.893	3.098	4.038	4.038	5.395	5.395	7.219	9.521
volume cm3/m		1587.5	1433.5	1313.400	1313.400	1140.100	1140.100	907.2	613.1
weight lbs/ft		1.298	2.125	2.770	2.770	3.701	3.701	4.952	6.531
volume in3/ft		29.430	26.575	24.349	24.349	21.136	21.136	16.818	11.366
NPS 2	2.375	0.065	0.109	0.154	0.154	0.218	0.218	0.344	0.436
DN 50	60.325	1.651	2.769	3.912	3.912	5.537	5.537	8.738	11.05
P/S ratio		0.04908	0.08372	0.12044447	0.12044447	0.17510555	0.17510555	0.29232	0.387522
weight kg/m		2.383	3.92	5.428	5.428	7.461	7.461	11.059	13.415
volume cm3/m		2553.8	2357.5	2164.800	2164.800	1905.100	1905.100	1445.5	1144.7
weight lbs/ft		1.635	2.689	3.723	3.723	5.118	5.118	7.586	9.202
volume in3/ft		47.344	43.705	40.132	40.132	35.318	35.318	26.798	21.221
NPS 2 1/2	2.875	0.083	0.12	0.203	0.203	0.276	0.276	0.375	0.532
DN65	73.025	2.108	3.048	5.156	5.156	7.010	7.010	9.525	14.021
P/S ratio		0.05184	0.07585	0.13189	0.13189	0.18392	0.18392	0.25911	0.39130
weight kg/m		3.677	5.246	8.607	8.607	11.382	11.382	14.876	20.348
volume cm3/m		3718.6	3518.2	3088.200	3088.200	3734.400	3734.200	2288.1	1589.2
weight lbs/ft		2.522	3.598	5.904	5.904	7.807	7.807	10.204	13.957
volume in3/ft		68.938	65.223	57.251	57.251	69.231	69.227	42.418	29.462
NPS 3	3.5	0.083	0.12	0.216	0.216	0.300	0.300	0.438	0.6
DN 80	88.9	2.108	3.048	5.486	5.486	7.620	7.620	11.1	15.24
P/S ratio		0.04239	0.06188	0.11429	0.11429	0.16252	0.16252	0.24718	0.35667
weight kg/m		4.5	6.436	11.255	11.255	15.233	15.233	21.24	27.61
volume cm3/m		5632.4	5385.1	4769.600	4769.600	4261.400	4261.400	3494.2	2680.5
weight lbs/ft		3.087	4.415	7.720	7.720	10.449	10.449	14.569	18.938
volume in3/ft		104.417	99.832	88.422	88.422	79.000	79.000	64.778	49.693

(Continued)

—cont'd

Pipe Size	OD	5S	10S	20	30	STD	40.000	60	XS	80.000	100	120	140	160	XXS
NPS 4	4.5	0.083	0.12			0.237	0.237		0.337	0.337		0.438		0.531	0.674
DN 100	114.3	2.108	3.048			6.020	6.020		9.525	9.525		12.7		15.875	19.05
P/S ratio		0.03281	0.04779			0.09669	0.09669		0.14048	0.14048		0.18673		0.23130	0.30396
weight kg/m		5.817	8.34			16.033	16.033		22.262	22.262		28.175		33.442	40.92
volume cm3/m		9517.9	9195.5			8213.000	8213.000		7417.300	7417.300		6662.1		5989.3	5034.1
weight lbs/ft		3.990	5.721			10.997	10.997		15.270	15.270		19.326		22.939	28.068
volume in3/ft		176.449	170.472			152.258	152.258		137.507	137.507		123.506		111.033	93.325
NPS 6	6.625	0.109	0.134			0.280	0.280		0.432	0.432		0.562		0.719	0.864
DN150	168.275	2.769	3.404			7.112	7.112		10.973	10.973		14.275		18.237	21.946
P/S ratio		0.02922	0.03604			0.07684	0.07684		0.12117	0.12117		0.16070		0.21063	0.25906
weight kg/m		11.272	13.804			28.191	28.191		42.454	42.454		54.07		67.3	78.985
volume cm3/m		20800	20477			18639	18639		16817	16817		15333		13644	12151
weight lbs/ft		7.732	9.468			19.337	19.337		29.120	29.120		37.088		46.162	54.177
volume in3/ft		385.603	379.615			345.541	345.541		311.764	311.764		284.253		252.941	225.263
NPS 8	8.625	0.109	0.148	0.25	0.277	0.322	0.322	0.406	0.500	0.500	0.594	0.719	0.812	0.906	0.875
DN 200	219.075	2.769	3.759	6.35	7.036	8.179	8.179	10.312	12.700	12.700	15.062	18.237	20.625	23.012	22.25
P/S ratio		0.07138	0.09820	0.17197	0.19245	0.22753	0.22753	0.29651	0.37980	0.37980	0.47065	0.60588	0.71986	0.85001	0.80519
weight kg/m		14.732	19.907	33.224	36.694	42.425	42.425	52.949	64.464	64.464	75.578	90.086	100.671	110.97	107.771
volume cm3/m		35813	35152	33451	33007	32275	32275	30931	29460	29460	28041	26188	24836	23520	13758.1
weight lbs/ft		10.105	13.655	22.789	25.169	29.100	29.100	36.319	44.217	44.217	51.841	61.792	69.052	76.117	73.922
volume in3/ft		663.924	651.670	620.135	611.904	598.334	598.334	573.418	546.148	546.148	519.841	485.489	460.425	436.028	255.056
NPS 10	10.75	0.134	0.155	0.25	0.307	0.365	0.365	0.5	0.500	0.594	0.719	0.844	1	1.125	1.125
DN 250	273.050	3.404	3.937	6.350	7.798	9.271	9.271	12.700	12.700	15.088	18.263	21.438	25.400	28.575	25.400
P/S ratio		0.02206	0.02556	0.04155	0.05127	0.06126	0.06126	0.08490	0.08490	0.10170	0.12448	0.14780	0.17768	0.20229	0.17768
weight kg/m		22.57	26.06	41.65	50.87	60.15	60.15	81.32	81.32	95.73	114.45	132.67	154.71	171.82	154.71
volume cm3/m		58538	58532	58493	58461	58422	58422	58303	58303	58199	58033	57835	57543	57274	57543
weight lbs/ft		15.483	17.874	28.571	34.895	41.257	41.257	55.782	55.782	65.661	78.500	90.999	106.121	117.856	106.121
volume in3/ft		1085.2	1085.1	1084.4	1083.8	1083.1	1083.1	1080.9	1080.9	1078.9	1075.8	1072.2	1066.8	1061.8	1066.8
NPS 12	12.75	0.156	0.18	0.25	0.33	0.375	0.406	0.562	0.500	0.688	0.844	1	1.125	1.312	1
DN 300	323.85	3.962	4.572	6.350	8.382	9.525	10.312	14.275	12.700	17.475	21.438	25.400	28.575	33.325	25.400
P/S ratio		0.02164	0.02502	0.03492	0.04635	0.05284	0.05734	0.08027	0.07110	0.09919	0.12312	0.14764	0.16772	0.19855	0.14764
weight kg/m		31.18	35.90	49.59	65.04	73.64	79.53	108.69	97.19	131.68	159.45	186.45	207.52	238.13	186.45
volume cm3/m		82347	82339	82308	82261	82229	82205	82052	82118	81892	81650	81358	81089	80627	81358
weight lbs/ft		21.384	24.627	34.013	44.610	50.510	54.548	74.553	66.666	90.325	109.372	127.890	142.345	163.336	127.890
volume in3/ft		1526.6	1526.5	1525.9	1525.0	1524.4	1524.0	1521.1	1522.4	1518.2	1513.7	1508.3	1503.3	1494.7	1508.3

NPS 14	14	0.156	0.188	0.312	0.375	0.375	0.438	0.594	0.500	0.750	0.938	1.094	1.25	1.406
DN 350	355.6	3.962	4.775	7.925	9.525	9.525	11.125	15.088	12.700	19.050	23.825	27.788	31.750	35.712
P/S ratio		0.01969	0.02378	0.03978	0.04801	0.04801	0.05631	0.07715	0.06454	0.09844	0.12471	0.14705	0.16990	0.19328
weight kg/m		34.27	41.20	67.77	81.08	81.08	94.26	126.36	107.11	157.69	194.42	224.04	252.90	280.98
volume cm3/m		99290	99279	99216	99172	99172	99120	98957	99062	98745	98423	98102	97731	97312
weight lbs/ft		23.506	28.263	46.483	55.612	55.612	64.654	86.673	73.469	108.162	133.355	153.676	173.467	192.729
volume in3/ft		1840.7	1840.5	1839.3	1838.5	1838.5	1837.6	1834.5	1836.5	1830.6	1824.6	1818.7	1811.8	1804.0
NPS 16	16	0.165	0.188	0.312	0.375	0.375	0.500	0.656	0.500	0.844	1.031	1.219	1.438	1.594
DN 400	406.4	4.191	4.775	7.925	9.525	9.525	12.700	16.662	12.700	21.438	26.187	30.963	36.525	40.488
P/S ratio		0.01821	0.02078	0.03472	0.04188	0.04188	0.05624	0.07445	0.05624	0.09686	0.11965	0.14309	0.17112	0.19158
weight kg/m		41.46	47.17	77.67	92.98	92.98	122.98	159.72	122.98	202.98	244.89	285.91	332.28	364.38
volume cm3/m		129690	129682	129619	129575	129575	129464	129281	129464	128996	128640	128212	127622	127142
weight lbs/ft		28.438	32.355	53.275	63.775	63.775	84.353	109.557	84.353	139.227	167.977	196.113	227.918	249.936
volume in3/ft		2404.3	2404.1	2403.0	2402.1	2402.1	2400.1	2396.7	2400.1	2391.4	2384.8	2376.9	2365.9	2357.0
NPS 18	18	0.165	0.188	0.312	0.375	0.438	0.562	0.750	0.500	0.938	1.156	1.375	1.562	1.781
DN 450	457.2	4.191	4.775	7.925	9.525	11.125	14.275	19.050	12.700	23.825	29.362	34.925	39.675	45.237
P/S ratio		0.01617	0.01845	0.03080	0.03714	0.04352	0.05619	0.07571	0.04983	0.09562	0.11922	0.14350	0.16471	0.19014
weight kg/m		46.70	53.14	87.57	104.88	122.06	155.51	205.29	138.85	253.96	308.98	362.73	407.43	458.37
volume cm3/m		164146	164138	164075	164031	163979	163854	163604	163920	163282	162819	162258	161701	160959
weight lbs/ft		32.030	36.448	60.066	71.938	83.723	106.667	140.815	95.237	174.193	211.934	248.807	279.465	314.402
volume in3/ft		3043.0	3042.9	3041.7	3040.9	3039.9	3037.6	3033.0	3038.9	3027.0	3018.4	3008.0	2997.7	2984.0
NPS 20	20	0.188	0.218	0.375	0.375	0.500	0.594	0.812	0.500	1.031	1.281	1.5	1.75	1.969
DN 500	508	4.775	5.537	9.525	9.525	12.700	15.088	20.625	12.700	26.187	32.537	38.100	44.450	50.013
P/S ratio		0.01659	0.01926	0.03336	0.03336	0.04474	0.05337	0.07370	0.04474	0.09454	0.11888	0.14070	0.16620	0.18909
weight kg/m		59.10	68.43	116.78	116.78	154.71	182.91	247.23	154.71	310.33	380.50	440.34	506.79	563.36
volume cm3/m		202648	202635	202541	202541	202430	202326	202015	202430	201606	201020	200403	199580	198754
weight lbs/ft		40.540	46.938	80.101	80.101	106.121	125.464	169.584	106.121	212.863	260.993	302.037	347.615	386.423
volume in3/ft		3756.8	3756.6	3754.8	3754.8	3752.8	3750.8	3745.1	3752.8	3737.5	3726.6	3715.2	3699.9	3684.6
NPS 24	24	0.218	0.25	0.375	0.375	0.562	0.688	0.969	0.500	1.219	1.531	1.812	2.062	2.344
DN 600	609.6	5.537	6.350	9.525	9.525	14.275	17.475	24.613	12.700	30.963	38.887	46.025	52.375	59.538
P/S ratio		0.01602	0.01840	0.02772	0.02772	0.04184	0.05147	0.07328	0.03714	0.09309	0.11837	0.14171	0.16294	0.18743
weight kg/m		82.27	94.22	140.58	140.58	209.02	254.50	354.13	186.45	440.66	545.86	637.97	717.81	805.49
volume cm3/m		291816	291801	291722	291722	291544	291384	290913	291611	290358	289489	288537	287555	286296
weight lbs/ft		56.429	64.625	96.428	96.428	143.369	174.568	242.904	127.890	302.256	374.418	437.597	492.361	552.502
volume in3/ft		5409.9	5409.6	5408.1	5408.1	5404.8	5401.9	5393.1	5406.1	5382.8	5366.7	5349.1	5330.9	5307.5

(Continued)

—cont'd

Pipe Size	OD	5S	10S	20	30	STD	40.000	60	XS	80.000	100	120	140	160	XXS
NPS 30	30	0.25	0.312	0.5	0.625	0.375			0.500						
DN 750	762	6.350	7.925	12.700	15.875	9.525			12.700						
P/S ratio		0.01469	0.01837	0.02960	0.03714	0.02212			0.02960						
weight kg/m		118.02	146.98	234.05	291.33	176.28			234.05						
volume cm3/m		455974	455939	455784	455642	455895			455784						
weight lbs/ft		80.951	100.817	160.542	199.828	120.917			160.542						
volume in3/ft		8453.1	8452.5	8449.6	8447.0	8451.7			8449.6						
NPS 34	34		0.312	0.500	0.625	0.375			0.500						
DN 850	863.6		7.925	12.700	15.875	9.525			12.700						
P/S ratio			0.01619	0.02607	0.03270	0.01949			0.02607						
weight kg/m			166.78	265.79	331.00	200.09			265.79						
volume cm3/m			585657	585502	585359	585613			585502						
weight lbs/ft			114.400	182.311	227.038	137.243			182.311						
volume in3/ft			10857.3	10854.4	10851.8	10856.5			10854.4						
NPS 36	36		0.312	0.5	0.625	0.375			0.500						
DN 900	914.4		7.925	12.700	15.875	9.525			12.700						
P/S ratio			0.01528	0.02461	0.03085	0.01840			0.02461						
weight kg/m			176.69	281.66	350.83	211.99			281.66						
volume cm3/m			656596	656441	656299	656552			656441						
weight lbs/ft			121.192	193.195	240.644	145.407			193.195						
volume in3/ft			12172.4	12169.5	12166.9	12171.6			12169.5						
NPS 42	42					0.375			0.500						
DN1050	1066.8					9.525			12.700						
P/S ratio						0.01575			0.02105						
weight kg/m						247.69			329.26						
volume cm3/m						893692			893581						
weight lbs/ft						169.896			225.848						
volume in3/ft						16567.8			16565.8						
NPS 48	48					0.375			0.500						
DN 1200	1219.2					9.525			12.700						
P/S ratio						0.01377			0.01840						
weight kg/m						283.39			376.87						
volume cm3/m						1167314			1167203						
weight lbs/ft						194.386			258.501						
volume in3/ft						21640.4			21638.3						

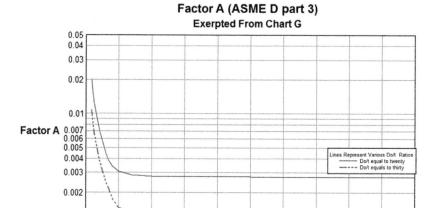

B Factor from ASME Charts
For Carbon Steel Yield of 24000 to 30000

Carbon Steel • Carbon Moly Steel • Low−Chrome Steel (Thru 3% Cr) (Ref 1) mm/meter	Temp, °C	Temp, °F	Carbon Steel • Carbon Moly Steel • Low−Chrome Steel (Thru 3% Cr) (Ref 1) Inches/Foot
	−198.33	−325	
	−195.56	−320	
	−190.00	−310	
	−184.44	−300	
−1.9166	−178.89	−290	−0.0230
−1.8749	−173.33	−280	−0.0225
−1.8333	−167.78	−270	−0.0220
−1.7916	−162.22	−260	−0.0215
−1.7416	−156.67	−250	−0.0209
−1.6916	−151.11	−240	−0.0203
−1.6499	−145.56	−230	−0.0198
−1.5999	−140.00	−220	−0.0192
−1.5583	−134.44	−210	−0.0187
−1.4999	−128.89	−200	−0.0180
−1.4249	−123.33	−190	−0.0171
−1.4083	−117.78	−180	−0.0169
−1.3583	−112.22	−170	−0.0163
−1.3166	−106.67	−160	−0.0158
−1.2666	−101.11	−150	−0.0152
−1.2166	−95.56	−140	−0.0146
−1.1666	−90.00	−130	−0.0140
−1.1083	−84.44	−120	−0.0133
−1.0583	−78.89	−110	−0.0127
−1.0083	−73.33	−100	−0.0121
−0.9666	−67.78	−90	−0.0116
−0.9083	−62.22	−80	−0.0109
−0.8583	−56.67	−70	−0.0103
−0.8000	−51.11	−60	−0.0096
−0.7250	−45.56	−50	−0.0087
−0.6666	−40.00	−40	−0.0080
−0.6083	−34.44	−30	−0.0073
−0.5416	−28.89	−20	−0.0065
−0.4833	−23.33	−10	−0.0058
−0.4250	−17.78	0	−0.0051
−0.3667	−12.22	10	−0.0044
−0.3083	−6.67	20	−0.0037
−0.2417	−1.11	30	−0.0029
−0.1833	4.44	40	−0.0022

(Continued)

—cont'd

Carbon Steel • Carbon Moly Steel • Low−Chrome Steel (Thru 3% Cr) (Ref 1) mm/meter	Temp, °C	Temp, °F	Carbon Steel • Carbon Moly Steel • Low−Chrome Steel (Thru 3% Cr) (Ref 1) Inches/Foot
−0.1250	10.00	50	−0.0015
−0.0583	15.56	60	−0.0007
0.0000	21.11	70	0.0000
0.0667	26.67	80	0.0008
0.1250	32.22	90	0.0015
0.1917	37.78	100	0.0023
0.2500	43.33	110	0.0030
0.3167	48.89	120	0.0038
0.3833	54.44	130	0.0046
0.4416	60.00	140	0.0053
0.5083	65.56	150	0.0061
0.5666	71.11	160	0.0068
0.6333	76.67	170	0.0076
0.7000	82.22	180	0.0084
0.7583	87.78	190	0.0091
0.8250	93.33	200	0.0099
0.8916	98.89	210	0.0107
0.9666	104.44	220	0.0116
1.0333	110.00	230	0.0124
1.1000	115.56	240	0.0132
1.1750	121.11	250	0.0141
1.2416	126.67	260	0.0149
1.3083	132.22	270	0.0157
1.3749	137.78	280	0.0165
1.4499	143.33	290	0.0174
1.5166	148.89	300	0.0182
1.5916	154.44	310	0.0191
1.6666	160.00	320	0.0200
1.7333	165.56	330	0.0208
1.8083	171.11	340	0.0217
1.8833	176.67	350	0.0226
1.9583	182.22	360	0.0235
2.0333	187.78	370	0.0244
2.0999	193.33	380	0.0252
2.1749	198.89	390	0.0261
2.2499	204.44	400	0.0270
2.3249	210.00	410	0.0279

(*Continued*)

—cont'd

Carbon Steel • Carbon Moly Steel • Low—Chrome Steel (Thru 3% Cr) (Ref 1) mm/meter	Temp, °C	Temp, °F	Carbon Steel • Carbon Moly Steel • Low—Chrome Steel (Thru 3% Cr) (Ref 1) Inches/Foot
2.3999	215.56	420	0.0288
2.4832	221.11	430	0.0298
2.5582	226.67	440	0.0307
2.6332	232.22	450	0.0316
2.7082	237.78	460	0.0325
2.8666	243.33	470	0.0344
2.8666	248.89	480	0.0344
2.9415	254.44	490	0.0353
3.0165	260.00	500	0.0362
3.0999	265.56	510	0.0372
3.1832	271.11	520	0.0382
3.2582	276.67	530	0.0391
3.3415	282.22	540	0.0401
3.4249	287.78	550	0.0411
3.5082	293.33	560	0.0421
3.5915	298.89	570	0.0431
3.6665	304.44	580	0.0440
3.7499	310.00	590	0.0450
3.8332	315.56	600	0.0460
3.9165	321.11	610	0.0470
4.0082	326.67	620	0.0481
4.0915	332.22	630	0.0491
4.1748	337.78	640	0.0501
4.2665	343.33	650	0.0512
4.3498	348.89	660	0.0522
4.4332	354.44	670	0.0532
4.5165	360.00	680	0.0542
4.6081	365.56	690	0.0553
4.6915	371.11	700	0.0563
4.7831	376.67	710	0.0574
4.8665	382.22	720	0.0584
4.9581	387.78	730	0.0595
5.0498	393.33	740	0.0606
5.1415	398.89	750	0.0617
5.2248	404.44	760	0.0627
5.3165	410.00	770	0.0638
5.4081	415.56	780	0.0649

(Continued)

—cont'd

Carbon Steel • Carbon Moly Steel • Low—Chrome Steel (Thru 3% Cr) (Ref 1) mm/meter	Temp, °C	Temp, °F	Carbon Steel • Carbon Moly Steel • Low—Chrome Steel (Thru 3% Cr) (Ref 1) Inches/Foot
5.4914	421.11	790	0.0659
5.5831	426.67	800	0.0670
5.6748	432.22	810	0.0681
5.7664	437.78	820	0.0692
5.8581	443.33	830	0.0703
5.9498	448.89	840	0.0714
6.0498	454.44	850	0.0726
6.1414	460.00	860	0.0737
6.2331	465.56	870	0.0748
6.3247	471.11	880	0.0759
6.4164	476.67	890	0.0770
6.5081	482.22	900	0.0781
6.5997	487.78	910	0.0792
6.6914	493.33	920	0.0803
6.7747	498.89	930	0.0813
6.8664	504.44	940	0.0824
6.9581	510.00	950	0.0835
7.0497	515.56	960	0.0846
7.1414	521.11	970	0.0857
7.2247	526.67	980	0.0867
7.3164	532.22	990	0.0878
7.4080	537.78	1,000	0.0889
7.5080	543.33	1010	0.0901
7.5997	548.89	1,020	0.0912
7.6997	554.44	1030	0.0924
77.9136	560.00	1,040	0.9350
7.8830	565.56	1050	0.0946
7.9830	571.11	1,060	0.0958
8.0830	576.67	1070	0.0970
8.1747	582.22	1,080	0.0981
33.2737	587.78	1090	0.3993
8.3663	593.33	1,100	0.1004
8.4580	598.89	1110	0.1015
8.5413	604.44	1,120	0.1025
8.6330	610.00	1130	0.1036
8.7163	615.56	1,140	0.1046
8.8080	621.11	1150	0.1057

(*Continued*)

—cont'd

Carbon Steel • Carbon Moly Steel • Low—Chrome Steel (Thru 3% Cr) (Ref 1) mm/meter	Temp, °C	Temp, °F	Carbon Steel • Carbon Moly Steel • Low—Chrome Steel (Thru 3% Cr) (Ref 1) Inches/Foot
8.8996	626.67	1,160	0.1068
8.9830	632.22	1170	0.1078
9.0746	637.78	1,180	0.1089
9.1580	643.33	1190	0.1099
9.2496	648.89	1,200	0.1110
9.3413	654.44	1210	0.1121
9.4330	660.00	1,220	0.1132
9.5330	665.56	1230	0.1144
9.6246	671.11	1,240	0.1155
9.7163	676.67	1250	0.1166
9.8079	682.22	1,260	0.1177
9.8996	687.78	1270	0.1188
9.9996	693.33	1,280	0.1200
10.0913	698.89	1290	0.1211
10.1829	704.44	1,300	0.1222
10.2746	710.00	1310	0.1233
10.3663	715.56	1,320	0.1244
10.4662	721.11	1330	0.1256
10.5579	726.67	1,340	0.1267
10.6496	732.22	1350	0.1278
10.8246	737.78	1,360	0.1299
10.9996	743.33	1370	0.1320
11.1829	748.89	1,380	0.1342
11.3579	754.44	1390	0.1363
11.1162	760.00	1,400	0.1334

Austenitic Stainless Steels (304, 316, 347) (Ret 1) mm/m	5Cr–1Mo Steel (Ret 2) mm/m	9Cr–1Mo Steel (Ref 2) mm/m	Temp, °C	Temp, °F	Austenitic Stainless Steels (304, 316, 347) (Ret 1) Inches/Foot	5Cr–1Mo Steel (Ret 2) Inches/Foot	9Cr–1Mo 2 Inches/Foot
0	−2.24991	−1.99992	−198.33	−325		−0.0270	−0.0240
0	−2.22372057	−1.97611143	−195.56	−320		−0.0267	−0.0237
0	−2.17134171	−1.92849429	−190.00	−310		−0.0261	−0.0231
0	−2.11896286	−1.88087714	−184.44	−300		−0.0254	−0.0226
−3.099876	−2.066584	−1.83326	−178.89	−290	−0.0372	−0.0248	−0.0220
−3.016546	−2.01420514	−1.78564286	−173.33	−280	−0.0362	−0.0242	−0.0214
−2.941549	−1.96182629	−1.73802571	−167.78	−270	−0.0353	−0.0235	−0.0209
−2.858219	−1.90944743	−1.69040857	−162.22	−260	−0.0343	−0.0229	−0.0203
−2.783222	−1.85706857	−1.64279143	−156.67	−250	−0.0334	−0.0223	−0.0197
−2.699892	−1.80468971	−1.59517429	−151.11	−240	−0.0324	−0.0217	−0.0191
−2.616562	−1.75231086	−1.54755714	−145.56	−230	−0.0314	−0.0210	−0.0186
−2.541565	−1.699932	−1.49994	−140.00	−220	−0.0305	−0.0204	−0.0180
−2.458235	−1.64755314	−1.45232286	−134.44	−210	−0.0295	−0.0198	−0.0174
−2.341573	−1.59517429	−1.40470571	−128.89	−200	−0.0281	−0.0191	−0.0169
−2.266576	−1.54279543	−1.35708857	−123.33	−190	−0.0272	−0.0185	−0.0163
−2.191579	−1.49041657	−1.30947143	−117.78	−180	−0.0263	−0.0179	−0.0157
−2.116582	−1.43803771	−1.26185429	−112.22	−170	−0.0254	−0.0173	−0.0151
−2.041585	−1.38565886	−1.21423714	−106.67	−160	−0.0245	−0.0166	−0.0146
−1.966588	−1.33328	−1.16662	−101.11	−150	−0.0236	−0.0160	−0.0140
−1.883258	−1.274949	−1.12116727	−95.56	−140	−0.0226	−0.0153	−0.0135
−1.799928	−1.216618	−1.07571455	−90.00	−130	−0.0216	−0.0146	−0.0129
−1.724931	−1.158287	−1.03026182	−84.44	−120	−0.0207	−0.0139	−0.0124
−1.641601	−1.099956	−0.98480909	−78.89	−110	−0.0197	−0.0132	−0.0118

(Continued)

—cont'd

Austenitic Stainless Steels (304, 316, 347) (Ret 1) mm/m	5Cr–1Mo Steel (Ret 2) mm/m	9Cr–1Mo Steel (Ref 2) mm/m	Temp, °C	Temp, °F	Austenitic Stainless Steels (304, 316, 347) (Ret 1) Inches/Foot	5Cr–1Mo Steel (Ret 2) Inches/Foot	9Cr–1Mo 2) Inches/Foot
-1.558271	-1.041625	-0.93935636	-73.33	-100	-0.0187	-0.0125	-0.0113
-1.466608	-0.983294	-0.89390364	-67.78	-90	-0.0176	-0.0118	-0.0107
-1.383278	-0.924963	-0.84845091	-62.22	-80	-0.0166	-0.0111	-0.0102
-1.291615	-0.866632	-0.80299818	-56.67	-70	-0.0155	-0.0104	-0.0096
-1.208285	-0.808301	-0.75754545	-51.11	-60	-0.0145	-0.0097	-0.0091
-1.116622	-0.74997	-0.66664	-45.56	-50	-0.0134	-0.0090	-0.0080
-1.024959	-0.6874725	-0.61108667	-40.00	-40	-0.0123	-0.0083	-0.0073
-0.933296	-0.624975	-0.55553333	-34.44	-30	-0.0112	-0.0075	-0.0067
-0.8333	-0.5624775	-0.49998	-28.89	-20	-0.0100	-0.0068	-0.0060
-0.741637	-0.49998	-0.44442667	-23.33	-10	-0.0089	-0.0060	-0.0053
-0.649974	-0.4374825	-0.38887333	-17.78	0	-0.0078	-0.0053	-0.0047
-0.558311	-0.49998	-0.33332	-12.22	10	-0.0067	-0.0060	-0.0040
-0.466648	-0.5624775	-0.27776667	-6.67	20	-0.0056	-0.0068	-0.0033
-0.366652	-0.624975	-0.22221333	-1.11	30	-0.0044	-0.0075	-0.0027
-0.274989	-0.6874725	-0.16666	4.44	40	-0.0033	-0.0083	-0.0020
-0.183326	-0.74997	-0.11110667	10.00	50	-0.0022	-0.0090	-0.0013
-0.091663	-0.8124675	-0.05555333	15.56	60	-0.0011	-0.0098	-0.0007
0	0	0	21.11	70	0.0000	0.0000	0.0000
0.099996	0.07051	0.5769	26.67	80	0.0012	0.0008	0.0069
0.191659	0.14102	1.1538	32.22	90	0.0023	0.0017	0.0138
0.283322	0.21153	1.7307	37.78	100	0.0034	0.0025	0.0208
0.374985	0.28204	2.3076	43.33	110	0.0045	0.0034	0.0277
0.466648	0.35255	2.8845	48.89	120	0.0056	0.0042	0.0346
0.566644	0.42306	3.4614	54.44	130	0.0068	0.0051	0.0415
0.658307	0.49357	4.0383	60.00	140	0.0079	0.0059	0.0485
0.74997	0.56408	4.6152	65.56	150	0.0090	0.0068	0.0554

0.841633	0.63459	5.1921	71.11	160	0.0101	0.0076	0.0623
0.933296	0.7051	5.769	76.67	170	0.0112	0.0085	0.0692
1.033292	0.77561	6.3459	82.22	180	0.0124	0.0093	0.0762
1.124955	0.84612	6.9228	87.78	190	0.0135	0.0102	0.0831
1.216618	0.91663	7.4997	93.33	200	0.0146	0.0110	0.0900
1.316614	0.983294	6.891391	98.89	210	0.0158	0.0118	0.0827
1.408277	1.049958	6.283082	104.44	220	0.0169	0.0126	0.0754
1.508273	1.116622	5.674773	110.00	230	0.0181	0.0134	0.0681
1.599936	1.183286	5.066464	115.56	240	0.0192	0.0142	0.0608
1.691599	1.24995	4.458155	121.11	250	0.0203	0.0150	0.0535
1.791595	1.316614	3.849846	126.67	260	0.0215	0.0158	0.0462
1.891591	1.383278	3.241537	132.22	270	0.0227	0.0166	0.0389
1.983254	1.449942	2.633228	137.78	280	0.0238	0.0174	0.0316
2.08325	1.516606	2.024919	143.33	290	0.0250	0.0182	0.0243
2.174913	1.58327	1.41661	148.89	300	0.0261	0.0190	0.0170
2.274909	1.658267	1.483274	154.44	310	0.0273	0.0199	0.0178
2.374905	1.733264	1.549938	160.00	320	0.0285	0.0208	0.0186
2.474901	1.808261	1.616602	165.56	330	0.0297	0.0217	0.0194
2.574897	1.883258	1.683266	171.11	340	0.0309	0.0226	0.0202
2.674893	1.958255	1.74993	176.67	350	0.0321	0.0235	0.0210
2.766556	2.033252	1.816594	182.22	360	0.0332	0.0244	0.0218
2.866552	2.108249	1.883258	187.78	370	0.0344	0.0253	0.0226
2.966548	2.183246	1.949922	193.33	380	0.0356	0.0262	0.0234
3.066544	2.258243	2.016586	198.89	390	0.0368	0.0271	0.0242
3.16654	2.33324	2.08325	204.44	400	0.0380	0.0280	0.0250
3.266536	2.399904	2.149914	210.00	410	0.0392	0.0288	0.0258
3.366532	2.466568	2.216578	215.56	420	0.0404	0.0296	0.0266
3.466528	2.533232	2.283242	221.11	430	0.0416	0.0304	0.0274

(*Continued*)

—cont'd

Austenitic Stainless Steels (304, 316, 347) (Ret 1) mm/m	5Cr–1Mo Steel (Ret 2) mm/m	9Cr–1Mo Steel (Ref 2) mm/m	Temp, °C	Temp, °F	Austenitic Stainless Steels (304, 316, 347) (Ret 1) Inches/Foot	5Cr–1Mo Steel (Ret 2) Inches/Foot	9Cr–1Mo 2) Inches/Foot
3.566524	2.599896	2.349906	226.67	440	0.0428	0.0312	0.0282
3.66652	2.66656	2.41657	232.22	450	0.0440	0.0320	0.0290
3.774849	2.733224	2.483234	237.78	460	0.0453	0.0328	0.0298
3.874845	2.799888	2.549898	243.33	470	0.0465	0.0336	0.0306
3.974841	2.866552	2.616562	248.89	480	0.0477	0.0344	0.0314
4.074837	2.933216	2.683226	254.44	490	0.0489	0.0352	0.0322
4.174833	2.99988	2.74989	260.00	500	0.0501	0.0360	0.0330
4.274829	3.08321	2.816554	265.56	510	0.0513	0.0370	0.0338
4.383158	3.16654	2.883218	271.11	520	0.0526	0.0380	0.0346
4.483154	3.24987	2.949882	276.67	530	0.0538	0.0390	0.0354
4.58315	3.3332	3.016546	282.22	540	0.0550	0.0400	0.0362
4.683146	3.41653	3.08321	287.78	550	0.0562	0.0410	0.0370
4.791475	3.49986	3.149874	293.33	560	0.0575	0.0420	0.0378
4.891471	3.58319	3.216538	298.89	570	0.0587	0.0430	0.0386
4.991467	3.66652	3.283202	304.44	580	0.0599	0.0440	0.0394
5.099796	3.74985	3.349866	310.00	590	0.0612	0.0450	0.0402
5.199792	3.83318	3.41653	315.56	600	0.0624	0.0460	0.0410
5.308121	3.908177	3.491527	321.11	610	0.0637	0.0469	0.0419
5.408117	3.983174	3.566524	326.67	620	0.0649	0.0478	0.0428
5.516446	4.058171	3.641521	332.22	630	0.0662	0.0487	0.0437
5.616442	4.133168	3.716518	337.78	640	0.0674	0.0496	0.0446
5.724771	4.208165	3.791515	343.33	650	0.0687	0.0505	0.0455
5.8331	4.283162	3.866512	348.89	660	0.0700	0.0514	0.0464
5.933096	4.358159	3.941509	354.44	670	0.0712	0.0523	0.0473
6.041425	4.433156	4.016506	360.00	680	0.0725	0.0532	0.0482
6.141421	4.508153	4.091503	365.56	690	0.0737	0.0541	0.0491

0.0500	0.0550	0.0750	700	371.11	4.1665	4.58315	6.24975
0.0509	0.0559	0.0763	710	376.67	4.241497	4.658147	6.358079
0.0518	0.0568	0.0776	720	382.22	4.316494	4.733144	6.466408
0.0527	0.0577	0.0789	730	387.78	4.391491	4.808141	6.574737
0.0536	0.0586	0.0802	740	393.33	4.466488	4.883138	6.683066
0.0545	0.0595	0.0815	750	398.89	4.541485	4.958135	6.791395
0.0554	0.0604	0.0828	760	404.44	4.616482	5.033132	6.899724
0.0563	0.0613	0.0841	770	410.00	4.691479	5.108129	7.008053
0.0572	0.0622	0.0854	780	415.56	4.766476	5.183126	7.116382
0.0581	0.0631	0.0867	790	421.11	4.841473	5.258123	7.224711
0.0590	0.0640	0.0880	800	426.67	4.91647	5.33312	7.33304
0.0599	0.0650	0.0893	810	432.22	4.991467	5.41645	7.441369
0.0608	0.0660	0.0906	820	437.78	5.066464	5.49978	7.549698
0.0617	0.0670	0.0920	830	443.33	5.141461	5.58311	7.66636
0.0626	0.0680	0.0933	840	448.89	5.216458	5.66644	7.774689
0.0635	0.0690	0.0946	850	454.44	5.291455	5.74977	7.883018
0.0644	0.0700	0.0959	860	460.00	5.366452	5.8331	7.991347
0.0653	0.0710	0.0972	870	465.56	5.441449	5.91643	8.099676
0.0662	0.0720	0.0986	880	471.11	5.516446	5.99976	8.216338
0.0671	0.0730	0.0999	890	476.67	5.591443	6.08309	8.324667
0.0680	0.0740	0.1012	900	482.22	5.66644	6.16642	8.432996
0.0689	0.0750	0.1026	910	487.78	5.741437	6.24975	8.549658
0.0698	0.0760	0.1039	920	493.33	5.816434	6.33308	8.657987
0.0707	0.0770	0.1053	930	498.89	5.891431	6.41641	8.774649
0.0716	0.0780	0.1066	940	504.44	5.966428	6.49974	8.882978
0.0725	0.0790	0.1080	950	510.00	6.041425	6.58307	8.99964
0.0734	0.0800	0.1094	960	515.56	6.116422	6.6664	9.116302
0.0743	0.0810	0.1107	970	521.11	6.191419	6.74973	9.224631
0.0752	0.0820	0.1121	980	526.67	6.266416	6.83306	9.341293

(Continued)

—cont'd

Austenitic Stainless Steels (304, 316, 347) (Ret 1) mm/m	5Cr–1Mo Steel (Ret 2) mm/m	9Cr–1Mo Steel (Ref 2) mm/m	Temp, °C	Temp, °F	Austenitic Stainless Steels (304, 316, 347) (Ret 1) Inches/Foot	5Cr–1Mo Steel (Ret Steel (Ret 2) Inches/Foot	9Cr–1Mo 2) Inches/Foot
9.449622	6.91639	6.341413	532.22	990	0.1134	0.0830	0.0761
9.566284	6.99972	6.41641	537.78	1,000	0.1148	0.0840	0.0770
9.682946	7.08305	6.49974	543.33	1010	0.1162	0.0850	0.0780
9.791275	7.16638	6.58307	548.89	1,020	0.1175	0.0860	0.0790
9.907937	7.24971	6.6664	554.44	1030	0.1189	0.0870	0.0800
10.016266	7.33304	6.74973	560.00	1,040	0.1202	0.0880	0.0810
10.132928	7.41637	6.83306	565.56	1050	0.1216	0.0890	0.0820
10.241257	7.4997	6.91639	571.11	1,060	0.1229	0.0900	0.0830
10.357919	7.58303	6.99972	576.67	1070	0.1243	0.0910	0.0840
10.474581	7.66636	7.08305	582.22	1,080	0.1257	0.0920	0.0850
10.58291	7.74969	7.16638	587.78	1090	0.1270	0.0930	0.0860
10.699572	7.83302	7.24971	593.33	1,100	0.1284	0.0940	0.0870
10.816234	7.91635	7.324707	598.89	1110	0.1298	0.0950	0.0879
10.924563	7.99968	7.399704	604.44	1,120	0.1311	0.0960	0.0888
11.041225	8.08301	7.474701	610.00	1130	0.1325	0.0970	0.0897
11.149554	8.16634	7.549698	615.56	1,140	0.1338	0.0980	0.0906
11.266216	8.24967	7.624695	621.11	1150	0.1352	0.0990	0.0915
11.382878	8.333	7.699692	626.67	1,160	0.1366	0.1000	0.0924
11.491207	8.41633	7.774689	632.22	1170	0.1379	0.1010	0.0933
11.607869	8.49966	7.849686	637.78	1,180	0.1393	0.1020	0.0942
11.716198	8.58299	7.924683	643.33	1190	0.1406	0.1030	0.0951
11.83286	8.66632	7.99968	648.89	1,200	0.1420	0.1040	0.0960
11.949522	8.74965	8.08301	654.44	1210	0.1434	0.1050	0.0970
12.057851	8.83298	8.16634	660.00	1,220	0.1447	0.1060	0.0980
12.174513	8.91631	8.24967	665.56	1230	0.1461	0.1070	0.0990
12.282842	8.99964	8.333	671.11	1,240	0.1474	0.1080	0.1000

			°C	°F			
12.399504	9.08297	8.41633	676.67	1250	0.1488	0.1090	0.1010
12.516166	9.1663	8.49966	682.22	1,260	0.1502	0.1100	0.1020
12.624495	9.24963	8.58299	687.78	1270	0.1515	0.1110	0.1030
12.741157	9.33296	8.66632	693.33	1,280	0.1529	0.1120	0.1040
12.849486	9.41629	8.74965	698.89	1290	0.1542	0.1130	0.1050
12.966148	9.49962	8.83298	704.44	1,300	0.1556	0.1140	0.1060
13.08281	9.58295	8.91631	710.00	1310	0.1570	0.1150	0.1070
13.191139	9.66628	8.99964	715.56	1,320	0.1583	0.1160	0.1080
13.307801	9.74961	9.08297	721.11	1330	0.1597	0.1170	0.1090
13.41613	9.83294	9.1663	726.67	1,340	0.1610	0.1180	0.1100
13.532792	9.91627	9.24963	732.22	1350	0.1624	0.1190	0.1110
13.649454	9.9996	9.33296	737.78	1,360	0.1638	0.1200	0.1120
13.757783	10.08293	9.41629	743.33	1370	0.1651	0.1210	0.1130
13.874445	10.16626	9.49962	748.89	1,380	0.1665	0.1220	0.1140
13.982774	10.24959	9.58295	754.44	1390	0.1678	0.1230	0.1150
14.099436	10.33292	9.66628	760.00	1,400	0.1692	**0.1240**	**0.1160**
14.199432			765.56	1410	0.1704		
14.33276			771.11	1,420	0.1720		
14.424423			776.67	1430	0.1731		
14.532752			782.22	1,440	0.1744		
14.641081			787.22	1450	0.1757		
14.757743			793.33	1,460	0.1771		
14.866072			798.89	1470	0.1784		
14.966068			804.44	1,480	0.1796		
15.091063			810.00	1490	0.1811		

Go down temperature chart °C or F to find final temperature; read inches per foot on the appropriate alloy to the right or mm per meter on the appropriate alloy to the left.

Note these are total expansion per ft/m from the reference temperature 70°F or 21°C to the temperature listed

Ref1: Grinnell Popong Engineering Handbook

Ref2: B31.1.1–2004

Values in bold excerpted others are linear interpolation

Approximate Weights of LR Elbows

Size NPS	Size DN	Sch 20		Sch 30		Std.		Sch 40		Sch 60	
		lbs	kg	lbs	kg	lbs	kg	lbs	kg	lbs	kg
1 1/4	32	—		—		0.58	0.18			0.58	0.18
1 1/2	40	—		—		0.87	0.27			0.87	0.27
2	50	—		—		1.5	0.46	1.5	0.46	—	
2 1/2	65	—		—		3	0.91	3.1	0.94	—	
3	80	—		—		4.8	1.46	4.8	1.5	—	
3 1/2	95	—		—		6.7	2.04	6.7	2.0	—	
4	100	—		—		9.2	2.80	9.2	3		
5	125	—		—		15	5	15	5		
6	150	—		—		24	7	24	7		
8	200	36	11	41	12	48	15	48	15	190	58
10	250	57	17	70	21	85	26	85	26	250	76
12	300	82	25	108	33	125	38	135	41	360	110
14	350	132	40	158	48	162	49	185	56	525	160
16	400	172	52	207	63	215	66	280	85	700	213
18	450	219	67	308	94	270	82	385	117		
20	500	323	98	428	130	330	101	520	158		
22	250	—	—	—		392	119	—			
24	600	468	143	702	214	480	146	850	259	1200	366
26	650	—	—	—		550	168	1300	396	—	
30	750	972	296	1215	370	733	223	—			
36	900	—		—		1061	323	—			
42	1050	—		—		1442	440	—			
48	1200	—		—		1883	574	—			

Wall Thickness

X-Stg.		Sch. 80		Sch. 100		Sch. 120		Sch. 140		Sch. 160		XX XXstrg	
lbs	kg	lbs	kg	lbs	kg	lbs	kg	lbs	kg	lbs	kg	lbs	kg
—	—	0.77	0.2	0.77	0.2	—	—	—	—	—	—	0.97	0.3
—	—	1.1	0.3	1.1	0.3	—	—	—	—	—	—	1.5	0.5
2.2	0.7	2.2	0.7	—	—	—	—	—	—	3.1	1	3.8	1.2
4	1.2	4	1.2	—	—	—	—	—	—	5.1	2	7	2
6.3	1.9	6.3	1.9	—	—	—	—	—	—	9	3	11.5	4
9.1	2.8	9.1	2.8	—	—	—	—	—	—	—	—	16	5
12.6	3.8	12.6	3.8	—	—	15.6	5	—	—	18.5	6	23	7
—	—	22	7	22	7	—	—	29	9	—	—	34	10
—	—	36	11	36	11	—	—	44	13	—	—	57	17
60	18	73	22	73	22	84	26	99	30	112	34	123	37
115	35	115	35	135	41	158	48	185	56	213	65	238	73
166	51	225	69	264	80	310	94	345	105	396	121	325	99
213	65	315	96	376	115	436	133	491	150	546	166	—	—
280	85	460	140	543	166	634	193	737	225	807	246	—	—
360	110	640	195	772	235	904	276	1016	310	1144	349	—	—
440	134	870	265	1054	321	1221	372	1404	428	1580	482	—	—
—	—	520	158	—	—	—	—	—	—	—	—	—	—
620	189	1470	448	1750	533	1810	552	2283	696	2510	765	—	—
—	—	—	—	—	—	—	—	—	—	—	—	—	—
720	219	972	296	—	—	—	—	—	—	—	—	—	—
—	—	1407	429	—	—	—	—	—	—	—	—	—	—
—	—	1916	584	—	—	—	—	—	—	—	—	—	—
—	—	2511	765	—	—	—	—	—	—	—	—	—	—

Approximate Weights – Reducers – Concentric and Eccentric

Nominal size		Schedule 20		Schedule 30		Standard		Schedule 40		Schedule 60	
NPS	DN	lbs	kg	lbs	kg	lbs	kg	lbs	kg	lbs	kg
2	50	—		—		0.9	0.4	0.9	0.4	—	
2 1/2	65	—		—		1.7	0.8	1.7	0.8	—	
3	80	—		—		2.2	1.0	2.2	1.0	—	
3 1/2	★	—		—		30	★	30	★	—	
4	100	—		—		3.6	1.6	3.6	1.6	—	
5	125	—		—		6.1	2.8	6.1	2.8	—	
6	150	—		—		8.7	3.9	8.7	3.9	—	
8	200	11	5.0	12	5.4	14	6.4	14	6.4	18	8.2
10	250	16	7.3	20	9.1	24	10.9	24	10.9	32	14.5
12	300	22	10.0	29	13.2	33	15.0	36	16.3	49	22.2
14	350	49	22.2	59	26.8	59	26.8	69	31.3	92	41.7
16	400	61	27.7	73	33.1	73	33.1	96	43.5	125	56.7
18	450	73	33.1	103	46.7	88	39.9	131	59.4	173	78.5
20	500	131	59.4	174	78.9	131	59.4	205	93.0	278	126.1
22	550	—		—		144	65.3	—		—	
24	600	158	71.7	236	107.0	158	71.7	285	129.3	397	180.1
26	650	—		—		205	93.0	—		—	
30	750	315	142.9	394	178.7	237	107.5	—		—	
36	900	—		—		285	129.3	—		—	
42	1050	—		—		334	151.5	—		—	
48	1200	—		—		395	179.2	—		—	

Note: Nominal size is based on large end weight and is estimated the same for different reductions because end to end dimensions are same.

Wall Thickness

X-STrong		Schedule 80		Schedule 100		Schedule 120		Schedule 140		Schedule 160		XX-STrong	
lbs	kg	lbs	kg	lbs	kg	lbs	kg	lbs	kg	lbs	kg	lbs	kg
1.2	0.5	1.2	0.5	—		—		—		1.9	0.9	2.3	1.0
2.2	1.0	2.2	1.0	—		—		—		2.9	1.3	4	1.8
3	1.4	3	1.4	—		—		—		4.2	1.9	5.4	2.4
4.2	★	4.2	★	—		6.3	2.9	—		5.8	2.6	7.6	3.4
5	2.3	5	2.3	—		12	5.4	—		7.5	3.4	9.2	4.2
8.6	3.9	8.6	3.9	—		17	7.7	—		14	6.4	16	7.3
13	5.9	13	5.9	25	11.3	30	13.6	34	15.4	21	9.5	24	10.9
22	10.0	22	10.0	45	20.4	52	23.6	61	27.7	37	16.8	36	16.3
32	14.5	38	17.2	71	32.2	84	38.1	93	42.2	67	30.4	61	27.7
44	20.0	59	26.8	142	64.4	164	74.4	185	83.9	107	48.5	84	38.1
78	35.4	115	52.2	194	88.0	224	101.6	262	118.8	208	94.3	—	
97	44.0	159	72.1	257	116.6	304	137.9	343	155.6	286	129.7	—	
117	53.1	213	96.6	425	192.8	494	224.1	567	257.2	358	162.4	—	
174	78.9	348	157.9	—		—		—		630	285.8	—	
191	86.6	—		—		—		—		—		—	
210	95.3	490	222.3	610	276.7	710	322.1	800	362.9	900	408.2	—	
272	123.4	—		—		—		—		—		—	
315	142.9	—		—		—		—		—		—	
379	171.9	—		—		—		—		—		—	
443	200.9	—		—		—		—		—		—	
515	233.6	—		—		—		—		—		—	

WEIGHT – TEES – STRAIGHT AND REDUCING

Nominal Size				Wall Thickness			
Run		Branch		Std.		Sched. 40	
DN	NPS	DN	NPS	Ibs	kg	Ibs	kg
20	3/4	20	3/4	0.45	0.20	0.45	0.20
20	3/4	15	1/2	0.5	0.23	0.5	0.23
25	1	25	1	0.63	0.29	0.63	0.29
25	1	20	3/4	0.58	0.26	0.58	0.26
25	1	15	1/2	0.57	0.26	0.57	0.26
32	11/4	32	11/4	13	0.54	1.2	0.54
32	1 1/4	25	1	1.1	0.50	1.1	0.50
32	1 1/4	20	3/4	1.1	0.50	1.1	0.50
32	11/4	15	1/2	1	0.45	1	0.45
40	11/2	40	11/2	1.7	0.77	1.7	0.77
40	11/2	32	11/4	1.6	0.73	1.6	0.73
40	11/2	25	1	1.6	0.73	1.6	0.73
40	11/2	20	3/4	1.5	0.68	1.5	0.68
40	11/2	15	1/2	1.5	0.68	1.5	0.68
50	2	50	2	4.2	1.91	4.2	1.91
50	2	40	11/2	4.2	1.91	4.2	1.91
50	2	32	11/4	4.2	13	4.2	1.91
50	2	25	1	4.2	1.91	4.2	1.91
50	2	20	34	4.2	1.91	4.2	1.91

APPROXIMATE WEIGHT – TEES – STRAIGHT AND REDUCING

X-Stg.		Sched. 80		Sched. 160		XX Stg.	
lbs	kg	lbs	kg	lbs	kg	lbs	kg
0.6	0.27	0.6	0.27	0.51	0.23	0.63	0.29
0.5	0.23	0.5	0.23	0.47	0.21	0.59	0.27
0.78	0.35	0.78	0.35	0.99	0.45	1.3	0.59
0.73	0.33	0.73	0.33	0.92	0.42	1.2	0.54
0.71	0.32	0.71	0.32	0.88	0.40	1.1	0.50
1.4	0.64	1.4	0.64	1.7	0.77	2.3	1.04
1.3	0.59	1.3	0.59	1.6	0.73	2.1	0.95
1.3	0.59	1.3	0.59	1.6	0.73	2.1	0.95
1.3	0.59	1.3	0.59	1.5	0.68	2	0.91
2.1	0.95	2.1	0.95	2.7	1.22	3.4	1.54

(Continued)

—cont'd

X-Stg.		Sched. 80		Sched. 160		XX Stg.	
lbs	kg	lbs	kg	lbs	kg	lbs	kg
2	0.91	2	0.91	2.5	1.13	3.3	1.50
1.9	0.86	1.9	0.86	2.5	1.13	3.1	1.41
1.9	0.86	1.9	0.86	2.4	1.09	3	1.36
1.9	0.86	1.9	0.86	—		3	1.36
4.1	1.86	4.1	1.86	5	2.27	5.9	2.68
4.1	1.86	4.1	1.86	4.6	2.09	5.5	2.49
4.1	1.86	4.1	1.86	4.5	2.04	5.4	2.45
4.1	1.86	4.1	1.86	4.4	2.00	5.3	2.40
4.1	1.86	4.1	1.86	4.3	1.95	5.1	2.31

Nominal Size				Wall Thickness			
Run		Branch		Std.		Schedule 40	
NPS	DN	NPS	Dn	Lbs	kg	Lbs	kg
2 1/2	65	21/2	65	5.9	2.7	5.9	2.7
2 1/2	65	2	50	5.9	2.7	5.9	2.7
2 1/2	65	1 1/2	40	6.8	3.1	6.8	3.1
2 1/2	65	1 1/4	32	6.8	3.1	6.8	3.1
2 1/2	65	1	25	6.8	3.1	6.8	3.1
3	80	3	80	8.4	3.8	8.4	3.8
3	80	21/2	65	8.4	3.8	8.4	3.8
3	80	2	50	8.4	3.8	8.4	3.8
3	80	1 1/2	40	8.4	3.8	8.4	3.8
3	80	1 1/4	32	8.4	3.8	8.4	3.8
31/2	★	31/2	★.	11		11	
31/2	★	3	★	11		11	
31/2	★	2 1 /2	★	11		11	
31/2	★	2	★	11		11	
31/2	★	1 1/2	★	11		11	
4	100	4	100	13	5.9	13	5.9
4	100	3 1/2	★	13	5.9	13	5.9
4	100	3	80	13	5.9	13	5.9
4	100	2 1/2	65	13	5.9	13	5.9
4	100	2	50	13	5.9	13	5.9
4	100	1 1/2	40	13	5.9	13	5.9

X-Stg.		Schedule 80		Schedule 120		Schedule 120		XX strong	
Lbs	kg	Lbs	kg	Lbs	kg	Lbs	kg	Lbs	kg
6.8	3.1	6.8	3.1	—		7.6	3.4	10	4.5
6.8	3.1	6.8	3.1	—		6.9	3.1	9.2	4.2
6.8	3.1	6.8	3.1	—		6.7	3.0	9	4.1

(Continued)

—cont'd

X-Stg.		Schedule 80		Schedule 120		Schedule 120		XX strong	
Lbs	kg	Lbs	kg	Lbs	kg	Lbs	kg	Lbs	kg
6.8	3.1	6.8	3.1	—		6.6	3.0	9	4.1
6.8	3.1	6.8	3.1	—		6.6	3.0	8.6	3.9
10	4.5	10	4.5	—		14	6.4	17	7.7
10	4.5	10	4.5	—		13	5.9	16	7.3
10	4.5	10	4.5	—		12	5.4	15	6.8
10	4.5	10	4.5	—		12	5.4	15	6.8
10	4.5	10	4.5	—		12	5.4	15	6.8
14		14		—		—		21	
14		14		—		—		20	
14		14		—		—		20	
14		14		—		—		19	
14		14		—		—		19	
19	8.6	19	8.6	23	10.4	34	15.4	34	15.4
19	8.6	19	8.6	23	10.4	28	12.7	34	15.4
19	8.6	19	8.6	23	10.4	28	12.7	34	15.4
19	8.6	19	8.6	23	10.4	28	12.7	34	15.4
19	8.6	19	8.6	23	10.4	28	12.7	34	15.4
19	8.6	19	8.6	23	10.4	28	12.7	34	15.4

Nominal Size						Wall Thickness					
Run		Branch		Schedule 20		Schedule 30		Standard		Schedule 40		Schedule 60	
NPS	DN	NPS	DN	lbs	kg	lbs	kg	lbs	kg	lbs	kg	lbs	kg
5	125	5	125	—		—		22	10.0	22	10.0	—	
5	125	4	100	—		—		22	10.0	22	10.0	—	
5★	31/2	★		—		—		22	10.0	22	10.0	—	
5	125	3	80	—		—		22	10.0	22	10.0	—	
5	125	2 1/2	65	—		—		22	10.1	22	10.0	—	
5	125	2	50	—		—		22	10.0	22	10.0	—	
6	150	6	150	—		—		36	16.3	36	16.3	—	
6	150	5	125	—		—		36	16.3	36	16.3	—	
6	150	4	100	—		—		33	15.0	33	15.0	—	
6★	31/2	★		—		—		33	15.0	33	15.0	—	
6	150	3	80	—		—		33	15.0	33	15.0	—	
6	150	21/2	65	—		—		33	15.0	33	15.0	—	
8	200	8	200	54	24.5	57	25.9	61	27.7	61	27.7	76	34.5
8	200	6	150	54	24.5	57	25.9	61	27.7	61	27.7	76	34.5
8	200	5	125	54	24.5	57	25.9	61	27.7	61	27.7	76	4.5
8	200	4	100	54	24.5	57	25.9	61	27.7	61	27.7	76	34.5
8★	31/2	★		54	24.5	57	25.9	61	27.7	61	27.7	76	34.5
10	250	10	250	73	33.1	81	36.7	91	41.3	91	41.3	129	58.5
10	250	8	200	70	31.8	78	35.4	91	41.3	91	41.3	116	52.6

(Continued)

—cont'd

Nominal Size								Wall Thickness						
Run		Branch		Schedule 20		Schedule 30		Standard		Schedule 40		Schedule 60	
NPS	DN	NPS	DN	lbs	kg	lbs	kg	lbs	kg	lbs	kg	lbs	kg
10	250	6	150	70	31.8	78	35.4	88	39.9	88	39.9	116	52.6
10	250	5	125	70	31.8	78	35.4	88	39.9	88	39.9	116	52.6
10	250	4	100	70	31.8	78	35.4	88	39.9	88	39.9	116	52.6
12	300	12	300	120	54.4	136	61.7	147	66.7	147	66.7	226	102.5
12	300	10	250	120	54.4	136	61.7	147	66.7	147	66.7	226	102.5
12	300	8	200	116	52.6	132	59.9	143	64.9	143	64.9	181	82.1
12	300	6	150	116	52.6	132	59.9	143	64.9	143	64.9	181	82.1
12	300	5	125	116	52.6	132	59.9	143	64.9	143	64.9	181	82.1
14	350	14	350	210	95.3	226	102.5	226	102.5	252	114.3	311	141.1
14	350	12	300	210	95.3	226	102.5	226	102.5	252	114.3	311	141.1
14	350	10	250	201	91.2	217	98.4	217	98.4	217	98.4	299	135.6
14	350	8	200	201	91.2	217	98.4	217	98.4	217	98.4	299	135.6
14	350	6	150	201	91.2	217	98.4	217	98.4	217	98.4	299	135.6
16	400	16	400	222	100.7	242	109.8	242	109.8	370	167.8	458	207.7
16	400	14	350	222	100.7	242	109.8	242	109.8	370	167.8	458	207.7
16	400	12	300	222	100.7	242	109.8	242	109.8	359	162.8	399	181.0
16	400	10	250	215	97.5	235	106.6	235	106.6	354	160.6	360	163.3
16	400	8	200	215	97.5	235	106.6	235	106.6	354	160.6	354	160.6
16	400	6	150	215	97.5	235	106.6	235	106.6	354	160.6	354	160.6
18	450	18	450	307	139.3	399	181.0	333	151.0	525	238.1	612	277.6
18	450	16	400	307	139.3	399	181.0	333	151.0	525	238.1	565	256.3
18	450	14	350	307	139.3	399	181.0	333	151.0	427	193.7	468	212.3
18	450	12	300	307	139.3	313	142.0	333	151.0	427	193.7	468	212.3
18	450	10	250	293	132.9	296	134.3	319	144.7	330	149.7	414	187.8
18	450	8	200	293	132.9	296	134.3	319	144.7	330	149.7	414	187.8
20	500	20	500	504	228.1	583	263.9	504	228.1	706	319.5	834	377.5
20	500	18	450	504	228.1	504	228.1	504	228.1	584	264.3	774	350.3
20	500	16	400	504	228.1	504	228.1	504	228.1	506	229.0	713	322.7
20	500	14	350	493	223.1	493	223.1	493	223.1	494	223.6	645	291.9
20	500	12	300	493	223.1	493	223.1	493	223.1	494	223.6	645	291.9
20	500	10	250	482	218.1	482	218.1	482	218.1	485	219.5	630	285.1
20	500	8	200	482	218.1	482	218.1	482	218.1	485	219.5	494	223.6
22	550	22	550	—		—		555	251.2	—		—	
22	550	20	500	—		—		555	251.2	—		—	
22	550	18	450	—		—		527	238.5	—		—	
22	550	16	400	—		—		527	238.5	—		—	
22	550	14	350	—		—		445	201.4	—		—	
22	550	12	300	—		—		445	201.4	—		—	
22	550	10	250	—		—		445	201.4	—		—	
24	600	24	600	765	346.2	977	442.2	765	346.2	1257	568.9	1446	654.4

(Continued)

—cont'd

Nominal Size				Wall Thickness										
Run		Branch		Schedule 20		Schedule 30		Standard		Schedule 40		Schedule 60	
NPS	DN	NPS	DN	lbs	kg	lbs	kg	lbs	kg	lbs	kg	lbs	kg
24	600	22	550	601	272.0	849	384.3	681	308.2	1130	511.4	1300	588.4
24	600	20	500	601	272.0	726	328.6	601	272.0	860	389.2	1200	543.1
24	600	18	450	601	272.0	726	328.6	601	272.0	860	389.2	1040	470.7
24	600	16	400	506	229.0	553	250.3	506	229.0	681	308.2	941	425.9
24	600	14	350	506	229.0	553	250.3	506	229.0	681	308.2	941	425.9
24	600	12	300	506	229.0	553	250.3	506	229.0	681	308.2	860	389.2
24	600	10	250	424	191.9	553	250.3	424	191.9	681	308.2	860	389.2

X-strong		Schedule 80		Schedule 100		Schedule 120		Schedule 140		Schedule 160		XX-Strong	
lbs	kg	lbs	kg	lbs	kg	lbs	kg	lbs	kg	lbs	kg	lbs	kg
28	12.7	28	12.7	—		44	20.0	—		53	24.0	53	24.0
28	12.7	28	12.7	—		44	20.0	—		44	20.0	53	24.0
28	12.7	28	12.7	—		44	20.0	—		44	20.0	53	24.0
28	12.7	28	12.7	—		44	20.0	—		44	20.0	53	24.0
28	12.7	28	12.7	—		44	20.0	—		44	20.0	44	20.0
28	12.7	28	12.7	—		44	20.0	—		44	20.0	44	20.0
42	19.1	42	19.1	—		64	29.0	—		44	20.0	85	38.6
42	19.1	42	19.1	—		64	29.0	—		85	38.6	85	38.6
42	19.1	42	19.1	—		64	29.0	—		67	30.4	64	29.0
42	19.1	42	19.1	—		64	29.0	—		67	30.4	64	29.0
42	19.1	42	19.1	—		64	29.0	—		67	30.4	64	29.0
42	19.1	42	19.1	—		64	29.0	—		64	29.0	64	29.0
76	34.5	76	34.5	97	44.0	115	52.2	133	60.3	152	68.9	152	68.9
76	34.5	76	34.5	97	44.0	115	52.2	115	52.2	115	52.2	114	51.7
76	34.5	76	34.5	97	44.0	97	44.0	115	52.2	115	52.2	114	51.7
76	34.5	76	34.5	97	44.0	97	44.0	109	49.4	109	49.4	114	51.7
76	34.5	76	34.5	97	44.0	97	44.0	109	49.4	109	49.4	114	51.7
129	58.5	161	73.0	180	81.6	215	97.5	241	109.3	280	127.0	241	109.3
116	52.6	157	71.2	161	73.0	197	89.4	219	99.3	241	109.3	219	99.3
116	52.6	120	54.4	161	73.0	180	81.6	201	91.2	223	101.2	201	91.2
116	52.6	116	52.6	120	54.4	157	71.2	177	80.3	197	89.4	177	80.3
116	52.6	116	52.6	116	52.6	157	71.2	177	80.3	188	85.3	177	80.3
187	84.8	245	111.1	304	137.9	353	160.1	404	183.3	429	194.6	353	160.1
187	84.8	226	102.5	279	126.6	329	149.2	353	160.1	377	171.0	329	149.2
180	81.6	180	81.6	269	122.0	294	133.4	318	144.2	341	154.7	294	133.4
180	81.6	180	81.6	245	111.1	245	111.1	270	122.5	318	144.2	245	111.1
280	81.6	180	81.6	226	102.5	226	102.5	270	122.5	318	144.2	226	102.5

(Continued)

—cont'd

X-strong		Schedule 80		Schedule 100		Schedule 120		Schedule 140		Schedule 160		XX-Strong	
lbs	kg	lbs	kg	lbs	kg	lbs	kg	lbs	kg	lbs	kg	lbs	kg
280	127.0	369	167.4	528	239.5	528	239.5	624	283.0	720	326.6	—	
280	127.0	315	142.9	528	239.5	528	239.5	648	293.9	696	315.7	—	
268	121.6	310	140.6	384	174.2	384	174.2	600	272.2	624	283.0	—	
268	121.6	268	121.6	360	163.3	456	206.8	480	217.7	528	239.5	—	
268	121.6	268	121.6	360	163.3	432	196.0	450	204.1	480	217.7	—	
369	167.4	548	248.6	780	353.8	826	374.7	962	436.4	1066	483.5	—	
369	167.4	440	199.6	676	306.6	728	330.2	806	365.6	910	412.8	—	
359	162.8	399	181.0	546	247.7	624	283.0	806	365.6	884	401.0	—	
352	159.7	360	163.3	468	212.3	572	259.5	806	365.6	884	401.0	—	
352	159.7	360	163.3	494	224.1	572	259.5	806	365.6	884	401.0	—	
352	159.7	360	163.3	494	224.1	572	259.5	806	365.6	884	401.0	—	
425	192.8	710	322.1	1131	513.0	1160	526.2	1189	539.3	1392	631.4	—	
425	192.8	615	279.0	928	420.9	957	434.1	1100	499.0	1160	526.2	—	
425	192.8	569	258.1	928	420.9	957	434.1	1100	499.0	1160	526.2	—	
339	153.8	516	234.1	928	420.9	957	434.1	1100	499.0	1160	526.2	—	
322	146.1	496	225.0	928	420.9	957	434.1	1100	499.0	1160	526.2	—	
322	146.1	449	203.7	928	420.9	957	434.1	1075	487.6	1150	521.6	—	

X-Strong		Schedule 80		Schedule 100		Schedule 120		Schedule 140		Schedule 160	
lbs	kg	lbs	kg	lbs	kg	lbs	kg	lbs	kg	lbs	kg
583	263.9	1021	462.1	1344	608.3	1504	680.7	1696	767.6	1792	811.0
504	228.1	903	408.7	1025	463.9	1312	593.8	1312	593.8	1536	695.2
504	228.1	782	353.9	928	420.0	1312	593.8	1312	593.8	1536	695.2
493	223.1	713	322.7	832	376.6	1206	545.8	1312	593.8	1536	695.2
493	223.1	713	322.7	832	376.6	1206	545.8	1312	593.8	1536	695.2
482	218.1	645	291.9	832	376.6	1206	545.8	1312	593.8	1536	695.2
482	218.1	645	291.9	832	376.6	1206	545.8	1312	593.8	1536	695.2
811	367.1	—		—		—		—		—	
811	367.1	—		—		—		—		—	
670	303.2	—		—		—		—		—	
670	303.2	—		—		—		—		—	
517	234.0	—		—		—		—		—	
517	234.0	—		—		—		—		—	
517	234.0	—		—		—		—		—	
934	422.7	1673	757.2	2592	1173.1	2808	1270.9	2950	1335.1	3096	1401.2
849	384.3	1673	757.2	2592	1173.1	2808	1270.9	2950	1335.1	3096	1401.2
683	309.1	1361	616.0	2100	950.4	2206	998.4	2320	1050.0	2800	1267.3
683	309.1	1200	543.1	1944	879.8	2206	998.4	2320	1050.0	2800	1267.3
509	230.4	1106	500.6	1944	879.8	2206	998.4	2320	1050.0	2440	1104.3

(*Continued*)

—cont'd

X-Strong		Schedule 80		Schedule 100		Schedule 120		Schedule 140		Schedule 160	
lbs	kg	lbs	kg	lbs	kg	lbs	kg	lbs	kg	lbs	kg
509	230.4	1106	500.6	1944	879.8	2206	998.4	2320	1050.0	2440	1104.3
509	230.4	1021	462.1	1800	814.7	2026	917.0	2160	977.6	2250	1018.3
509	230.4	1021	462.1	1800	814.7	2026	917.0	2160	977.6	2250	1018.3

Nominal Size							
Run		Branch		Schedule 20		Schedule 30	
NPS	DN	NPS	DN	lbs	kg	lbs	kg
26	650	26	650	—		—	
26	650	24	600	—		—	
26	650	22	550	—		—	
26	650	20	500	—		—	
26	650	18	450	—		—	
26	650	16	400	—		—	
26	650	14	350	—		—	
26	650	12	300	—		—	
30	750	30	750	1375	623.7	1517	688.1
30	750	26	650	1257	570.2	1517	688.1
30	750	24	600	1090	494.4	1232	558.8
30	750	22	550	1090	494.4	1232	558.8
30	750	20	500	1090	494.4	1232	558.8
30	750	18	450	1090	494.4	1232	558.8
30	750	16	400	1090	494.4	1232	558.8
30	750	14	350	1090	494.4	1232	558.8
36	900	36	900	2165	982.0	2700	1224.7
36	900	30	750	1893	858.6	2700	1224.7
36	900	26	650	1504	682.2	2280	1034.2
36	900	24	600	1504	682.2	2280	1034.2
36	900	20	500	1504	682.2	2280	1034.2
36	900	18	450	1321	599.2	2280	1034.2
36	900	16	400	1321	599.2	1900	861.8
42	1050	42	1050	—		—	
42	1050	36	900	—			
42	1050	30	750	—		—	
42	1050	24	600	—		—	
42	1050	20	500	—		—	
48	1200	48	1200	—		—	
48	1200	42	1050	—		—	
48	1200	36	900	—		—	
48	1200	24	600	—		—	

*means no DN equivalent

Wall Thickness

Standard		Schedule 40		X- Strong	
lbs	kg	lbs	kg	lbs	kg
826	374.7	—		1121	508.5
826	374.7	—		1121	508.5
727	329.8	—		925	419.6
727	329.8	—		925	419.6
614	278.5	—		713	323.4
614	278.5	—		713	323.4
614	278.5	—		713	323.4
614	278.5	—		713	323.4
1130	512.6	—		1510	684.9
1065	483.1	—		1257	570.2
1065	483.1	—		1257	570.2
921	417.8	—		1048	475.4
921	921	—		1048	475.4
921	417.8	—		1048	475.4
792	359.2	—		921	921
792	359.2	—		921	417.8
1617	733.5	2925	1326.8	2165	982.0
1524	691.3	2925	1326.8	1893	858.6
1321	599.2	2380	1079.5	1504	682.2
1321	599.2	2380	1079.5	1504	682.2
1321	599.2	2380	1079.5	1504	682.2
1321	599.2	2380	1079.5	1321	599.2
1136	515.3	2010	911.7	1321	599.2
2900	1315.4	—		2900	1315.4
2755	1249.6	—		2755	1249.6
2755	1249.6	—		2755	1249.6
2755	1249.6	—		2755	1249.6
2755	1249.6	—		2755	1249.6
3215	1458.3	—		4080	1850.7
2420	1097.7	—		3535	1603.4
2420	1097.7	—		3535	1603.4
2420	1097.7	—		3535	1603.4

* means no DN equivalent

Suggested Starting Torque Values

ASTM A307

Bolt Size	TPI	Proof Load (lbs)	Clamp Load (lbs)	Tightening Torque (ft lbs)		
				Waxed	Galv	Plain
1/4	20	1145	859	2	4	4
5/16	18	1886	1415	4	9	7
3/8	16	2790	2093	7	16	13
7/16	14	3827	2870	10	26	21
1/2	13	5108	3831	16	40	32
9/16	12	6552	4914	23	58	46
5/8	11	8136	6102	32	79	64
3/4	10	12024	9018	56	141	113
7/8	9	15200	11400	83	208	166
1	8	20000	15000	125	313	250
1 1/8	7	25200	18900	177	443	354
1 1/4	7	32000	24000	250	625	500
1 3/8	6	38100	28575	327	819	655
1 1/2	6	46400	34800	435	1088	870
1 3/4	5	68400	51300	748	1870	1496
2	4 ½	90000	67500	1125	2813	2250
2 1/4	4 ½	117000	87750	1645	4113	3291
2 1/2	4	144000	108000	2250	5625	4500
2 3/4	4	177480	133110	3050	7626	6101
3	4	214920	161190	4030	10074	8060
3 1/4	4	255600	191700	5192	12980	10384
3 1/2	4	299880	224910	6560	16400	13120
3 3/4	4	347760	260820	8151	20377	16301
4	4	398880	299160	9972	24930	19944

SAE Grade 2

Bolt Size	TPI	Proof Load (lbs)	Clamp Load (lbs)	Tightening Torque (ft lbs)		
				Waxed	Galv	Plain
1/4	20	1750	1313	3	7	5
5/16	18	2900	2175	6	14	11
3/8	16	4250	3188	10	25	20
7/16	14	5850	4388	16	40	32
1/2	13	7800	5850	24	61	49
9/16	12	10000	7500	35	88	70
5/8	11	12400	9300	48	121	97
3/4	10	18400	13800	86	216	173
7/8	9	15200	11400	83	208	166
1	8	20000	15000	125	313	250
1 1/8	7	25200	18900	177	443	354
1 1/4	7	32000	24000	250	625	500
1 3/8	6	38100	28575	327	819	655
1 1/2	6	46400	34800	435	1088	870

ASTM A325 / ASTM A449 / SAE Grade 5

Bolt Size	TPI	Proof Load (lbs)	Clamp Load (lbs)	Tightening Torque (ft lbs)		
				Waxed	Galv	Plain
1/4	20	2700	2025	4	11	8
5/16	18	4450	3338	9	22	17
3/8	16	6600	4950	15	39	31
7/16	14	9050	6788	25	62	49
1/2	13	12050	9038	38	94	75
9/16	12	15450	11588	54	136	109
5/8	11	19200	14400	75	188	150
3/4	10	28400	21300	133	333	266
7/8	9	39250	29438	215	537	429
1	8	51500	38625	322	805	644
1 1/8	7	56450	42338	397	992	794
1 1/4	7	71700	53775	560	1400	1120
1 3/8	6	85450	64088	734	1836	1469
1 1/2	6	104000	78000	975	2438	1950
1 3/4	5	104500	78375	1143	2857	2286
2	$4^1/_2$	137500	103125	1719	4297	3438

(Continued)

—cont'd

Bolt Size	TPI	Proof Load (lbs)	Clamp Load (lbs)	Tightening Torque (ft lbs)		
				Waxed	Galv	Plain
2 1/4	$4^1/_2$	178750	134063	2514	6284	5027
2 1/2	4	220000	165000	3438	8594	6875
2 3/4	4	271150	203363	4660	11651	9321
3	4	328350	246263	6157	15391	12313

ASTM A193 B7

Bolt Size	TPI	Proof Load (lbs)	Clamp Load (lbs)	Tightening Torque (ft lbs)		
				Waxed	Galv	Plain
1/4	20	3350	2513	5	13	10
5/16	18	5500	4125	11	27	21
3/8	16	8150	6113	19	48	38
7/16	14	11150	8363	30	76	61
1/2	13	14900	11175	47	116	93
9/16	12	19100	14325	67	168	134
5/8	11	23750	17813	93	232	186
3/4	10	35050	25288	164	411	329
7/8	9	48500	36375	265	663	530
1	8	63650	47738	398	995	796
1 1/8	7	80100	60075	563	1408	1126
1 1/4	7	101750	76313	795	1987	1590
1 3/8	6	121300	90975	1042	2606	2085
1 1/2	6	147550	110663	1383	3458	2767
1 3/4	5	199500	149625	2182	5455	4364
2	$4^1/_2$	262500	196875	3281	8203	6563
2 1/4	$4^1/_2$	341250	255938	4799	11997	9598
2 1/2	4	420000	315000	6563	16406	13125
2 3/4	4	468500	351263	8050	20124	16100
3	4	567150	425363	10634	26585	21268
3 1/4	4	674500	505875	13701	34252	27402
3 1/2	4	791350	593513	17311	43277	34622
3 3/4	4	917700	688275	21509	53771	43017
4	4	1052600	7894505	26315	65788	52630

ASTM A354–BD / ASTM A490 / SAE Grade 8

Bolt Size	TPI	Proof Load (lbs)	Clamp Load (lbs)	Tightening Torque	
				Waxed	Plain
1/4	20	3800	2850	6	12
5/16	18	6300	4725	12	25
3/8	16	9300	6975	22	44
7/16	14	12750	9563	35	70
1/2	13	17050	12788	53	107
9/16	12	21850	16388	77	154
5/8	11	27100	20325	106	212
3/4	10	40100	30075	188	376
7/8	9	55450	41588	303	606
1	8	72700	54525	454	909
1 1/8	7	91550	68663	644	1287
1 1/4	7	120000	90000	938	1875
1 3/8	6	138600	103950	1191	2382
1 1/2	6	168600	126450	1581	3161
1 3/4	5	228000	171000	2494	4988
2	4½	300000	225000	3750	7500
2 1/4	4½	390000	292500	5484	10969
2 1/2	4	480000	360000	7500	15000
2 3/4	4	517650	388238	8897	17794
3	4	626850	470138	11753	23507
3 1/4	4	745500	559125	15143	30286
3 1/2	4	874650	655988	19133	38266
3 3/4	4	1014300	760725	23773	47545
4	4	1052600	789450	26315	52630

Notes:

1. Values calculated using industry accepted formula $T = KDP$ where T = Torque, K = torque coefficient (dimensionless), D = nominal diameter (inches), P = bolt clamp load, Ib.
2. K values: **waxed** (e.g. pressure wax as supplied on high strength nuts) = .10, hot dip galvanized = .25, and plain non–plated bolts(as received) = .20
3. Torque has been converted into ft/Ibs by dividing the result of the formula by 12.
4. All calculations are for Coarse Thread Series(UNC).
5. Grade 2 calculation only cover fasteners 1/4̋–3/4̋ in diameter up to 6̋ long; for longer fasteners the torque is reduced significantly.
6. Clamp loads are based on 75% of the minimum proof loads for each grade and size.
7. Proof load, stress area, yield strength, and other data is based on IFI 7th Edition(2003) Technical Data N-68,SAE J429, ASTM A307,A325,A449, and A490.

The above estimated torque calculations are only offered as a guide. Use if its content by anyone is the sole responsibility of that person and they assume all risk. Due to many variables that affect the torque-tension relationship like, human error, surface texture, lubrication etc, the only way to determine the correct torque is through experimentation undet actual joint and assembly conditions.

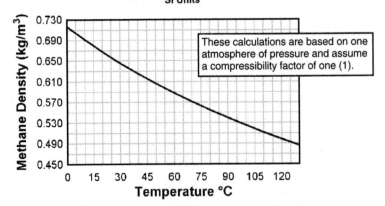

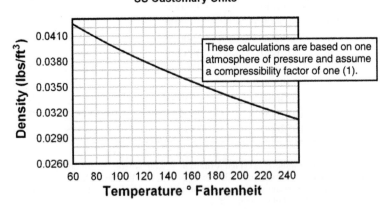

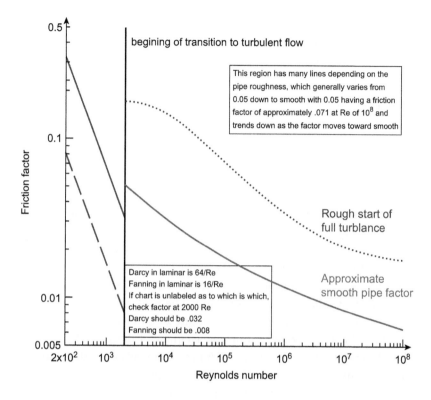

This chart is a comparison chart. The lines on the right represent how the two different factors vary with the two factors in laminar flow. The right side shows the region for the factor as the flow moves through the transition region from the beginning of turbulence to full turbulence as it varies with the friction factor and Reynolds number. The solid line represents smooth pipe and the dotted line represents an approximate full turbulence point.

Sample layout for Colebrook equation

Variables		
f	0.016032	
Reynolds#	3.85E+05	
E	0.00015	
D (Ft)	0.665	

$$\frac{1}{\sqrt{f}} = -2\log_{10}\left[\left(\frac{1}{3.7}\right)\left(\frac{\varepsilon}{D}\right) + \frac{2.51}{Re\sqrt{f}}\right]$$

	E/D	0.000226
	2.51/(Re*f^.5)	5.15E-05
	1/3.7	0.27027
	inside bracket	1.12E-04
	"-2log*bracket	7.90E+00

e factors to use				
new clean pipe		left side	7.897773	right side 7.90E+00
smooth like brass, glass, plastic	0.000005			
new steel	0.00015		set to 0	-1.05E-05
asphalt dipped cast iron	0.0004			use goal seek to set cell to 0 by changing f
galvanized iron	0.0005,			then f will be the calculated friction factor
uncoated cast iron	0.00085			
concrete	0.001to0.01			

This is a suggested layout for the solution to the Colebrook equation, which normally requires iterative solutions to find the answer. The development of spreadsheets with a goal seek function in them will allow one to set up cells for the opposite side of the equation and then set up a cell with the formula of subtracting one side from the other. The goal seek for that cell is to set that subtraction equal to zero. The reader will note that in this case the subtraction does not result in absolute zero but a very small number. This is most likely sufficient within the accuracy of the variable data.

For those efficient in the use of VBA in such spreadsheets, the worksheet can be made much more elegant and more automatic. This layout is to give the reader a picture of how to set up the calculation. As noted in the text in Chapter 4, the Swamee-Jain equation is a direct solution equation and is well within the accuracy of the variable data.

Gas or Vapor	Formula	Specific Heat Capacity				Ratio of Specific Heats	Individual Gas constant -R-	
		C_p (kJ/kg K)	C_v (kJ/kg K)	C_p (Btu/lb$_m$ °F)	C_v (Btu/lb$_m$ °F)	$K = C_p/C_v$	$C_p - C_v$ (kJ/kg K)	$C_p - C_v$ (ft lb$_f$/lb$_m$ °R)
Acetone		1.47	1.32	0.35	0.32	1.11	0.15	
Acetylene	C_2H_2	1.69	1.37	0.35	0.27	1.232	0.319	59.34
Air		1.01	0.718	0.24	0.17	1.4	0.287	53.34
Alcohol	C_2H_5OH	1.88	1.67	0.45	0.4	1.13	0.22	
Alcohol	CH_3OH	1.93	1.53	0.46	0.37	1.26	0.39	
Ammonia	NH_3	2.19	1.66	0.52	0.4	1.31	0.53	96.5
Argon	Ar	0.52	0.312	0.12	0.07	1.667	0.208	
Benzene	C_6H_6	1.09	0.99	0.26	0.24	1.12	0.1	
Blast furnace gas		1.03	0.73	0.25	0.17	1.41	0.3	55.05
Bromine		0.25	0.2	0.06	0.05	1.28	0.05	
Butatiene						1.12		
Butane	C_4H_{10}	1.67	1.53	0.395	0.356	1.094	0.143	26.5
Carbon Dioxide	CO_2	0.844	0.655	0.21	0.16	1.289	0.189	38.86
Carbon Monoxide	CO	1.02	0.72	0.24	0.17	1.4	0.297	55.14
Carbon disulphide		0.67	0.55	0.16	0.13	1.21	0.12	
Chlorine	CL_2	0.48	0.36	0.12	0.09	1.34	0.12	
Chloroform		0.63	0.55	0.15	0.13	1.15	0.08	
Ethane	C_2H_6	1.75	1.48	0.39	0.32	1.187	0.276	51.5
Ether		2.01	1.95	0.48	0.47	1.03	0.06	
Ethylene	C_2H_4	1.53	1.23	0.4	0.33	1.24	0.296	55.08
Freon 22						1.18		
Helium	HE	5.19	3.12	1.25	0.75	1.667	2.08	386.3
Hexane						1.06		
Hydrogen	H_2	14.32	10.16	3.42	2.43	1.405	4.12	765.9
Hydrogen Chloride	HCL	0.8	0.57	0.191	0.135	1.41	0.23	42.4

(Continued)

—cont'd

Gas or Vapor	Formula	Specific Heat Capacity				Ratio of Specific Heats	Individual Gas constant -R-	
		C_p (kJ/kg K)	C_v (kJ/kg K)	C_p (Btu/lb$_m$ °F)	C_v (Btu/lb$_m$ °F)	$K = C_p/C_v$	$C_p - C_v$ (kJ/kg K)	$C_p - C_v$ (ft lb$_f$/lb$_m$°R)
Hydrogen Sulfide	H_2S	1.76	1.27	0.243	0.187	1.32		45.2
Hydroxyl	OH					1.384	0.489	96.4
Methane	CH_4	2.22	1.7	0.59	0.45	1.304	0.518	30.6
Methyl Chloride	CH_3Cl	2.34	1.85	0.24	0.2	1.2		79.1
Natural Gas				0.56	0.44	1.27	0.5	
Neon		1.03	0.618			1.667	0.412	
Nitric Oxide	NO	0.995	0.718	0.23	0.17	1.386	0.277	
Nitrogen	N_2	1.04	0.743	0.25	0.18	1.4	0.297	54.99
Nitrogen tetroxide		4.69	4.6	1.12	1.1	1.02	0.09	
Nitrous oxide	N_2O	0.88	0.69	0.21	0.17	1.27	0.18	35.1
Oxygen	O_2	0.919	0.659	0.22	0.16	1.395	0.26	48.24
Pentane						1.07		
Propane	C_3H_8	1.67	1.48	0.39	0.34	1.127	0.189	35
Propene (propylene)	C_3H_6	1.5	1.31	0.36	0.31	1.15	0.18	36.8
Water Vapor Steam 1 psia. 120 - 600 °F		1.93	1.46	0.46	0.35	1.32	0.462	
Steam 14.7 psia. 220 - 600 °F		1.97	1.5	0.47	0.36	1.31	0.46	
Steam 150 psia. 360 - 600 °F		2.26	1.76	0.54	0.42	1.28	0.5	
Sulfur dioxide (Sulphur dioxide)	SO_2	0.64	0.51	0.15	0.12	1.29	0.13	24.1

k = cp / cv - the specific heat capacity ratio courtesy enginneringtoolbox.com
cp = specific heat in a constant pressure process
cv = specific heat in a constant volume process
R- Individual Gas constant

MOLECULAR WEIGHT OF COMMON GASES

Molecular Weight sorted alphabeticly		Molecular Weight sorted numerically	
Acetylene, C2H2	26.04	Deuterium	2.014
Air	28.966	Hydrogen, H2	2.016
Ammonia	17.02	Helium, He	4.02
Argon, Ar	39.948	Methane, CH4	16.044
Benzene	78.11	Hydroxyl, OH	17.01
N-Butane, C4H10	58.12	Ammonia	17.02
Iso-Butane (2-Metyl propane)	58.12	Water Vapor, H2O	18.02
Butadiene	54.09	Natural Gas	19
1-Butene	56.108	Neon, Ne	20.179
cis -2-Butene	56.108	Acetylene, C2H2	26.04
trans-2-Butene	56.108	Carbon Monoxide, CO	28.011
Isobutene	56.108	Nitrogen, N2	28.0134
Carbon Dioxide, CO2	44.01	Ethylene, C2H4	28.054
Carbon Disulphide	76.13	Air	28.966
Carbon Monoxide, CO	28.011	Nitric Oxide, NO2	30.006
Chlorine	70.906	Ethane, C2H6	30.07
Cyclohexane	84.16	Oxygen, O2	31.9988
Deuterium	2.014	Sulfur	32.02
Ethane, C2H6	30.07	Methyl Alcohol	32.04
Ethyl Alcohol	46.07	Hydrogen Sulfide	34.076
Ethyl Chloride	64.515	Hydrogen Chloride	36.461
Ethylene, C2H4	28.054	Hydrochloric Acid	36.47
Fluorine	37.996	Fluorine	37.996
Helium, He	4.02	Argon, Ar	39.948
N-Heptane	100.2	Propylene	42.08
Hexane	86.17	Carbon Dioxide, CO2	44.01
Hydrochloric Acid	36.47	Nitrous Oxide	44.012
Hydrogen, H2	2.016	Propane, C3H8	44.097
Hydrogen Chloride	36.461	Ethyl Alcohol	46.07
Hydrogen Sulfide	34.076	Ozone	47.998
HydroxyI, OH	17.01	Sulfuric Oxide	48.1
Krypton	83.8	Methyl Chloride	50.488
Methane, CH4	16.044	Butadiene	54.09
Methyl Alcohol	32.04	1 -Butene	56.108
Methyl Butane	72.15	cis-2-Butene	56.108
Methyl Chloride	50.488	trans-2-Butene	56.108
Natural Gas	19	Isobutene	56.108
Neon, Ne	20.179	N-Butane, C4H10	58.12
Nitric Oxide, NO2	30.006	Iso-Butane (2-Metyl propane)	58.12

(Continued)

—cont'd

Molecular Weight sorted alphabeticly		Molecular Weight sorted numerically	
Nitrogen, N2	28.0134	R-611	60.05
Nitrous Oxide	44.012	Sulfur Dioxide	64.06
N-Octane	114.22	Ethyl Chloride	64.515
Oxygen, O2	31.9988	Chlorine	70.906
Ozone	47.998	Methyl Butane	72.15
N-Pentane	72.15	N-Pentane	72.15
Iso-Pentane	72.15	Iso-Pentane	72.15
Propane, C3H8	44.097	Carbon Disulphide	76.13
Propylene	42.08	Benzene	78.11
R-11	137.37	Krypton	83.8
R-12	120.92	Cyclohexane	84.16
R-22	86.48	Hexane	86.17
R-114	170.93	R-22	86.48
R-123	152.93	Toluene	92.13
R-134a	102.03	N-Heptane	100.2
R-611	60.05	R-134a	102.03
Sulfur	32.02	N-Octane	114.22
Sulfur Dioxide	64.06	R-12	120.92
Sulfuric Oxide	48.1	Xenon	131.3
Toluene	92.13	R-11	137.37
Xenon	131.3	R-123	152.93
Water Vapor, H2O	18.02	R-114	170.93

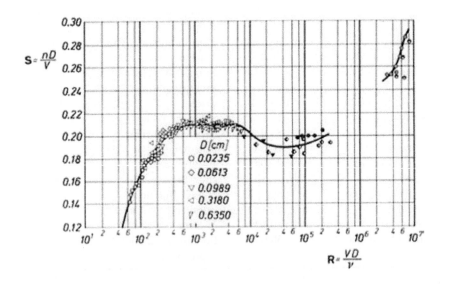

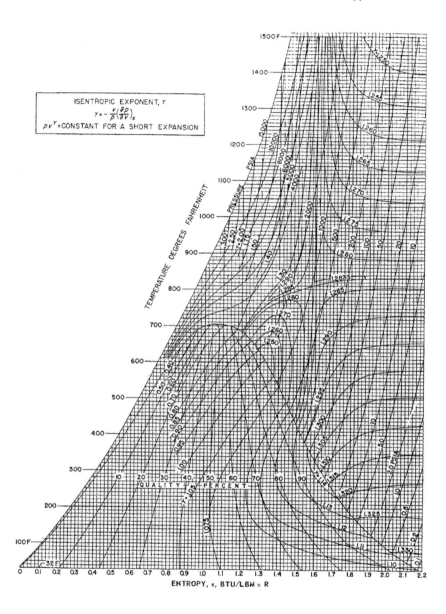

ISENTROPIC EXPONENT, γ

$$\gamma = -\frac{v}{p}\left(\frac{\partial p}{\partial v}\right)_s$$

pv^γ =CONSTANT FOR A SHORT EXPANSION

ENTROPY, s, BTU/LBM × R

Non-asbestos calcium silicate insulation board and pipe insulation features light weight, low thermal conductivity, and high temperature and chemical resistance.

Calcium silicate is a rigid, high density material used for high temprature applications ranging from *250° F(121° C)-1000° F (540°C).*

The relationship between temperature and thermal conductivity is indicated in the diagram below:

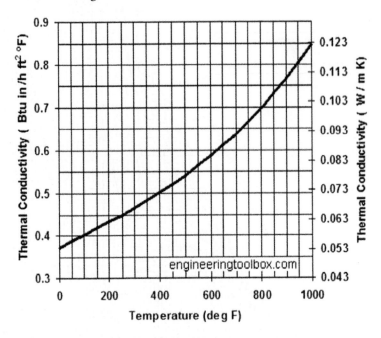

Thermal conductivity of some common metals

Thermal conductivity - k - is the quantity of heat transmitted, due to unit temperature gradient, in unit time under steady conditions in a direction normal to a surface of unit area. Thermal conductivity - k - is used in Fourier's equation.

- 1 Btu/(hr °F ft^2/ft) = 1 Btu/(hr °F ft) = 1.731 W/(m K) = 1.488 kcal/(h m °C)

Metal	Temperature - t - (°F)	Thermal Conductivity - k - (Btu/(hr °F ft))
Admiralty Brass	68	64
Aluminum, pure	68	118
	200	124
	400	144
Aluminum Bronze	68	44
Antimony	68	10.7
Beryllium	68	126

(Continued)

—cont'd

Metal	Temperature - t - (°F)	Thermal Conductivity - k - (Btu/(hr °F ft))
Beryllium Copper	68	38
Bismuth	68	4.9
Cadmium	68	54
Carbon Steel, max 0.5% C	68	31
Carbon Steel, max 1.5% C	68	21
	752	19
	2192	17
Cast Iron, gray	70	27–46
Chromium	68	52
Cobalt	68	40
Copper, pure	68	223
	572	213
	1112	204
Copper bronze (75% Cu, 25% Zi)	68	15
Copper brass (70% Cu, 30% Zi)	68	64
Cupronickel	68	17
Gold	68	182
Hastelloy B		6
Hastelloy C	70	5
Inconel	70-212	8.4
Incoloy	32-212	6.8
Iridium	68	85
Iron, nodular pearlitic	212	18
Iron, pure	68	42
	572	32
	1832	20
Iron, wrought	68	34
Lead	68	20
	572	17.2
Manganese Bronze	68	61
Magnesium	68	91.9
Mercury	68	4.85
Molybdenum	68	81
Monel	32-212	15
Nickel	68	52
Nickel Wrought	32-212	35–52
Niobium (Columbium)	68	30
Osmium	68	35
Platinum	68	42
Plutonium	68	4.6

(Continued)

—cont'd

Metal	Temperature - t - (°F)	Thermal Conductivity - k - (Btu/(hr °F ft))
Potassium	68	57.8
Red Brass	68	92
Rhodium	68	86.7
Selenium	68	0.3
Silicon	68	48.3
Silver, pure	68	235
Sodium	68	77.5
Stainless Steel	68	7-26
Tantalum	68	31
Thorium	68	24
Tin	32	36-39
Titanium	68	11–13
Tungsten	68	94–100
Uranium	68	14
Vanadium	68	35
Wrought Carbon Steel	32	34
Yellow Brass	68	67
Zinc	—	67
Zirconium	—	145

- $T(^{0}C) = 5/9[T(^{0}F) - 32]$
- $1\ Btu/(hr\ ^{0}F\ ft^{2}/ft) = 1\ Btu/(hr\ ^{0}F\ ft) = 1.731\ W/(m\ K) = 1.488\ kcal/(h\ m\ ^{0}C)$

Thermal conductivity is the quantity of heat transmitted through a unit thickness in a direction normal to a surface of unit area, due to a unit temperature gradient under steady state conditions.

Thermal conductivity, or heat transfer coefficients, of some common materials and products are indicated in the table below.

- $W/(mK) = 1\ W/(m^{0}C) = 0.85984\ kcal/(hr\ m^{0}C) = 0.5779\ Btu/(ft\ hr\ ^{0}F)$

Thermal Conductivity - k - (W/mK)

Material/Substance	Temperature		
	25	125	225
Acetone	0.16		
Acrylic	0.2		
Air	0.024		

(Continued)

—cont'd

Thermal Conductivity - k - (W/mK)

Material/Substance	Temperature		
	25	125	225
Alcohol	0.17		
Aluminum	250	255	250
Aluminum Oxide	30		
Ammonia	0.022		
Antimony	18.5		
Argon	0.016		
Asbestos–cement board	0.744		
Asbestos–cement sheets	0.166		
Asbestos–cement	2.07		
Asbestos, loosely packed	0.15		
Asbestos mill board	0.14		
Asphalt	0.75		
Balsa	0.048		
Bitumen	0.17		
Benzene	0.16		
Beryllium	218		
Brass	109		
Brick dense	1.31		
Brick work	0.69		
Cadmium	92		
Carbon	1.7		
Carbon dioxide	0.0146		
Cement, portland	0.29		
Cement, mortar	1.73		
Chalk	0.09		
Chrome Nickel Steel (18% Cr, 8% Ni)	16.3		
Clay, dry to moist	0.15–1.8		
Clay, saturated	0.6–2.5		
Cobalt	69		
Concrete, light	0.42		
Concrete, stone	1.7		
Constantan	22		
Copper	401	400	398
Corian (ceramic filled)	1.06		
Corkboard	0.043		
Cork, regranulated	0.044		
Cork	0.07		
Cotton	0.03		

(*Continued*)

—cont'd

Thermal Conductivity - k - (W/mK)

Material/Substance	Temperature		
	25	125	225
Carbon Steel	54	51	47
Cotton Wool insulation	0.029		
Diatomaceous earth (Sil-o-cel)	0.06		
Earth, dry	1.5		
Ether	0.14		
Epoxy	0.35		
Felt insulation	0.04		
Fiberglass	0.04		
Fiber insulating board	0.048		
Fiber hardboard	0.2		
Fireclay brick 500°C	1.4		
Foam glass	0.045		
Freon 12	0.073		
Gasoline	0.15		
Glass	1.05		
Glass, Pearls, dry	0.18		
Glass, Pearls, saturated	0.76		
Glass, window	0.96		
Glass, wool Insulation	0.04		
Glycerol	0.28		
Gold	310	312	310
Granite	1.7-4.0		
Gypsum or plaster board	0.17		
Hairfelt	0.05		
Hardboard high density	0.15		
Hardwoods (oak, maple . . .)	0.16		
Helium	0.142		
Hydrogen	0.168		
Ice (0°C, 32°F)	2.18		
Insulation materials	0.035-0.16		
Iridium	147		
Iron	80	68	60
Iron, wrought	59		
Iron, cast	55		
Kapok insulation	0.034		
Kerosene	0.15		
Lead Pb	35		
Leather, dry	0.14		

(Continued)

—cont'd

Thermal Conductivity - k - (W/mK)

Material/Substance	25	125	225
		Temperature	
Limestone	1.26–1.33		
Magnesia insulation (85%)	0.07		
Magnesite	4.15		
Magnesium	156		
Marble	2.08–2.94		
Mercury	8		
Methane	0.030		
Methanol	0.21		
Mica	0.71		
Mineral insulation materials, wool blankets . . .	0.04		
Molybdenum	138		
Monel	26		
Nickel	91		
Nitrogen	0.024		
Nylon 6	0.25		
Oil, machine lubricating SAE 50	0.15		
Olive oil	0.17		
Oxygen	0.024		
Paper	0.05		
Paraffin Wax	0.25		
Perlite, atmospheric pressure	0.031		
Perlite, vacuum	0.00137		
Plaster, gypsum	0.48		
Plaster, metal lath	0.47		
Plaster, wood lath	0.28		
Plastics, foamed (insulation materials)	0.03		
Plastics, solid			
Platinum	70	71	72
Plywood	0.13		
Polyethylene HD	0.42–0.51		
Polypropylene	0.1–0.22		
Polystyrene expanded	0.03		
Porcelain	1.5		
PTFE	0.25		
PVC	0.19		
Pyrex glass	1.005		
Quartz mineral	3		
Rock, solid	2–7		

(Continued)

—cont'd

Thermal Conductivity - k - (W/mK)

Material/Substance	25	125	225
Rock, porous volcanic (Tuff)	0.5–2.5		
Rock Wool insulation	0.045		
Sand, dry	0.15–0.25		
Sand, moist	0.25–2		
Sand, saturated	2–4		
Sandstone	1.7		
Sawdust	0.08		
Silica aerogel	0.02		
Silicone oil	0.1		
Silver	429		
Snow (temp < 0°C)	0.05–0.25		
Sodium	84		
Softwoods (fir, pine . . .)	0.12		
Soil, with organic matter	0.15–2		
Soil, saturated	0.6–4		
Steel, Carbon 1%	43		
Stainless Steel	16	17	19
Straw insulation	0.09		
Styrofoam	0.033		
Tin Sn	67		
Zinc Zn	116		
Urethane foam	0.021		
Vermiculite	0.058		
Vinyl ester	0.25		
Water	0.58		
Water, vapor (steam)		0.016	
Wood across the grain, white pine	0.12		
Wood across the grain, balsa	0.055		
Wood across the grain, yellow pine	0.147		
Wood, oak	0.17		
Wool, felt	0.07		

- $1\ W/(m\ K) = 1\ W/(m°\ C) = 0.85984\ kcal/(h\ m\ °C) = 0.5779\ Btu/(ft\ h\ °F)$

A COLLECTION OF FORMULAS

I. Linear Spring Constants

("Load" per inch deflection)

Coil dia. D; wire dia. d; n turns
$$k = \frac{Gd^4}{8nD^3}$$ (1)

Cantilever
$$k = \frac{3EI}{l^3}$$ (2)

Cantilever
$$k = \frac{2EI}{l^3}$$ (3)

Beam on two supports; centrally loaded
$$k = \frac{48EI}{l^3}$$ (4)

Beam on two supports; load off center
$$k = \frac{3EIl}{l_1^2 l_2^2}$$ (5)

Clamped-clamped beam; centrally loaded
$$k = \frac{192EI}{l^3}$$ (6)

Circular plate, thickness t; centrally loaded; circumferential edge simply supported
$$k = \frac{16\pi D}{R^2}\frac{1+\mu}{3+\mu}$$ (7)

in which the plate constant is
$$D = \frac{Et^3}{12(1-\mu^2)}$$ (7a)

μ = Poisson's ratio ≈ 0.3

Circular plate; circumferential edge clamped
$$k = \frac{16\pi D}{R^2}$$ (8)

Two springs in series
$$k = \frac{1}{1/k_1 + 1/k_2}$$ (9)

II. Rotational Spring Constants

("Load" per radian rotation)

Twist of coil spring; wire dia. d; coil dia. D; n turns
$$k = \frac{Ed^4}{64nD}$$ (10)

Bending of coil spring $\quad k = \dfrac{Ed^4}{32nD} \cdot \dfrac{1}{1 + E/2G}$ $\qquad$ (11)

Spiral spring; total length l; moment of inertia of cross section I $\qquad k = \dfrac{EI}{l}$ $\qquad$ (12)

Twist of hollow circular shaft, outer dia. D, inner dia. d, length l $\qquad k = \dfrac{GI_p}{l} = \dfrac{\pi}{32} \dfrac{G(D^4 - d^4)}{l}$ $\qquad$ (13)

$\qquad\qquad$ For steel $\quad k = 1.18 \times 10^6 \times \dfrac{D^4 - d^4}{l}$

Cantilever $\qquad k = \dfrac{EI}{l}$ $\qquad$ (14)

Cantilever $\qquad k = \dfrac{2EI}{l^2}$ $\qquad$ (15)

Beam on two simple supports; couple at center $\qquad k = \dfrac{12EI}{l}$ $\qquad$ (16)

Clamped-clamped beam; couple at center $\qquad k = \dfrac{16EI}{l}$ $\qquad$ (17)

III. Natural Frequencies of Simple Systems

End mass M; spring mass m, spring stiffness k $\qquad \omega_n = \sqrt{k/(M + m/3)}$ $\qquad$ (18)

End inertia I; shaft inertia I_s, shaft stiffness k $\qquad \omega_n = \sqrt{k/(I + I_s/3)}$ $\qquad$ (19)

Two disks on a shaft $\qquad \omega_n = \sqrt{\dfrac{k(I_1 + I_2)}{I_1 I_2}}$ $\qquad$ (20)

Cantilever; end mass M; beam mass m, stiffness by formula (2) $\qquad \omega_n = \sqrt{\dfrac{k}{M + 0.23m}}$ $\qquad$ (21)

Simply supported beam; central mass M; beam mass m; stiffness by formula (4) $\qquad \omega_n = \sqrt{\dfrac{k}{M + 0.5m}}$ $\qquad$ (22)

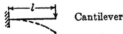 Massless gears, speed of I_2 n times as large as speed of I_1 $\qquad \omega_n = \sqrt{\dfrac{1}{\dfrac{1}{k_1} + \dfrac{1}{n^2 k_2}} \times \dfrac{I_1 + n^2 I_2}{I_1 \cdot n^2 I_2}}$ $\qquad$ (23)

Longitudinal vibration of cantilever: A = cross section, E = modulus of elasticity.

μ_1 = mass per unit length, $n = 0,1,2,3$ = number of nodes

$$\omega_n = \left(n + \frac{1}{2}\right)\pi\sqrt{\frac{AE}{\mu_1 l^2}} \qquad (25)$$

For steel and l in inches this becomes

$$f = \frac{\omega_n}{2\pi} = (1 + 2n)\frac{51,000}{l}$$

cycles per second $(25a)$

Longitudinal vibration of beam clamped (or free) at both ends

For steel, l in inches

$$\omega_n = n\pi\sqrt{\frac{AE}{\mu_1 l^2}} \quad n = 1, 2, 3, \ldots (26)$$

$$f = \frac{\omega_n}{2\pi} = \frac{102,000}{l} \text{ cycles/sec.} \quad (26a)$$

Torsional vibration of beams

Same as (25) and (26); replace tensional stiffness AE by torsional stiffness C ($= GI_p$ for circular cross section); replace μ_1 by the moment of inertia per unit length $i = \dfrac{I_{bar}}{l}$

Organ pipe open at one end, closed at the other

For air at 60°F., l in inches:

$$f = \frac{\omega_n}{2\pi} = (1 + 2n)\frac{3,300}{l}$$

cycles/sec. $(27a)$

$$n = 0, 1, 2, 3, \ldots$$

Water column in rigid pipe closed at one end (l in inches)

$$f = \frac{\omega_n}{2\pi} = (1 + 2n)\frac{14,200}{l}$$

cycles/sec. $(27b)$

$$n = 0, 1, 2, 3, \ldots$$

Organ pipe closed (or open) at both ends (air at 60°F.)

$$f = \frac{n6,600}{l} \text{ cycles/sec.} \qquad (28a)$$

$$n = 1, 2, 3, \ldots$$

Water column in rigid pipe closed (or open) at both ends

$$f = \frac{n28,400}{l} \text{ cycles/sec.} \qquad (28b)$$

$$n = 1, 2, 3, \ldots$$

For water columns in non-rigid pipes

$$\frac{f_{non-rigid}}{f_{rigid}} = \frac{1}{\sqrt{1 + \dfrac{300,000D}{E_{pipe}t}}}$$

E_{pipe} = elastic modulus of pipe, lb./in.^{2}

D, t = pipe diameter and wall thickness, same units

Membrane of any shape of area A roughly of equal dimensions in all directions, fundamental mode:

$$\omega_n = \text{const.} \sqrt{\frac{T}{\mu_1 A}} \tag{33}$$

circle.......................... const. $= 2.40\pi = 4.26$
square......................... const. $= 4.44$
quarter circle.................. const. $= 4.55$
2×1 rectangle.............. const. $= 4.97$

Circular plate of radius r, mass per unit area μ_1; the "plate constant D" defined by Eq. (7a),

$$\omega_n = a \sqrt{\frac{D}{\mu_1 r^4}} \tag{34}$$

For free edges, 2 perp. nodal diameters............. $a = 5.25$
For free edges, one nodal circle, no diameters....... $a = 9.07$
Clamped edges, fundamental mode................ $a = 10.21$
Free edges, clamped at center, umbrella mode....... $a = 3.75$

Rectangular plate, all edges simply supported, dimensions l_1 and l_2:

$$\omega_n = \pi^2 \left(\frac{m^2}{l_1^2} + \frac{n^2}{l_2^2} \right) \sqrt{\frac{D}{\mu_1}} \qquad m = 1, 2, 3, \ldots ; n = 1, 2, 3, \ldots \tag{35}$$

Square plate, all edges clamped, length of side l, fundamental mode:

$$\omega_n = \frac{36}{l^2} \sqrt{\frac{D}{\mu_1}} \tag{36}$$

Uniform Beams

(Transverse or bending vibrations)

The same general formula holds for all the following cases,

$$\omega_n = a_n \sqrt{\frac{EI}{\mu_1 l^4}}$$

where EI is the bending stiffness of the section, l is the length of the beam, μ_1 is the mass per unit length $= W/gl$, and a_n is a numerical constant, different for each case and listed below

Cantilever or "clamped-free" beam

$a_1 = 3.52$
$a_2 = 22.0$
$a_3 = 61.7$
$a_4 = 121.0$
$a_5 = 200.0$

Simply supported or "hinged-hinged" beam

$a_1 = \pi^2 = 9.8$
$a_2 = 4\pi^2 = 39.5$
$a_3 = 9\pi^2 = 88.9$
$a_4 = 16\pi^2 = 158.$
$a_5 = 25\pi^2 = 247.$

"Free-free" beam or floating ship

$a_1 = 22.0$
$a_2 = 61.7$
$a_3 = 121.0$
$a_4 = 200.0$
$a_5 = 298.2$

"Clamped-clamped" beam has same frequencies as "free-free"

$a_1 = 22.0$
$a_2 = 61.7$
$a_3 = 121.0$
$a_4 = 200.0$
$a_5 = 298.2$

"Clamped-hinged" beam may be considered as half a "clamped-clamped" beam for even a-numbers

$a_1 = 15.4$
$a_2 = 50.0$
$a_3 = 104.$
$a_4 = 178.$
$a_5 = 272.$

"Hinged-free" beam or wing of autogyro may be considered as half a "free-free" beam for even a-numbers

$a_1 = 0$
$a_2 = 15.4$
$a_3 = 50.0$
$a_4 = 104.$
$a_5 = 178.$

K Factors for Equivalent Length Calculations

Tee, Flanged, Dividing Line Flow	0.2
Tee, Threaded, Dividing Line Flow	0.9
Tee, Flanged, Dividing Branched Flow	1.0
Tee, Threaded, Dividing Branch Flow	2.0
Union, Threaded	0.08
Elbow, Flanged Regular 90°	0.3
Elbow, Threaded Regular 90°	1.5
Elbow, Threaded Regular 45°	0.4
Elbow, Flanged Long Radius 90°	0.2
Elbow, Threaded Long Radius 90°	0.7
Elbow, Flanged Long Radius 45°	0.2
Return Bend, Flanged 180°	0.2
Return Bend, Threaded 180°	1.5
Globe Valve, Fully Open	10
Angle Valve, Fully Open	2
Gate Valve, Fully Open	0.15
Gate Valve, 1/4 Closed	0.26
Gate Valve, 1/2 Closed	2.1
Gate Valve, 3/4 Closed	17
Swing Check Valve, Forward Flow	2
Ball Valve, Fully Open	0.05
Ball Valve, 1/3 Closed	5.5
Ball Valve, 2/3 Closed	200
Diaphragm Valve, Open	2.3
Diaphragm Valve, Half Open	4.3
Diaphragm Valve, 1/4 Open	21
Water Meter	7

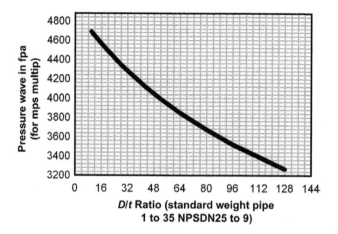

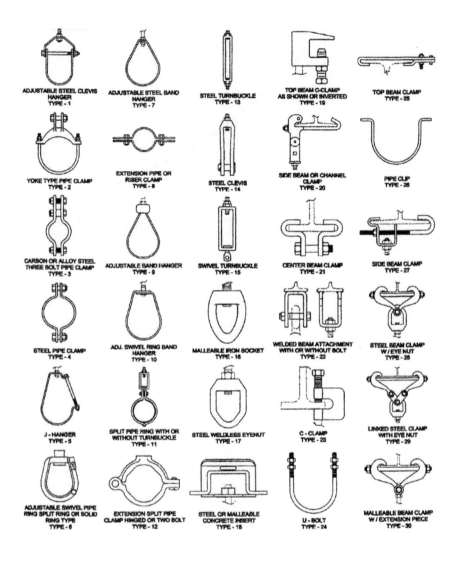

ADJUSTABLE STEEL CLEVIS HANGER
TYPE - 1

ADJUSTABLE STEEL BAND HANGER
TYPE - 7

STEEL TURNBUCKLE
TYPE - 13

TOP BEAM C-CLAMP AS SHOWN OR INVERTED
TYPE - 19

TOP BEAM CLAMP
TYPE - 25

YOKE TYPE PIPE CLAMP
TYPE - 2

EXTENSION PIPE OR RISER CLAMP
TYPE - 8

STEEL CLEVIS
TYPE - 14

SIDE BEAM OR CHANNEL CLAMP
TYPE - 20

PIPE CLIP
TYPE - 26

CARSON OR ALLOY STEEL THREE BOLT PIPE CLAMP
TYPE - 3

ADJUSTABLE BAND HANGER
TYPE - 9

SWIVEL TURNBUCKLE
TYPE - 15

CENTER BEAM CLAMP
TYPE - 21

SIDE BEAM CLAMP
TYPE - 27

STEEL PIPE CLAMP
TYPE - 4

ADJ. SWIVEL RING BAND HANGER
TYPE - 10

MALLEABLE IRON SOCKET
TYPE - 16

WELDED BEAM ATTACHMENT WITH OR WITHOUT BOLT
TYPE - 22

STEEL BEAM CLAMP W / EYE NUT
TYPE - 28

J - HANGER
TYPE - 5

SPLIT PIPE RING WITH OR WITHOUT TURNBUCKLE
TYPE - 11

STEEL WELDLESS EYENUT
TYPE - 17

C - CLAMP
TYPE - 23

LINKED STEEL CLAMP WITH EYE NUT
TYPE - 29

ADJUSTABLE SWIVEL PIPE RING SPLIT RING OR SOLID RING TYPE
TYPE - 6

EXTENSION SPLIT PIPE CLAMP HINGED OR TWO BOLT
TYPE - 12

STEEL OR MALLEABLE CONCRETE INSERT
TYPE - 18

U - BOLT
TYPE - 24

MALLEABLE BEAM CLAMP W / EXTENSION PIECE
TYPE - 30

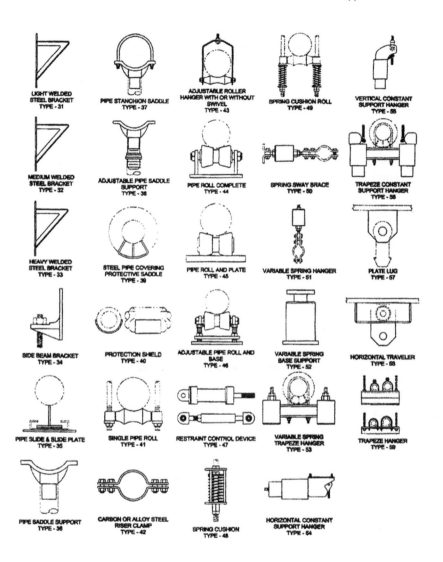

LIGHT WELDED
STEEL BRACKET
TYPE - 31

PIPE STANCHION SADDLE
TYPE - 37

ADJUSTABLE ROLLER
HANGER WITH OR WITHOUT
SWIVEL
TYPE - 43

SPRING CUSHION ROLL
TYPE - 49

VERTICAL CONSTANT
SUPPORT HANGER
TYPE - 55

MEDIUM WELDED
STEEL BRACKET
TYPE - 32

ADJUSTABLE PIPE SADDLE
SUPPORT
TYPE - 38

PIPE ROLL COMPLETE
TYPE - 44

SPRING SWAY BRACE
TYPE - 50

TRAPEZE CONSTANT
SUPPORT HANGER
TYPE - 56

HEAVY WELDED
STEEL BRACKET
TYPE - 33

STEEL PIPE COVERING
PROTECTIVE SADDLE
TYPE - 39

PIPE ROLL AND PLATE
TYPE - 45

VARIABLE SPRING HANGER
TYPE - 51

PLATE LUG
TYPE - 57

SIDE BEAM BRACKET
TYPE - 34

PROTECTION SHIELD
TYPE - 40

ADJUSTABLE PIPE ROLL AND
BASE
TYPE - 46

VARIABLE SPRING
BASE SUPPORT
TYPE - 52

HORIZONTAL TRAVELER
TYPE - 58

PIPE SLIDE & SLIDE PLATE
TYPE - 35

SINGLE PIPE ROLL
TYPE - 41

RESTRAINT CONTROL DEVICE
TYPE - 47

VARIABLE SPRING
TRAPEZE HANGER
TYPE - 53

TRAPEZE HANGER
TYPE - 59

PIPE SADDLE SUPPORT
TYPE - 36

CARBON OR ALLOY STEEL
RISER CLAMP
TYPE - 42

SPRING CUSHION
TYPE - 48

HORIZONTAL CONSTANT
SUPPORT HANGER
TYPE - 54

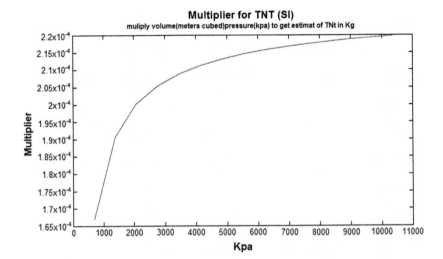

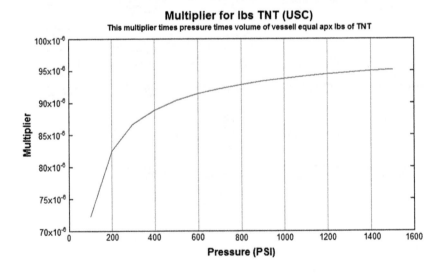

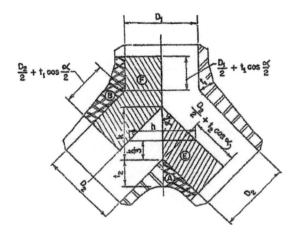

$$S_A \gtreqless \frac{p\left(E + \frac{1}{2}A\right)}{A}$$

$$S_B \gtreqless \frac{p\left(F + \frac{1}{2}B\right)}{B} \quad \text{USE ALSO FOR } 45° \text{ ELBOW}$$

WYE OR 45° ELBOW

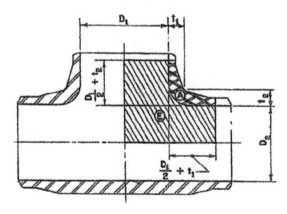

$$S_A \gtreqless \frac{p\left(E + \frac{1}{2}A\right)}{A}$$

TEE

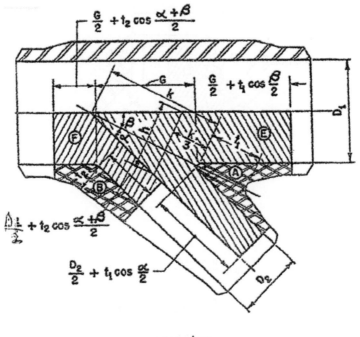

$$S_A \gtreqqless \frac{p(E + \frac{1}{2}A)}{A}$$

$$S_B \gtreqqless \frac{p(F + \frac{1}{2}B)}{B}$$

LATERAL

COMPANY IDENTITY

PROJECT NUMBER xxxx

TEST NUMBER xx RESULTS

On mm/dd/yyyy, Company conducted a burst test on a 6″. NPS schedule 120 Buttweld Tee, having dimensions in accordance with ANSI B16.9xxxx Edition.

The test procedure of Company specification xxxx, Revision x, dated mm/dd/yyyy was followed. The pressure to exceed 100% of P(adj) per ANSI B16.9, to successfully complete the burst test.

This pressure is:

FITTING DESCRIPTION	PRESSURE TO EXCEED 100% P(adj) PER ANSI B16.9
6″. Sch 120 Buttweld Tee	13,262 psig

The actual pressure achieved at burst for this fittings was 14,440 psig; 108.8% of the pressure required by ANSI B16.9.

The results show the fitting design meets the Section 9 of ANSI B16.9–1986 Edition for the Buttweld Tees.

Name and title of the signer of report

Project Number xxxx

Burst Test Number xx

BURST TEST PRESSURE CALCULATIONS

I. PURPOSE

The purpose of these calculations is to determine the required burst pressure of the test sample of this program.

The calculated pressures will be based on the pipe which the fitting tested is intended.

The actual pipe used in the this test will be a heavier schedule than that for which the fitting is designed. The purpose is to test fitting, not the pipe.

II. CALCULATIONS

The following calculations are based on the rules and guidlines of Company specification xxxx, Revidion x, dated mm/dd/yyyy.

(a) The adjusted proof test pressure shall be determind as follows:

$$P(adj.) = Px \underline{S(act.)} S$$

Where:

P(adj.) Adjusted proof test pressure, psig

P Computed bursting pressure of pipe that the fitting marking identifies, psig

S Minimum specified tensile strength of the pipe that the fitting marking identifies, psi

S(act.) Actual tensile strength of the test fitting material, psi

(b) The computed proof test pressure shall be determined as follows:

$$P = \frac{2St}{D}$$

Where:

P Computed burst pressure of the pipe that the fitting marking identifies, psig

PROJECT NUMBER xxxx
BURST TEST NUMBER xx

ASSEMBLY FABRICATION DRAWING

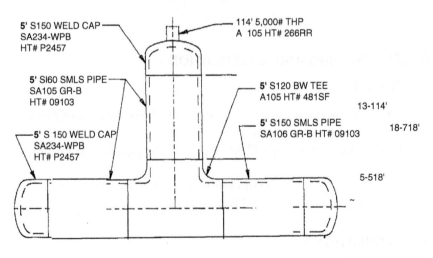

5' S150 WELD CAP
SA234-WPB
HT# P2457

5' SI60 SMLS PIPE
SA105 GR-B
HT# 09103

5' S 150 WELD CAP
SA234-WPB
HT# P2457

114' 5,000# THP
A 105 HT# 266RR

5' S120 BW TEE
A105 HT# 481SF

13-114'

5' S150 SMLS PIPE
SA106 GR-B HT# 09103

18-718'

5-518'

5' S150 WELD CAP
SA234-WPB
HT# P2457

S Minimum specified tensile strength of the pipe, psi
t Nominal pipe wall thickness, inches
D Specified outside diameter of pipe, inches

III. HEAT CODE DESCRIPTION OF THE FITTINGS AND PIPE

6" S120 Tee	No. 4818F
6" S160 Pipe	No. 09103
6" S160 End Cap	No. P2457
1/4" NPT 6,000# Threaded Pipet	No. 266RR

IV. FITTING QUALIFICATIONS (BASED ON THE BRANCH)

The following sizes are qualified by company Burst Test Number xx in accordance with section 9 of ANSI B16.9-1986 Edition.

SIZE AND LIMITING WALL THICKNESS	STANDARD SCHEDULES
3" NPS (0.148" W) thru (0.891 W)	Sch 40 thru Sch XXS
3-1/2" NP S (0.170" W) thru (1.018 W)	Sch 40 thru Sch 80
4" NP S (0.191" W) thru (1.145 W)	Sch 40 thru 8ch XX8
5" NP S (0.236" W) thru (1.416 W)	Sch 40 thru 8ch XX8
6" NP S (0.281" W) thru (1.686 W)	8ch 80 thru Sch XX8
8" NPS (0.366" W) thru (2.195 W)	8ch 60 thru 8ch 160
10" NPS (0.456" W) thru (2.736 W)	Sch 60 thru 8ch 160
12" NPS (0.541" W) thru (3.245 W)	8ch 60 thru Sch 160

Nuclear Products, 'NC

Project Number xxxx
Burst Test Number xx

V. MINIMUM SPECIFIED PROPERTIES OF THE PIPE

MATERIAL	O.D. IN	NOMINAL WALL IN.	MIN. SPEC. YLD. STG. psi	MIN. SPEC. TEN. STG. psi
A106 GR–B	6.625	0.562	35,000	60,000

VI. ACTUAL PROPERTIES AND REQUIRED PRESSURES

DESCRIPTION FITTING	S(act) psi	COMPUTED BURST (psi)	P(adj) (psi)	105% P(adj) (psi)	110% P(adj) (psi)
6″ S120 BW TEE	71,063	10,180	12,057	12,659	13,262
6″ S160 Pipe	64,000	★★			
6″ S160 Cap	65,000	★★			
1/4″ 6,000# THP	86,363	★★			

★★ These pieces are not being qualified by this test program, they were required to complete the test assembly

VII. SUMMARY

For ASME B16.9-xxxx, the test assembly must withstand at least 100% of P(adj). If the pipe ruptures or if sufficient pressure to rupture any part of the assembly can not be attained, the test pressure is acceptable if a final test pressure is at least 105% of the adjusted proof test pressure

The test has shown that the company design philosophy for these type fittings proved successful. These results satisfy the code requirement of 100% strength replacement for the design of pipe fittings, and futher reinforce the statement that the company design procedure for these type fittings is safe and conservative.

Conversions Table One

Metric Base Unit and Comparable USC Units

Base	Name	Symbol	UsCom	Name	Symbol
Length	meter	m	Length	foot	ft
Mass	kilogram	kg	Mass[1,2]	slug	slug
Time	Second	S	Time	second	second
Electric current	ampere	A	Electric current	ampere	A
Thermodynamic temperature	Kelvin	K	Thermodynamic temperature	Rankine	R
Amount of subatance	mole	mole	Amount of substance	mole	mole
Luminous intensity	candela	Cd	luminous intensity	candela	cd

Notes:

[1]The weight in U.S. customary is 1 pound fource and requires the slug for those calculations that require inertial units such as $F = ma$ from Newton's laws. One way the slug is avioded is whenever one see a formula, such as Newton's, using mass, substitute this with the ratio of weight/gravitational force in the engineering system. This keeps the units compatible because weight is defined as mass(g).

[2]In general, all other units are derived by diamentional analysis or definition. The other concern might be the prefixes, which in the SI system are the simple movement of the decimal places with prefixes that are by name.

Growing Larger			Growing Smaller		
Factor	Name	Symbol	Factor	Name	Symbol
10^1	deka	da	10^{-1}	deci	d
10^2	hecto	h	10^{-2}	centi	c
10^3	kilo	k	trun -110^{-3}	milli	m
10^6	mega	M	10^{-6}	micro	^
10^9	giga	G	10^{-9}	nano	n
10^{12}	tera	T	10^{-12}	pico	p
10^{15}	peta	P	10^{-15}	femto	f
10^{18}	exa	E	10^{-18}	atto	a
10^{21}	zeta	Z	10^{-21}	zepto	z
10^{24}	yotta	Y	10^{-24}	yocto	y

SI Derived Units Relating to Piping

Derived Quantity	SI Derived Unit	Expressed in Terms
Plane angle	Radian	$180/\pi$
Frequency	Hertz	$1s^{-1}$
Force	Newton	$1kg/m^2$
Pressure, Stress	Pascal Pa	$1N/m^2$
Energy	Joule	$1N/m$
Power	Watt	$1J/s$

Conversion Factor

Multiply	By	To Obtain
Absolute viscosity (poise)	1	Gram/second centimeter
Absolute viscosity (centipoise)	0.01	Poise
Acceleration due to gravity (g)	32.174	Feet/second2
	980.6	Centimeters/second2
Acres	0.4047	Hectares
	10	Square Chains
	43,560	Square Feet
	4047	Square Meters
	0.001562	Square Miles
	4840	Square Yards
	160	Square Rods
Acre-feet	43,560	Cubic Feet
	325,851	Gallons (US)
	1233.49	Cubic Meters
	1,233,490	Liters
Acre-feet/hr	726	Cubic feet/Minute
	5430.86	Gallons Minute
Angstroms	1.00E-09	Meters
Ares	0.01	Hectares
	1076.39	Square Feet
	0.02471	Acres
Atmospheres	76	Cms of Hg at 32°F
	29.921	Inches of Hg at 32° F
	33.94	Feet of Water at 62- F
	10,333	Kgs Square meter
	14.6963	Pounds Square inch
	1.058	Tons/Square foot
	1013.15	Millibars
	235.1408	Ounces/Square inch
Bags of cement	94	Pounds of cement
Barrels of oil	42	Gallons of oil (US)
Barrels of cement	376	Pounds of cement
Barrels (not legal)	31	Gallons (US)
or	31.5	Gallons (US)
Board feet	144 x 1 in.*	Cubic inches
Boiler horsepower	33,479	BTU/hour
	9.803	Kilowatts
	34.5	Pounds of water evaporated/ hour at 2120 F

(Continued)

—cont'd

Multiply	By	To Obtain
BTU	252.016	Calories (gm)
	0.252	Calonss (Kg)
	778.26	Foot pounds
	0.0003927	Horsepower hours
	1055.1	Joules
	107.5	Kilogram meters
	0.0002928	Kilowatt hours
BTU/cubic foot	8.89	Calories (Kg)/Cu meter at 32- F
Btu/hr/ft2/°F/ft	0.00413	Cal (gm)/Sec/cm2/"C/cm
	1.49	Cal (Kg)/Hr/M2/"C/Meter
Btu/minutes	12.96	Foot pounds/second
	0.02356	Horsepower
	0.01757	Kilowatts
BTU/minute	17.57	Watts
BTU/pound	0.556	Calories (Kg)!Kilogram
Bushels	2150.4	Cubic inches
	35.24	Liters
	4	Pecks
	32	Quarts (dry)
Cables	120	Fathoms
Calories (gm)	0.003968	BTU
	0.001	Calories (Kg)
	3.088	Foot pounds
	1.558×10^{-6}	Horsepower hours
	4.185	Joules
	0.4265	Kilogram meters
	1.1628×10^{-6}	Kilowatt hours
	0.0011628	Watt hours
Cal (gm)/sec/cm2/°C/ cm	242.13	BTU/Hr/ft2/°F/ft
Calories (Kg)	3.968	BTU
	1000	Calories (gm)
	3088	Foot pounds
	0.001558	Horse power hours
	4185	Joules
	426.5	Kilograms meters
	0.0011628	Kilowatt hours
	1.1628	Watt hours
Calories (Kg) cubic meter	0.1124	BTUlCu foot at O- C
Cal (Kg)/Hr/M2rC/M	0.671	BTU Hr/ft2/"F/foot

(Continued)

—cont'd

Multiply	By	To Obtain
Calories (Kg)/Kg	1.8	BTU pound
Calories (Kg) minute	51.43	Foot pounds/second
	0.09351	Horsepower
	0.06972	Kilowatts
Carats (diamond)	200	Milligrams
Centares (Centiares)	1	Square meters
Centigram	0.01	Grams
Centiliters	0.01	Liters
	0.3937	Inches
Centimeters	0.032808	Feet
	0.01	Meters
	10	Millimeters
Centimeters of Hg at 32°F	0.01316	Atmospheres
	0.4461	Feet of water at 62° F
	136	Kgs/Square meter
	27.85	Pounds/Square foot
	0.1934	Pounds/Square inch
Centimeters/second	1.969	Feet/minute
	0.03281	Feet/second
	0.036	Kilometers/hour
	0.6	Meters/minute
	0.02237	Miles/hour
	0.0003728	Miles/minute
Centimeters/second2	0.03281	Feet/second2
Centipoise	0.000672	Pounds/sec foot
	2.42	Pounds/hour foot
	0.01	Poise
Chains (Gunter's)	4	Rods
	66	Feet
	100	Links
Cheval-vapeur	1	Metric horsepower
	75	Kilogram meters/second
	0.98632	Horsepower
Circular inches	106	Circular mils
	0.7854	Square inches
	785,400	Square mils
Circular mils	0.7854	Square mils
	10	Circular inches
	7.854 x 10.5	Square inches

(*Continued*)

—cont'd

Multiply	By	To Obtain
Cubic centimeters	3.531 x 10.5	Cubic feet
	0.06102	Cubic inches
	1.00E-05	Cubic meters
	1.308e10-6	Cubic yards
	0.0002642	Gallons (US)
	0.001	Liters
	0.002113	Pints (liq, US)
	0.001057	Quarts (Liq. US)
	0.0391	Ounces (fluid)
Cubic feet	28.32	Cubic centimeters
	1728	Cubic inches
	0.02832	Cubic meters
	0.03704	Cubic yards
	7.48052	Gallons (US)
	28.32	Liters
	59.84	Pints (liq, US)
	29.92	Quarts (Iiq. US)
	2.30E-05	Acre feet
	0.803564	Bushels
Cubic feet of water	62.4266 F	Pounds at 39.2°
	62.3352	Pounds at 62° F
Cubic feet/minute	472	Cubic centimeters/sec
	0.1247	Gallons (US)/second
	0.472	Liters/second
	62.34	Pounds water/min at 62°F
	7.4805	Gallons (US)/minute
	10,772	Gallons/24 hours
	0.033058	Acre feet/24 hours
Cubic feet/second	646.317	Gallons (US)/24 hours
	448.831	Gallons/minute
	1.98347	Acre feet/24 hours
Cubic inches	16.387	Cubic centimeters
	0.0005787	Cubic feet
	1.64E-05	Cubic meters
	2.14E-05	Cubic yards
	0.004329	Gallons (US)
	0.01639	Liters
	0.03463	Pints (liq. US)
	0.01732	Quarts (liq. US)

(*Continued*)

—cont'd

Multiply	By	To Obtain
Cubic meters	1.00E+07	Cubic centimeters
	35.31	Cubic feet
	61.023	Cubic inches
	1.308	Cubic yards
	264.2	Gallons (US)
	1000	Liters
	2113	Pints (liq. US)
	1057	Quarts (Iiq. US)
Cubic yards	764,600	Cubic centimeters
	27	Cubic feet
	46,656	Cubic inches
	0.7646	Cubic meters
	202	Gallons (US)
	764.6	Liters
	1616	Pints (liq. US)
	807.9	Quarts (iiq. US)
Cubic yards/minute	0.45	Cubic feet/second
	3.367	Gallons (US)/second
	12.74	Liters/second
Cubit	18	Inches
Days (mean)	1440	Minutes
	24	Hours
	86,400	Seconds
Days (sidereal)	86,164.10	Solar seconds
Decigrams	0.1	Grams
Deciliters	0.1	Liters
Decimeters	0.1	Meters
Degrees (angle)	60	Minutes
	0.01745	Radians
	3600	Seconds
Degrees F (Less 32)	0.5556	Degrees C
Degrees F	1 [Plus 460)	Degrees F above absolute 0 (R)
Degrees C	1.8 [Plus 32]	Degrees F
	1 [Plus 273]	Degrees C above absolute 0 (K)
Degrees/second	0.01745	Radians/second
	0.1667	Revolutions/min
	0.002778	Revolutions/sec
Dekagrams	10	Grams

(Continued)

—cont'd

Multiply	By	To Obtain
Dekaliters	10	Liters
Dekameters	10	Meters
Diameter (circle)	3.14159(π)	Circumference
(approx)	3.1416	
(approx)	3.14	
Diameter (circle)	0.88623	Side of equal square
	0.7071	Side of inscribed square
Diametercubed (sphere)	0.5236	Volume (sphere)
Diam (major) x diam (minor)	0.7854	Area of ellipse
Diametersquared (circle)	0.7854	Area (circle)
Diametersquared (sphere)	3.1416	Surface (sphere)
Diam (inches) x RPM	0.262	Belt speed ft/mi
Digits	0.75	Inches
Drams (avoirdupois)	27.34375	Grains
	0.0625	Ounces (avoir.)
	1.771845	Grams
Fathoms	0.16667	Feet
Feet	30.48	Centimeters
	12	Inches
	0.3048	Meters
	0.333	Yards
	0.06061	Rods
Feet of water at 62° F	0.029465	Atmospheres
	0.88162	Inches of Hg at 32° F
	62.3554	Pounds/square meter
	0.43302	Pounds/square meter
	304.44	Kilogram/square meter
Feet/minute	0.508	Centimeters/second
	0.01667	Feet/second
	0.01829	Kilometers/hour
Feet/minute	0.3048	Meters/minute
	0.01136	Miles/hour
Feet/second	30.48	Centimeters/second
	1.097	Kilometers/hour
	0.5921	Knots
	18.29	Meters/minute
	0.6818	Miles/hour
	0.01136	Miles/minute
Feet/second2	30.48	Centimeters/second^2
	0.3048	0.3048 Meters/second^2

(*Continued*)

—cont'd

Multiply	By	To Obtain
Flat of a hexagon	1.155	Distance across comers
Flat of a square	1.414	Distance across corners
Foot Pounds	0.00128492	BTU
	0.32383	Calories (gm)
	0.0003238	Calories (Kg)
	5.05 x 10.7	Horsepower hours
	1.3558	Joules
	0.13826	Kilogram meters
	3.766 x 10-7	Kilowatt hours
	0.0003766	Watt hours
Foot pounds/minute	0.001286	BTU/minute
	0.01667	Foot pounds/second
	3.03 x 10.5	Horsepower
	0.0003241	Calories (Kg)/minute
	2.26 x 10-5	Kilowatts
Foot pounds/second	0.07717	BTU/minute
	0.001818	Horsepower
	0.01945	Calories (Kg)/minute
	0.001356	Kilowatts
Furlong	40	Rods
	220	Yards
	660	Feet
	0.125	Miles
	0.2042	Kilometers
Gallons (Imperial)	277.42	Cubic inches
	4.543	Liters
	1.20095	Gallons (US)
Gallons (US)	3785	Cubic centimeters
	0.13368	Cubic feet
	231	Cubic inches
	0.003785	Cubic meters
	0.004951	Cubic yards
	3.785	liters
	8	Pints (liq, US)
	4	Quarts (liq. US)
	0.83267	Gallons (Imperial)
	3.069 x 10-6	Acre feet
Gallons (US) of water at 62°F		
	8.333	Pounds of water

(*Continued*)

—cont'd

Multiply	By	To Obtain
@Gallons (US) of water/minute	6.0086	Tons of water/24 hours
Gallons (US)/minute	0.002228	Cubic feet/second
	0.13368	Cubic feet/minute
	8.0208	Cubic feet/hour
	0.06309	Liters/second
	3.78533	Liters/minute
	0.0044192	Acre feet/24 hours
Grains	1	Grains (avoirdupois)
	1	Grains (apothecary)
	1	Grains (troy)
	0.0648	Grams
	0.0020833	Ounces (troy)
	0.0022857	Ounces (avoir.)
Grains/gallon (US)	17.128	Parts/million
	142.86	Pounds/million gallons (US)
Grams	980.7	Dynes
	15.43	Grains
	0.001	Kilograms
	1000	Milligrams
	0.03527	Ounces (avoir.)
	0.03215	Ounces (troy)
	0.002205	Pounds
Grams/centimeters	0.00521	Pounds/inch
Grams/cubic centimeter	62.45	Pounds/cubic foot
	0.03613	Pounds/cubic inch
	4.37	Grains/100 cubic ft
Grams/liter	58.405	Grains/galion (US)
	8.345	Pounds/lOG gallons (US)
	0.062427	Pounds/cubic foot
	1000	Parts/million
Gravity (g)	32.174	Feet/second
	980.6	Centimeters/second^2
Hand	4	Inches
	10.16	Centimeters
Hectares	2.471	Acres
	107,639	Square feet
	100	Ares
Hectograms	100	Grams
Hectoliters	100	Liters
Hectometers	100	Meters

(Continued)

—cont'd

Multiply	By	To Obtain
Hectowatts	100	Watts
Hogshead	63	Gallons (US)
	238.4759	Liters
Horsepower	42.44	BTU/minute
	33,000	Foot pounds/minute
	550	Foot pounds/second
	1.014	Metric Horsepower (Cheval vapeur)
	10.7	Calories {Kg}/min
	0.7457	Kilowatts
	745.7	Watts
Horsepower (boiler)	33,479	BTU/hour
	9.803	Kilowatts
	34.5	Pounds of water evaporated/hour at 2120 F
Horsepower hours	2546.5	BTU
	641,700	Calories (gm)
	641.7	Calories (Kg)
	1980198	Foot pounds
	2688172	Joules
	273,740	Kilogram meters
	0.7455	Kilowatt hours
	745.5	Watt hours
Inches	2.54	Centimeters
	0.08333	Feet
	1000	1000 Mils
	12	Lines
	72	Points
Inches of Hg at 32° F	0.03342	Atmospheres
	345.3	Kilograms/square meter
	70.73	Pounds/square foot
	0.49117	Pounds/square inch
	1.1343	Feet of water at 62° F
Inches of Hg at 32° F	13.6114	Inches of water at 62° F
	7.85872	Ounces/square inch
Inches of water at 62° F	0.002455	Atmospheres
	25.37	Kilograms/square meter
	0.5771	Ounces/square inch
	5.1963	Pounds/square foot
	0.03609	Pounds/square inch
	0.07347	Inches of Hg at 32° F

(Continued)

—cont'd

Multiply	By	To Obtain
Joules	0.00094869	BTU
	0.239	Calories (gm)
	0.000239	Calories (Kg)
	0.73756	Foot pounds
	3.72 x 10.7	Horsepower hours
	0.10197	Kilogram meters
	2.778 x 10.7	Kilowatt hours
	0.0002778	Watt hours
	1	Watt second
Kilograms	980,665	Dynes
	2.205	Pounds
	0.001102	Tons (short)
	1000	Grams
	35.274	Ounces (avoir.)
	32.1507	Ounces (troy)
Kilogram meters	0.009302	BTU
	2.344	Calories (gm)
	0.002344	Calories (Kg)
	7.233	Foot pounds
	3.653 x 10-6	Horse power hours
	9.806	Joules
	2.724 x 10'-	Kilowatt hours
	0.002724	Watt hours
Kilograms/cubic meter	0.06243	Pounds! cubic foot
Kilograms/meter	0.672	Pounds/foot
Kilograms/sq centimeter	14.223	Pounds/sq inch
	1	Metric atmosphere
Kilograms/sq meter	9.68E-05	Atmospheres
	0.003285	Feet of water at 62° F
	0.002896	Inches of Hg at 32° F
	0.2048	Pounds/square foot
	0.001422	Pounds/square inch
	0.007356	Centimeters of Hg at 32° F
Kiloliters	1000	Liters
Kilometers	100,000	Centimeters
	1000	Meters
	3281	Feet
	0.6214	Miles
	1094	Yards

(Continued)

—cont'd

Multiply	By	To Obtain
Kilometers/hour	27.78	Centimeters/second
	54.68	Feet/minute
	0.9113	Feet/second
	16.67	Meters/minute
	0.6214	Miles/hour
	0.5396	Knots
Kilometers/hr/sec	27.78	Centimeters/sec/sec
	0.9113	Feet/sec/sec
	0.2778	Meters/sec/sec
Kilowatts	56.92	BTU/minute
	44,250	Foot pounds/minute
	737.6	Foot pounds/second
	1.341	Horsepower
	14.34	Calories (Kg)/min 10e-3 Watts
Kilowatt hours	3413	BTU
Kilowatt hours	859,999	Calories (gm)
	858.99	Calories (Kg)
	2,655,200	Foot pounds
	1.341	Horsepower
	3,600,000	Joules
	367,100	Kilogram meters
	1000	Watt hours
Knots	1	Nautical miles/hr
	1.1516	Miles/hr
	1.8532	Kilometers/hr
Leagues	3	Miles
Lines	0.08333	Inches
Links	7.92	Inches
Liters	1000	Cubic centimeters
	0.03531	Cubic feet
	61.02	Cubic inches
	0.001	Cubic meters
	0.001308	Cubic yards
	0.2642	Gallons (US)
	0.22	Gallons (Imp)
	2.113	Pints (liq. US)
	1.057	Quarts (liq. US)
	8.11E-07	Acre feet
	2.2018	Pounds of water

(Continued)

—cont'd

Multiply	By	To Obtain
Liters/minute	0.0005886	Cubic feet/second
	0.004403	Gallons (US)/second
	0.26418	Gallons (USD)/minute
Meters	100	Centimeters
	3.281	Feet
	39.37	Inches
	1.094	Yards
	0.001	Kilometers
	1000	Milimeters
Meters/minute	1.667	Centimeters/second
	3.281	Feet/minute
	0.05468	Feet/second
	0.06	Kilometers/minute
	0.03728	Miles/hr
Meters/second	196.8	Feet/minute
	3.281	Feet/second
	3.6	Kilometers/hour
	0.06	Kilometers/minute
	2.237	Miles/hour
	0.03728	Miles/minute
Microns	1000000	Meters
	0.001	Millimeters
	0.03937	Mils
Mils	0.001	Inches
	0.0254	Millimeters
	25.4	Microns
Miles	160,934	Centimeters
	5280	Feet
	63,360	Inches
	1.609	Kilometers
	1760	Yards
	80	Chains
	320	Rods
	0.8684	Nautical miles
Miles/hour	44.7	Centimeters/second
	88	Feet/minute
	1.467	Feet/second
	1.609	Kilometers/hours
	0.8684	Knots
	26.82	Meters/minute

(Continued)

—cont'd

Multiply	By	To Obtain
Miles/minute	2682	Centimeters/second
	88	Feet/second
	1,609	Kilometers/minute
	60	Miles/hour
Milibars	0.000987	Atmosphere
1Milliers	1000	Kilograms
Milligrams	0.01	Grams
	0.01543	Grains
Milligrams/liter	1	Parts/million
Kiloliters	0.001	Liters
Million gals/24 hours	1.54723	Cubic feet/second
Millimeters	0.1	Centimeters
	0.03937	Inches
	39.37	Mils
	1000	Microns
Miner's inches	1.5	Cubic feet/minute
Minutes (angle)	0.0002909	Radians
Nautical miles	6080.2 1.516	Feet Miles
Ounces (avoirdupois)	16	Drams (avoir.)
	437.5	Grains
	0.0625	Pounds (avoir.)
	28.349527	Grams
	0.9115	Ounces (troy)
Ounces(fluid)	1.805	Cubic inches
	0.02957	Liters
	29.57	Cubic centimeters
	0.25	Gills
Ounces (troy)	480	Grains
	20	Pennyweights (troy)
	0.08333	Pounds (troy)
	31.103481	Grams
	1.09714	Ounces (avoir.)
Ounces/square inch	0.0625	Pounds/square inch
	1.732	Inches of water at 62° F
	4.39	Centimeters of water at 62° F
	0.12725	Inches of Hg at 32° F
	0.004253	Atmospheres
Palms	3	Inches
Parts/million	0.0584	Grains/gallon (US)
	0.07016	Grains/gallon (Imp)
	8.345	Pounds/million gal (US)

(Continued)

—cont'd

Multiply	By	To Obtain
Pennyweights	24	Grains
	1.55517	Grams
	0.05	Ounces (troy)
	0.0041667	Pounds (troy)
Pints (liq. US)	4	Gills
	16	Ounces (fluid)
	0.5	Quarts (liq. US)
	28.875	Cubic inches
	473.1	Cubic centimeters
Pipe	126	Gallons (US)
Points	0.01389	Inches
	0.0672	Pounds/sec foot
Poise		
	242	Pounds/hour foot
	100	Centipoise
Poncelots	100 1.315	Kilogram meters/second Horsepower
Pounds (avoirdupois)	16	Ounces (avoir.)
	256	Drams (avoir.)
	7000	Grains
	0.0005	Tons (short)
	453.5924	Grams
	1.21528	Pounds (troy)
	14.5833	Ounces (troy)
Pounds (troy)	5760	Grains
	240	Pennyweights (troy)
	12	Ounces (troy)
	373.24177	Grams
	0.822857	Pounds (avoir.)
	13.1657	Ounces (avoir.)
	0.00036735	Tons (long)
	0.00041143	Tons (short)
	0.00037324	Tons (metric)
Pounds of water at 62° F	0.01604	Cubic feet
	27.72	Cubic inches
Pounds of water/min.	0.12	Gallons (US)
at 62° F	0.0002673	Cubic feet/second
Pounds/cubic foot	0.01602	Grams/cubic centimeter
	16.02	Kilograms/cubic meter
	0.0005787	Pounds/cubic inch
Pounds/cubic inch	27.68	Grams/cubic centimeter
	27.68	Kilograms/cubic meter
	1728	Pounds/cubic foot

(*Continued*)

—cont'd

Multiply	By	To Obtain
Pounds/foot	1.488	Kilograms/meter
Pounds/inch	178.6	Grams/centimeter
Pounds/hour foot	0.4132	Centipoise
	0.004132	Poise grams/sec cm
Pounds/sec foot	14.881	Poise grams/sec cm
	1488.1	Centipoise
Pounds/square foot	0.016037	Feet of water at 62° F
	4.882	Kilograms/square meter
	0.006944	Pounds/square inch
	0.014139	Inches of Hg at 32° F
	0.0004725	Atmospheres
Pounds/square inch	0.068044	Atmospheres
	2.30934	Feet of water at 62° F
	2.036	Inches of Hg at 32° F
	703.067	Kilograms/square meter
	27.912	Inches of water at 62° F
Quadrants (angular)	90	Degrees
	5400	Minutes
	324,000	Seconds
	1.751	Radians
Quarts (dry)	67.2	Cubic inches
Quarts (liq.US)	2	Pints (liq. US)
	0.9463	Liters
	32	Ounces (fluid)
	57.75	Cubic inches
	946.3	Cubic centimeters
Quintal, Argentine	101.28	Pounds
Brazil	129.54	Pounds
Castile, Peru	101.43	Pounds
Chile	101.41	Pounds
Metric	220.46	Pounds
Mexico	101.47	Pounds
Quires	25	Sheets
Radians	57.3	Degrees
	3438	Minutes
	206.186	Seconds
	0.637	Quadrants
Radians/second	57.3	Degrees/second
	0.1592	Revolutions/second
	9.549	Revolutions/minute

(*Continued*)

—cont'd

Multiply	By	To Obtain
Radians/second	573	Revolutions/minute'
	0.1592	Revolutions/second
Reams	500	Sheets
Revolutions	360	Degrees
	4	Quadrants
	6.283	Radians
Revolutions/minute	6	Degrees/second
	0.1047	Radians/second
	0.01667	Revolutions/second
Revolutions/minute^2	0.001745	Hadiana/second
	0.0002778	Revolutions/second
Revolutions/second	360	Degrees/second
	6.283	Radians/second
	60	Revolutions/minute
Revolutions/second^2	6.283	Radians/seconds
	3600	Revolutions/minute
Rods	16.5	Feet
	5.5	Yards
Seconds (angle)	4.848 x 10"	Radians
Sections	1	Square miles
Side of a square	1.4142	Diameter of inscribed circle
	1.1284	Diameter of circle with equal area
Span	9	Inches
Square centimeters	0.001076	Square feet
	0.155	Square inches
	0.0001	Square meters
	100	Square millimeters
Square feet	2.296 x 10.5	Acres
	929	Square centimeters
	144	Square inches
	0.0929	Square meters
	3.587 x 10.8	Square miles
	0.1111	Square yards
Sqaure inches	6.452	Square centimeters
	0.006944	Square feet
	645.2	Square millimeters
	1.27324	Circular inches
	1,273,239	Circular mils
	1,000,000	Square mils

(*Continued*)

—cont'd

Multiply	By	To Obtain
Square kilometers	247.1	Acres
	10,760,000	Square feet
	1,000,000	Square meters
	0.3861	Square miles
	1,196,000	Square yards
Square meters	0.0002471	Acres
	10.764	Square feet
	1.196	Square yards
	1	Centares
Square miles	640	Acres
	27,878,400	Square feet
Square miles	2.59	Square kilometers
	259	Hectares
	3,097,600	Square yards
	102,400	Square rods
	1	Sections
Square millimeters	0.01	Square centimeters
	0.00155	Square inches
	1550	Square mils
	1973	Circular mils
Square mils	1.27324	Circular mils
	0.0006452	Square millimeters
	10"	Square inches
Square yards	0.0002066	Acres
	9	Square feet
	0.8361	Square meters
	3.228 x 10.7	Square miles
Stere	1	Cubic meters
Stone	14	Pounds
	6.35029	Kilograms
Tons (long)	1016	Kilograms
	2240	Pounds
	1.12	Tons (short)
Tons (metric)	1000	Kilograms
	2205	Pounds
	1.1023	Tons (short)
Tons (short)	2000	Pounds
	32,000	Ounces
	907.185	Kilograms
	0.90718	Tons (metric)
	0.89286	Tons (long)

(Continued)

—cont'd

Multiply	By	To Obtain
Tons of Refrigeration	12000	BTU/hour
	288,000	BTU/24 hours
Tons of water/24 hours at 62° F	83.33	Pounds of water/hour
	0.1651	Gallons (US)/minute
	1.3263	Cubic feet/hour
Watts	0.05692	BTU/minute
	44.26	Foot pounds/minute
	0.7376	Foot pounds/second
	0.001341	Horsepower
	0.01434	Calories (Kg)/minute
	0.001	Kilowatts
	1	Joule/second
Watt hours	3.413	BTU
	860.5	Calories (gm)
	0.8605	Calories (Kg)
	2655	Foot pounds
	0.001341	Horsepower hours
	3600	Joules
	367.1	Kilogram meters
	0.001	Kilowatt hours
Watts/square inch	8.2	BTU/sq ft/minute
	6373	Foot pounds/sq ft/minute
	0.1931	Horsepower/square foot
Yards	91.44	Centimeters
	3	Feet
	36	Inches
	0.9144	Meters
	0.1818	Rods
Year (365 days)	8760	Hours

BIBLIOGRAPHY

Some of the following books are older than others. This is for two reasons. First, there is a relative shortage of modern books on pipe stress and related subjects. Second, many modern books on pipe stress tend to point to computer science solutions along with computer programs, which is not necessarily helpful for the way this book is intended to be used.

Title	Author, Publisher, or Sponsor
ASCE 7-05 Minimum Design Loads	ASCE
B16 Code Books	ASME
B31 Code Books	ASME
Design of Piping Systems	M. W. Kellogg Co.
Flow Measurement Engineering Handbook	R. W. Miller
Flow of Fluids Technical Paper 410	Crane Company
Fluid Transients in Systems	E. Benjamin Wylie and Victor C. Streeter
Handbook of Valves	Philip A. Schweitzer
Introduction to Heat Transfer	A. I. Brown and S. M. Marco
Introduction to Pipe Stress Analysis	Sam Kannappan
Mark's Standard Handbook for Mechanical Engineers	E. A. Avallone, T. Baumeister, and Ali Sadegh
Mechanical Vibrations	J. P. Den Hartog
MSS Standard Practices	Manufacturers Standardization Society
Perry's Chemical Engineers' Handbook	D. W. Green and R. H. Perry
Piping and Pipe Support Systems	Paul R. Smith and Tom J. Van Laan
Piping Design and Engineering	ITT/Grinnell
Piping Engineering	Tube Turns Inc.
Piping Stress Calculations Simplified	S. W. Spielvogel
Piping Systems and Pipeline	J. Phillip Ellenberger
Principals of Engineering Heat Transfer	W. H. Giedt
Theory and Design of Pressure Vessels	John F. Harvey P.E.
The Wiley Engineers Desk Reference	Sanford I. Heisler
Roark's Formulas for Stress and Strain, 2nd, 4th, and 6th editions	Raymond J. Roark and Warren C. Young, 6th edition
Wikepedia.com	Internet
Google.com	Internet
EngineeringToolBox.com	Internet

Note: Page numbers followed by "f" denote figures; "t" tables; "b" boxes.

Printed and bound by CPI Group (UK) Ltd, Croydon, CR0 4YY

03/10/2024

01040421-0004